Mathematics for the Green Industry

Essential Calculations for Horticulture and Landscape Professionals

Michael L. Agnew

Nancy H. Agnew

Nick E. Christians

Ann Marie VanDerZanden

John Wiley & Sons, Inc.

This book is printed on acid-free paper. ∞

Copyright © 2008 by John Wiley & Sons, Inc. All rights reserved

Published by John Wiley & Sons, Inc., Hoboken, New Jersey
Published simultaneously in Canada

No part of this publication may be reproduced, stored in a retrieval system, or transmitted in any form or by any means, electronic, mechanical, photocopying, recording, scanning, or otherwise, except as permitted under Section 107 or 108 of the 1976 United States Copyright Act, without either the prior written permission of the Publisher, or authorization through payment of the appropriate per-copy fee to the Copyright Clearance Center, 222 Rosewood Drive, Danvers, MA 01923, (978) 750-8400, fax (978) 646-8600, or on the web at www.copyright.com. Requests to the Publisher for permission should be addressed to the Permissions Department, John Wiley & Sons, Inc., 111 River Street, Hoboken, NJ 07030, (201) 748-6011, fax (201) 748-6008, or online at www.wiley.com/go/permissions.

Limit of Liability/Disclaimer of Warranty: While the publisher and the author have used their best efforts in preparing this book, they make no representations or warranties with respect to the accuracy or completeness of the contents of this book and specifically disclaim any implied warranties of merchantability or fitness for a particular purpose. No warranty may be created or extended by sales representatives or written sales materials. The advice and strategies contained herein may not be suitable for your situation. You should consult with a professional where appropriate. Neither the publisher nor the author shall be liable for any loss of profit or any other commercial damages, including but not limited to special, incidental, consequential, or other damages.

For general information about our other products and services, please contact our Customer Care Department within the United States at (800) 762-2974, outside the United States at (317) 572-3993 or fax (317) 572-4002.

Wiley also publishes its books in a variety of electronic formats. Some content that appears in print may not be available in electronic books. For more information about Wiley products, visit our web site at www.wiley.com.

Library of Congress Cataloging-in-Publication Data:

Mathematics for the green industry / Michael Agnew ... [et al.].
   p. cm.
  Includes bibliographical references and index.
  ISBN 978-0-470-13672-0 (pbk. : alk. paper)
  1. Landscaping industry—Mathematics. I. Agnew, Michael Lewis, 1949-
SB472.5.M38 2008
635.9—dc222
                                                    2007050459

TABLE OF CONTENTS

Preface   v
About the Authors   vii

PART 1   MATHEMATICAL PRINCIPLES   1

Chapter 1   Basic Math Skills   1
Chapter 2   Measurement and Calculations with Measured Values   31
Chapter 3   Geometry   47

PART 2   GREEN INDUSTRY APPLICATIONS   79

Chapter 4   Calculating the Area of Landscape Features   79
Chapter 5   Fertilizer Calculations   95
Chapter 6   Pesticide and Plant Growth Regulator Calculations   127
Chapter 7   Calibration of Application Equipment   151
Chapter 8   Mathematical Applications for the Turfgrass Industry   169
Chapter 9   Mathematical Applications for the Landscape Industry   191
Chapter 10 Mathematical Applications for the Greenhouse, Nursery, and Interior Landscape Industries   223

Appendix A: Metric System Prefixes   285
Appendix B: Tables of Equivalents   287
Appendix C: Table of Conversion Factors   297
Appendix D: Squaring-Up Gardens and Garden Structures   307
Appendix E: Solutions to Practice Problems   309

Index   395

PREFACE

The green industry is comprised of a diverse group of companies and individuals involved in the production, sales, and service of ornamental plants. On a daily basis, these professionals perform mathematical tasks that allow them to estimate area and volume; apply fertilizers, pesticides, and plant growth regulators; calibrate application equipment; plan, price, and execute design/build projects; manage either greenhouse or nursery crops and the growing environment; and establish and maintain turf areas. It is essential to complete these mathematical tasks accurately. Errors can negatively affect profitability. Further, when the calculation relates to the use of fertilizers, pesticides, or plant growth regulators, crop quality and the environment in general is impacted. The purpose of *Mathematics for the Green Industry* is to provide an overview of correct procedures for common mathematical tasks within each of the major industry segments. The authors designed this text to be a reference for professionals, a college-level text for a course in mathematics about horticulture, and an adjunct text for college-level, commodity-based courses in horticulture.

The introductory chapters refresh the mathematical skills of industry professionals and horticulture students. Proper expression and manipulation of numerical values are reviewed in Chapter 1. Chapter 2 reviews measurement, accuracy, and precision of measured values; the use of significant digits and rules for rounding measured and calculated values; units of measure; and conversion between units of measure. The final introductory chapter refreshes geometry skills and provides the formulae for determining area and volume of geometric figures.

In Chapters 4 through 10, specific green industry mathematical applications are presented. Chapter 4 continues the presentation of geometry skills with examples of how green industry professionals can measure and calculate the area of landscape features. Chapters 5 and 6 review the basics of fertilizer and pesticide calculations. Calculations to calibrate fertilizer and pesticide application equipment are presented in Chapter 7. Chapter 8 provides examples of calculations unique to the turfgrass industry, such as lawn care, sports turf management, and sod production. Landscape maintenance and landscape design/build industry calculations are discussed in Chapter 9. Finally, Chapter 10 reviews calculations common to the greenhouse, nursery, and interior landscape industries.

Throughout this text, the authors provide appropriate industry examples for each mathematical task. In addition, a systematic approach to each mathematical problem shows students and professionals how to calculate accurate results. Individuals who use this text are able to practice their mathematical skills through the numerous practice problems and solutions the authors provide in each chapter.

Ultimately, the authors hope *Mathematics for the Green Industry* will facilitate the development of confidence and a "sense of ease" for horticulture students and industry professionals when they perform the common, and sometimes quite complex, mathematical tasks required of green industry professionals.

ABOUT THE AUTHORS

Michael L. Agnew Michael L. Agnew, PhD, is a senior field technical manager for Syngenta Professional Products, Greensboro, North Carolina, and former associate professor and extension turfgrass specialist at Iowa State University. He earned a BS and a MS in horticulture and earned a PhD in horticulture with specialization in turfgrass science and plant stress physiology from Kansas State University. Dr. Agnew has been providing service to the turfgrass and golf industries locally, nationally, and internationally for 25 years. He currently provides technical support for Syngenta customers and staff; provides leadership in turfgrass fungicide product development; serves several trade and professional organizations; and is a frequently requested speaker and author within the turfgrass and green industries. For 20 years, he has co-taught calculations and practical math for golf courses at the annual meeting of the Golf Course Superintendents Association of America. During his tenure at Iowa State University, he served as state Extension leader for horticulture and held board positions on the Iowa Turfgrass Institute, Iowa Horticulture Society, Iowa Arboretum, and the City of Ames Parks and Recreation Department.

Nancy H. Agnew Nancy H. Agnew, PhD, is an instructor with the Professional Gardener Training Program at Longwood Gardens, Inc. and former associate professor of horticulture at Iowa State University. She earned a BS and MS in horticulture and earned a PhD in horticulture with specialization in floriculture and plant stress physiology from Kansas State University. Dr. Agnew has been teaching, advising, and writing in the field of horticulture for 25 years. She currently provides curriculum development services for Longwood Gardens, Inc. in Kennett Square, Pennsylvania, and teaches horticulture math, greenhouse management, and greenhouse crop production courses in the Longwood Gardens Professional Gardener Training Program. During her tenure at Iowa State University, she instructed courses in herbaceous ornamentals, tropical plants and interior landscaping, home horticulture, and floriculture crop production; advised students and student organizations; conducted research that served the greenhouse and landscape horticulture industries; authored refereed and non-refereed journal publications; provided service and education to the Iowa greenhouse and landscape horticulture industries; and served as the professor in charge of the campus horticulture garden.

Nick E. Christians Nick E. Christians, PhD, is a university professor of horticulture at Iowa State University, where he has been involved in teaching and research of turfgrass management since 1979. He received his BS from Colorado State University and both MS and PhD degrees from Ohio State University. He is the author of *Fundamentals of Turfgrass Management* and co-author with Dr. Mike Agnew of *The Mathematics of Turfgrass Maintenance*, which are both publications of John Wiley & Sons, Inc. He is also co-author of *Scotts Lawns: Your Guide to a Beautiful Yard*. He has authored or co-authored more than 900 refereed papers, popular publications, abstracts, and published research reports. He has received a number of national and local awards for teaching and advising, including the American Society for Horticultural Science Outstanding Undergraduate Educator Award in 1991.

Ann Marie VanDerZanden   Ann Marie VanDerZanden, PhD is an associate professor of horticulture and a member of the Iowa State University horticulture department. She has taught landscape horticulture courses, including herbaceous plant identification, landscape design, landscape construction, and landscape contracting and estimating for 13 years. In addition to her teaching responsibilities, she is an extension specialist for the nursery and landscape industry and previously served as an extension specialist in consumer horticulture. Her research interests include undergraduate pedagogy and using new technology to enhance the learning experiences of students and nursery/landscape professionals.

She is an avid writer and has published numerous manuscripts on teaching and extension outreach projects in peer reviewed journals, as well as general interest articles in garden magazines including *American Nurseryman*, *Iowa Gardening*, and *Northwest Woman*. In 2007, her first college textbook, *Landscape Design: Theory and Application*, was published. In addition to writing, Dr. VanDerZanden has a number of speaking engagements each year reaching members of nursery and landscape industry and home gardeners.

Dr. VanDerZanden completed her academic training in horticulture science and earned a BS and PhD from Washington State University and a MS from Cornell University. Prior to joining the faculty at ISU, she was a faculty member at Oregon State University and Illinois State University.

PART ONE — MATHEMATICAL PRINCIPLES
Chapter 1 — Basic Math Skills

Introduction

Numbers used to express counted, measured, or calculated values can be found in various forms. This chapter is designed to review the various ways numbers can be expressed and manipulated. A review of proper order of operations, the isolation of x, and a review of the use of ratios and proportions in problem solving also are included.

Whole Numbers, Fractions, and Mixed Numbers

❧ Definition ❧ Whole Numbers

Whole numbers are sometimes referred to as **natural numbers** or **counting numbers**. This set of numbers includes all of the positive integers.

$$1, 2, 3, 4, 5 \ldots$$

These numbers are **exact** and have a fractional part of zero. ❧

Counted numbers in horticulture can be used to express the number of plants to be planted in a landscaping project, the number of potted plants or cut flowers to be grown in a greenhouse, or the number of paving units required for a paving project. Paving units are purchased as a whole unit and plants are not useful unless they are whole.

❧ Definition ❧ Fraction or Rational Number

A *fraction* or *rational number* is a number expressed in a ratio format.

$$\frac{a}{b}, \text{ where } b \neq 0$$

❧

A fraction represents a part of a whole. The word rational number comes from the fact that a rational number or fraction is expressed as a ratio. The number a is called the numerator, and the number b is called the denominator. The number b may not be equal to zero.

1

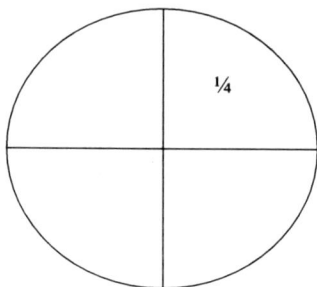

Figure 1-1

$\frac{1}{4}$ is a fraction. The one is the numerator and four is the denominator, which is greater than zero.

$\frac{1}{4}$ also can be described as one of four parts, as illustrated in Figure 1-1.

Fractions can be used and defined in many ways. The preferred way to write a fraction is in the form of a proper fraction, which is defined below. Improper fractions often are the result of performing mathematical operations with fractions. Mixed numbers combine whole numbers and fractions to express a number. Horticulturists are called on to manipulate fractions especially when performing mathematical operations on measures of length.

◂ Definition ▸ Proper Fraction

A *proper fraction* is one in which:

$$\frac{a}{b} < 1, \text{ and } a, b > 0$$

$\frac{1}{8}$ is a proper fraction because it is less than 1 and 8, the denominator, is greater than zero.

$\frac{24}{32}$ also is a proper fraction, but it could be written in the more efficient reduced or simplified form. Reducing a fraction requires dividing both the numerator and denominator by the greatest common factor. This results in a fraction that is expressed in lowest terms.

◂ Definition ▸ Greatest Common Factor (GCF)

The *greatest common factor* is the largest number that divides two numbers evenly. The *greatest common factor* is used to reduce fractions to lowest terms.

EXAMPLE 1-1 Reducing Fractions Using the GCF

Reduce $\frac{24}{32}$ to lowest terms by using the GCF.

Step 1

Factor the numerator and the denominator.

$$\frac{24}{32} = \frac{8 \times 3}{4 \times 8}$$

Step 2

Find the fraction that equals 1 and isolate it.

$$\frac{8 \times 3}{4 \times 8} = \frac{3}{4} \times \frac{8}{8}$$

Step 3

Reduce the fraction.

$$\frac{3}{4} \times \frac{8}{8} = \frac{3}{4} \times 1 = \frac{3}{4}$$

or

$$\frac{24}{32} \div \frac{8}{8} = \frac{3}{4}$$

$$\frac{24}{32} = \frac{3}{4}$$

Step 4

Check using division.

$$\frac{24}{32} = 24 \div 32 = 0.75$$

$$\frac{3}{4} = 3 \div 4 = 0.75$$

> **◀ Definition ▶ Improper Fraction**
>
> An *improper fraction* is one in which:
>
> $$\frac{a}{b} \geq 1, \text{ where } a \geq b \text{ and } a, b > 0$$

$\frac{9}{8}$ is an improper fraction because it is greater than 1 and the numerator is greater than or equal to the denominator.

An improper fraction can be rewritten in the form of a mixed number. For example, the improper fraction $\frac{9}{8}$ may be written as the mixed number $1\frac{1}{8}$.

> **◀ Definition ▶ Mixed Numbers**
>
> *Mixed numbers* are a combination of a whole number and a fractional part. It may be written as the sum of a whole number and a proper fraction.
>
> $$n + \frac{a}{b}$$

Mixed numbers are more accurate than whole numbers when estimating a value. For example, a garden bed may measure $61\frac{3}{4}$ inches (in.) deep. This is a more accurate estimate than 61 or 62 in. Mixed numbers are commonly used when estimating length using the U.S. Customary units of measurement.

EXAMPLE 1-2 *Converting Improper Fractions to Mixed Numbers and vice versa*

Convert $\frac{134}{25}$ to a mixed number.

Step 1

Divide the numerator by the denominator.

$$134 \div 25 = 5.64$$

Step 2

Write down the whole number.

$$5$$

Step 3

Calculate the remainder by subtracting the product of the whole number times the denominator from the numerator.

$$134 - (5 \times 25) = 134 - 125 = 9$$

Step 4

Combine the whole number with the remainder over the denominator.

$$5\frac{9}{25}$$

TO REVERSE THE PROCESS:

Convert $5\frac{9}{25}$ to an improper fraction.

Step 1

Multiply the whole number times the denominator.

$$5 \times 25 = 125$$

Step 2

Add the numerator to the solution in step one.

$$9 + 125 = 134$$

Step 3

Place the solution in step two in the fraction as the numerator.

$$\frac{134}{25}$$

Practice Problem Set 1-1 — *Whole Numbers, Fractions, and Mixed Numbers*

Identify each of the following numbers as a whole number, proper fraction, improper fraction, or mixed number:

1. $\frac{9}{8}$

2. $\dfrac{3}{4}$

3. 7

4. $1\dfrac{5}{8}$

5. $\dfrac{25}{25}$

Convert each of the following improper fractions to a mixed number:

6. $\dfrac{11}{8}$

7. $\dfrac{67}{32}$

8. $\dfrac{128}{5}$

9. $\dfrac{13}{4}$

10. $\dfrac{117}{16}$

Adding, Subtracting, Multiplying, and Dividing Fractions

Algebraic manipulation of fractions can be challenging. Here is a review of the proper ways to combine fractions when adding, subtracting, multiplying, and dividing.

$$\frac{a}{b} + \frac{c}{d} = \frac{ad + bc}{bd}$$

$$\frac{a}{b} - \frac{c}{d} = \frac{ad - bc}{bd}$$

$$\frac{a}{b} \times \frac{c}{d} = \frac{ac}{bd}$$

$$\frac{a}{b} \div \frac{c}{d} = \frac{ad}{bc}$$

EXAMPLE 1-3 Adding Fractions

$$\frac{a}{b} + \frac{c}{d} = \frac{ad + bc}{bd}$$

$$\frac{3}{4} + \frac{1}{2} = \frac{(3)(2) + (4)(1)}{(4)(2)} = \frac{10}{8} = \frac{5}{4} = 1\frac{1}{4}$$

Note that the fraction is reduced at the end of the computation.

$$\frac{10}{8} \div \frac{2}{2} = \frac{5}{4}$$

Then the improper fraction is changed to a mixed number.

$$\frac{5}{4} = 1\frac{1}{4}$$

EXAMPLE 1-4 Subtracting Fractions

$$\frac{a}{b} - \frac{c}{d} = \frac{ad - bc}{bd}$$

$$\frac{15}{16} - \frac{1}{4} = \frac{(15)(4) - (16)(1)}{(16)(4)}$$

$$\frac{60 - 16}{64} = \frac{44}{64} = \frac{11}{16}$$

EXAMPLE 1-5 Multiplying Fractions

$$\frac{a}{b} \times \frac{c}{d} = \frac{ac}{bd}$$

$$\frac{5}{8} \times \frac{2}{3} = \frac{(5)(2)}{(8)(3)} = \frac{10}{24} = \frac{5}{12}$$

EXAMPLE 1-6 Dividing Fractions

$$\frac{a}{b} \div \frac{c}{d} = \frac{ad}{bc}$$

$$\frac{5}{6} \div \frac{2}{4} = \frac{(5)(4)}{(6)(2)} = \frac{20}{12} = \frac{5}{3} = 1\frac{2}{3}$$

Practice Problem Set 1-2 Adding, Subtracting, Multiplying, and Dividing Fractions

Solve the following problems involving fractions:

1. $\dfrac{3}{32} + \dfrac{15}{16}$

2. $\dfrac{3}{8} + \dfrac{1}{2}$

3. $\dfrac{10}{32} - \dfrac{1}{4}$

4. $\dfrac{7}{8} - \dfrac{1}{3}$

5. $\dfrac{3}{8} \times \dfrac{1}{4}$

6. $\dfrac{2}{16} \times \dfrac{1}{10}$

7. $\dfrac{10}{16} \div \dfrac{3}{8}$

8. $\dfrac{1}{2} \div \dfrac{1}{3}$

Decimal Numbers, Place Value, and Decimal Fractions

✒ Definition ✒ Decimal Number
A *decimal number* is the base-10 system used for expressing a mixed number. In other words, it is a way of naming the values that lie between whole numbers. The whole number is separated from the fractional portion of the number with a decimal point.

An example of a decimal number is five and four-tenths, and it is written as 5.4. Five is the whole number, and four-tenths is the fractional part of the number. The number five and four-tenths lies between the whole numbers five and six. A decimal point, written as a dot or a period, separates the whole number from the fractional part of the number.

✒ Definition ✒ Decimal Point
A *decimal point* is a period or dot in a decimal number that serves to separate the whole number portion of a number from the fractional part of the number.

When writing decimal numbers, it is important to understand the concept of place value. The benchmark for place value is the decimal point because it separates the whole number portion of the decimal number from the fractional portion of the decimal number (see Table 1-1). Notice that place value changes by a magnitude of ten as digit placement moves to the left or the right of the decimal point.

TABLE 1-1 • PLACE VALUE TABLE

1,000	100	10	1	.	$\frac{1}{10}$	$\frac{1}{100}$	$\frac{1}{1,000}$	$\frac{1}{10,000}$
Thousands	Hundreds	Tens	Ones	Decimal Point	One Tenth	One Hundredth	One Thousandth	One Ten-Thousandth

> **Definition** *Place Value*
>
> **Place value** is the value of a digit in a number. A digit's value depends on its position in relation to the decimal point.

Place value aids in understanding and comparing the value of numbers. It also explains the relationship between digits in a number. For example, place value can be used to compare the whole numbers 302 and 320. Although the numbers are similar as written, they are very different in value. By identifying that the 2 in 302 is 2 ones, you can see that it is a smaller number than 320, since the 2 in 320 is 2 tens. Notice that, as the placement of a digit moves to the left, its value increases by a magnitude of ten, and as the placement of the digit moves to the right, its value decreases by a magnitude of ten. Comparing decimal numbers can be more difficult because a number of digits can be located to the left *and* to the right of the decimal point. Writing a decimal number in an expanded form using a place value table can aid in comparing one number to another and in comparing digits within a number.

Table 1-2 demonstrates how the number 3,924.1256 is written in expanded form. Here is the expanded form of 3,924.1256 without the use of a table:

$$(3 \times 1{,}000) + (9 \times 100) + (2 \times 10) + (4 \times 1) + \left(1 \times \frac{1}{10}\right) + \left(2 \times \frac{1}{100}\right) + \left(5 \times \frac{1}{1{,}000}\right) + \left(6 \times \frac{1}{10{,}000}\right)$$

TABLE 1-2 • PLACE VALUE TABLE FOR THE NUMBER 3,924.1256

1,000	100	10	1	.	$\frac{1}{10}$	$\frac{1}{100}$	$\frac{1}{1,000}$	$\frac{1}{10,000}$
Thousands	Hundreds	Tens	Ones	Decimal Point	One Tenth	One Hundredth	One Thousandth	One Ten-Thousandth
3	9	2	4	.	1	2	5	6

Definition ▸ Decimal Fraction

A *decimal fraction* is a fraction in which the denominator is a power of ten.

$$\frac{a}{10^p}$$

Fractions can be converted to decimal fractions in the following way:

$$\frac{1}{2} \times \frac{5}{5} = \frac{5}{10} \text{ and is read as five-tenths}$$

$$\frac{1}{2} \times \frac{50}{50} = \frac{50}{100} \text{ and is read as fifty-one-hundredths}$$

Note that the denominator is always a power of ten, and the numerator and denominator are always multiplied by the same number.

Practice Problem Set 1-3 ▸ Decimal Numbers and Place Value

Write each of the following numbers in expanded form:

1. 43,560
2. 5,280.5
3. 3.14
4. 0.05
5. 24.175

Rewrite each of the following fractions in the form of a decimal fraction:

6. $\frac{1}{5}$
7. $\frac{4}{20}$
8. $\frac{3}{4}$
9. $\frac{1}{25}$
10. $\frac{7}{5}$

Converting Decimals into Fractions and Fractions into Decimals

Converting decimal numbers into fractions is a simple process, once the concept of place value is understood.

EXAMPLE 1-7

$$0.1 = \frac{1}{10}$$

This equivalent can be explained mathematically this way:

$$0.1 = \frac{0.1}{1} \times \frac{10}{10} = \frac{1}{10}$$

EXAMPLE 1-8

$$0.12 = \frac{0.12}{1} \times \frac{100}{100} = \frac{12}{100}$$

EXAMPLE 1-9

$$0.125 = \frac{0.125}{1} \times \frac{1,000}{1,000} = \frac{125}{1,000}$$

EXAMPLE 1-10

$$0.1256 = \frac{0.1256}{1} \times \frac{10,000}{10,000} = \frac{1256}{10,000}$$

Converting fractions into decimal numbers is also a simple process. Divide the numerator (top number) by the denominator (bottom number).

EXAMPLE 1-11

$$\frac{1}{2} = 1 \div 2 = 0.50$$

EXAMPLE 1-12

$$\frac{12}{60} = 12 \div 60 = 0.20$$

EXAMPLE 1-13

$$\frac{15}{16} = 15 \div 16 = 0.9375$$

Certain fraction to decimal conversions should be committed to memory. Common equivalents are listed in Table 1-3.

Some of the decimal numbers listed in Table 1-3 are written in a format that includes the use of a line over one of the numbers ($0.\overline{3}$). This line is called a vinculum and is used to indicate a repeating number or number sequence. Converting fractions to decimals result in either an even number or in a number with the last number or

TABLE 1-3 • DECIMAL EQUIVALENTS OF COMMONLY USED FRACTIONS

Fraction	Decimal Equivalent
$\frac{1}{2}$	0.50
$\frac{1}{3}$	$0.\overline{3}$
$\frac{2}{3}$	$0.\overline{6}$
$\frac{1}{4}$	0.25
$\frac{3}{4}$	0.75
$\frac{1}{8}$	0.125

sequence of numbers repeating. Some additional examples of decimal numbers with repeating numbers or repeating number sequences follow.

$$\frac{1}{6} = 0.1\overline{6}$$

$$\frac{3}{11} = 0.\overline{27}$$

$$\frac{2}{7} = 0.\overline{285714}$$

Practice Problem Set 1-4 Converting Decimals into Fractions and Fractions into Decimals

Convert the following decimals into decimal fractions:

1. 0.50

2. 0.88

3. 0.125

4. 0.3456

5. 0.75896

Convert the following fractions into decimal form:

6. $\frac{3}{8}$

7. $\frac{25}{75}$

8. $\frac{75}{100}$

9. $\dfrac{13}{16}$

10. $\dfrac{9}{12}$

Exponents, Scientific Notation, and Square Root

Exponents

The process of exponentiation is a multiplication process. The exponent (or power) in a mathematical expression indicates the number of times a number or mathematical expression is multiplied by itself. For example:

$$2^3 = 2 \times 2 \times 2 = 8$$

> **◄ Definition ►** *Exponent*
>
> An **exponent** is a number in a mathematical expression that is found to the upper right of a number or expression that indicates how many times the number or expression is multiplied by itself. An exponent is sometimes called a power.
>
> $$n^p$$
>
> where n equals a number or mathematical expression and p is the exponent. ►

Exponents can be either negative or positive integers. The definition of a negative exponent and Table 1-4 below show the impact of both negative and positive exponents on a base.

TABLE 1-4 • THE EXPONENTIATION OF BASE 2

$2^3 = 2 \times 2 \times 2 = 8$
$2^2 = 2 \times 2 = 4$
$2^1 = 2$
$2^0 = 1$
$2^{-1} = \frac{1}{2} = 0.05$
$2^{-2} = \frac{1}{2} \times \frac{1}{2} = \frac{1}{4} = 0.25$
$2^{-3} = \frac{1}{2} \times \frac{1}{2} \times \frac{1}{2} = \frac{1}{8} = 0.125$

Exponents, Scientific Notation, and Square Root

> **◆ Definition ◆ Negative Exponent**
>
> A *negative exponent* means that the reciprocal of a number or expression is multiplied by itself *p* number of times.
>
> $$n^{-p} = \frac{1}{n^p}$$

The primary use of exponents in horticulture is when units of square measure and units of cubic measure are expressed. Units of square measure represent area, and units of cubic measure represent volume.

For example:

144 in.² are equivalent to 1 square foot (ft²), when 12 in. = 1 ft

$$12^2 = 12 \times 12 = 144 \text{ in.}^2/\text{ft}^2$$

27 cubic feet (ft³) are equivalent to 1 cubic yard (yd³), when 3 ft = 1 yd

$$3^3 = 3 \times 3 \times 3 = 27 \text{ ft}^3/\text{yd}^3$$

Scientific Notation

Occasionally horticulturists need to work with either very small or very large numbers, which can be cumbersome to write out. Scientific notation is mathematical shorthand for writing these types of numbers in a more compact, or manageable, form.

> **◆ Definition ◆ Scientific Notation**
>
> *Scientific notation* is a method used to express a number in a form in which:
>
> $$n = a \times 10^p$$
>
> where *a* is the coefficient, $a \geq 1$, $a < 10$; and p is the power or the exponent.

The exponent or the power (*p*) indicates how many times the base (10 is the base in scientific notation) is multiplied by itself. The operation of raising the base to a specific power is called exponentiation. The exponent can be a positive number or a negative number. When an exponent is positive, the number is large. When the exponent is negative, the number is small. Table 1-5 demonstrates this principle for base 10.

Here are some examples of how to use scientific notation to express very large or very small numbers.

324,000,000,000,000 can be expressed as: 3.24×10^{14}

3[24,000,000,000,000]

TABLE 1-5 • THE EXPONENTIATION OF BASE 10

$10^4 = 10 \times 10 \times 10 \times 10 = 10{,}000 = 1.0 \times 10^4$
$10^3 = 10 \times 10 \times 10 = 1{,}000$
$10^2 = 10 \times 10 = 100$
$10^1 = 10$
$10^0 = 1$
$10^{-1} = 0.1$
$10^{-2} = 0.1 \times 0.1 = 0.01$
$10^{-3} = 0.1 \times 0.1 \times 0.1 = 0.001$
$10^{-4} = 0.1 \times 0.1 \times 0.1 \times 0.1 = 0.0001$

The bracketed digits in the number 3[24, 000, 000, 000, 000] indicate the number of spaces the decimal point has been moved. In this case, the decimal point has moved 14 spaces to the left and the coefficient (3.24) is multiplied by 10^{14}.

0.0000000000324 can be expressed as: 3.24×10^{-11}

0[00000000003]24

The bracketed digits in the number above indicate the number of spaces the decimal point has been moved. In this case, the decimal point has moved 11 spaces to the right and the coefficient (3.24) is multiplied by 10^{-11}.

324,125,600 can be expressed as: 3.241256×10^8

or

$$3.24 \times 10^8$$

The decimal point has moved eight spaces to the left, and the coefficient is multiplied by 10^8. In the alternate expression of the number, the coefficient has been reduced to two decimal places by rounding.

0.0003241256 can be expressed as: 3.241256×10^{-4}

or

$$3.24 \times 10^{-4}$$

The decimal point has moved four spaces to the right, and the coefficient is multiplied by 10^{-4}. In the alternate expression of the number, the coefficient has been reduced to two decimal places by rounding.

Square Root

> ◄ Definition ► *Square Root*
> The *square root* of x^2 is x.
> $$\sqrt{x^2} = x$$

Exponents, Scientific Notation, and Square Root 15

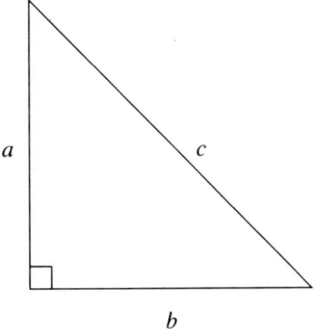

Figure 1-2

Square root examples:

$$\sqrt{25} = 5$$
$$\sqrt{144} = 12$$
$$\sqrt{49} = 7$$

Determining the square root of a number comes up occasionally in horticultural operations. One common use of square root is to find out the length of the long side, or hypotenuse, of a right triangle (Figure 1-2) using the Pythagorean Theorem.

EXAMPLE 1-14

Pythagorean Theorem: $a^2 + b^2 = c^2$

If $a = 9$ ft and $b = 7$ ft, then how long is c?

$$a^2 + b^2 = c^2$$
$$9^2 + 7^2 = c^2$$
$$81 + 49 = 130$$
$$c^2 = 130$$
$$c = \sqrt{130}$$
$$c = 11.4 \text{ ft}$$

✎ Practice Problem Set 1-5 ✎ Scientific Notation, Exponential Notation, and Square Root

Express the following numbers using scientific notation:

1. 56,200,000
2. 0.0000302
3. 775,000
4. 310,000,000,000
5. 0.000000000922

Provide the numerical equivalent to each of the following numbers expressed as exponents:

6. 12^2
7. 4^4
8. 3^{-1}
9. 7^0
10. 4^{-2}

Complete the following problems involving square roots:

11. $\sqrt{625}$
12. $\sqrt{2,500}$
13. $\sqrt{562}$
14. What is the length of the hypotenuse of a right triangle when one side is 15 ft and the other side is 20 ft?
15. What is the length of the hypotenuse of a right triangle when one side is 9 ft and the other side is 12 ft?

Order of Mathematical Operations

More complex mathematical equations need some special attention to the order in which the mathematical operations are performed. The acronym **PEMDAS** or the mnemonic device ***Please Excuse My Dear Aunt Sally*** are employed to help remember the proper order of mathematical operations. If the mathematical operations are not performed in the proper order, an incorrect answer is the result.

> **◆ Definition ◆** *Proper Order of Mathematical Operations or PEMDAS*
> 1. **P**—**P**arentheses
> 2. **E**—**E**xponents (powers, square roots)
> 3. **MD**—**M**ultiplication and **D**ivision (performed left to right)
> 4. **AS**—**A**ddition and **S**ubtraction (performed left to right)

P—Parentheses

Perform operations within the parentheses first.

$$3 \times (4 - 2) = 3 \times 2 = 6 \quad \text{CORRECT}$$
$$3 \times (4 - 2) = 12 - 2 = 10 \quad \text{INCORRECT}$$

E—Exponents (Powers, Square Roots)

Perform operations involving exponents before multiplying, dividing, adding, or subtracting.

$$3 \times 4^2 = 3 \times 16 = 48 \quad \text{CORRECT}$$
$$3 \times 4^2 = 12^2 = 144 \quad \text{INCORRECT}$$

MD—Multiplication and Division

Perform multiplication and division prior to addition or subtraction. Always work from left to right.

$$3 + 4 \times 2 = 3 + 8 = 11 \quad \text{CORRECT}$$
$$3 + 4 \times 2 = 7 \times 2 = 14 \quad \text{INCORRECT}$$

Perform multiplication and division in order from left to right.

$$10 \div 2 \times 5 = 5 \times 5 = 25 \quad \text{CORRECT}$$
$$10 \div 2 \times 5 = 10 \div 10 = 1 \quad \text{INCORRECT}$$

AS—Addition and Subtraction

Perform addition and subtraction in order from left to right.

$$10 - 3 + 2 = 7 + 2 = 9 \quad \text{CORRECT}$$
$$10 - 3 + 2 = 10 - 5 = 5 \quad \text{INCORRECT}$$

EXAMPLE 1-15

Perform mathematic operations using PEMDAS in a more complex equation.

$$2 \times (5 + 3)^2 - (2 \times 3) \div 2$$
$$= 2 \times 8^2 - 6 \div 2$$
$$= 2 \times 64 - 6 \div 2$$
$$= 128 - 3 = 125$$

Practice Problem Set 1-6 — *Order of Mathematical Operations*

Complete the following problems using PEMDAS:

1. $\dfrac{5}{3}[(3 + 4) + 2(8 + 3 + 5) + 4(5 + 6 + 5 + 4)]$
2. $\sqrt{6(6 - 4)(6 - 3)(6 - 5)}$
3. $32(6^2 + 5^3)$

4. $7(3 + 4^2) + 8(25 + 7^3)$
5. $2(21 \div 3) \div (3 + 4)$

Solving for x

Many mathematical problems in horticulture involve solving for an unknown value. The unknown value in a mathematical equation is often identified by the letter x. Equations are mathematical statements of equality. An equation always has an equal sign, and the value of what is on the left side of the equal sign is the same as the value of what is on the right side of the equal sign.

> **◄ Definition ►** *Equation*
> An *equation* is a mathematical statement of equality. For example:
> $$x = a + b$$

Solving a mathematical problem when there is an unknown (x) begins with the isolation of x. This means the equation needs to be manipulated so that x is by itself on one side of the equal sign and the other elements of the equation are on the opposite side of the equal sign. When manipulating equations, keep in mind that whatever changes are made on one side of the equal sign must be made on the opposite side to maintain equality.

Solving for x in Equations with Addition and Subtraction

Solving for x in equations with addition and subtraction involves simple technique for the isolation of x.

EXAMPLE 1-16

$$x + 4 = 12$$
$$x + 4(-4) = 12 - 4$$
$$x = 8$$

EXAMPLE 1-17

$$x - 5 = 10$$
$$x - 5(+5) = 10 + 5$$
$$x = 15$$

Solving for x in Equations with Multiplication and Division

Solving for *x* in equations with multiplication and division is a little more complex.

EXAMPLE 1-18

$$5x = 20$$
$$\frac{5x}{5} = \frac{20}{5}$$
$$x = 4$$

EXAMPLE 1-19

$$x \div 4 = 2$$
$$\frac{x}{4} = 2$$
$$\frac{x}{4} \times 4 = 2 \times 4$$
$$x = 8$$

Solving for x in Equations with Combined Operations

When solving for *x* in equations with combined operations, it is best to complete the operations involving addition and subtraction first and then perform the operations involving multiplication and division last.

EXAMPLE 1-20

$$3x + 6 = 36$$
$$3x + 6(-6) = 36 - 6$$
$$3x = 30$$
$$\frac{3x}{3} = \frac{30}{3}$$
$$x = 10$$

EXAMPLE 1-21

$$\frac{x}{5} + 3 = 5$$
$$\frac{x}{5} + 3(-3) = 5 - 3$$

$$\frac{x}{5} = 2$$
$$\frac{x}{5} \times 5 = 2 \times 5$$
$$x = 10$$

✐ Practice Problem Set 1-7 ✐ *Solving for x*

Isolate and solve for x.

1. $x + 25 = 100$
2. $\frac{x}{7} = 49$
3. $x^2 + 16 = 25$
4. $2x + (7 - x) = 10$
5. $x(4 + 20) = 384$

---

*R*atios and Proportions

A ratio is a comparison of the relative size of two quantities. It is commonly written as a fraction or as two numbers separated by a colon (:). For example:

$$\frac{1}{2} \text{ or } 1:2$$

These ratios are read as a ratio of 1 to 2.

> **◄ Definition ► Ratio**
> A *ratio* is the relative size of two quantities expressed as the quotient of one divided by the other: $\frac{a}{b}$ or $a:b$ ✣

A proportion is a mathematical statement that two ratios are equal. A proportion can be expressed in the following ways:

$$\frac{1}{2} = \frac{4}{8} \text{ or } 1:2 = 4:8$$

These proportions are read as: one is to two as four is to eight.

In a proportion, the product of the *means* is equal to the product of the *extremes*. For example, review the following:

$$1:2 = 4:8$$

The word *means* refers to the two *inner* numbers in the proportion (2 and 4), and the word *extremes* refers to the two *outer* numbers in the proportion (1 and 8). The word *product* refers to the mathematical operation of multiplication. Translating this statement into a mathematical equation results in the following:

$$2 \times 4 = 1 \times 8$$

or

$$8 = 8$$

> **◆ Definition ◆ *Proportion***
> A *proportion* is a relation of equality between two ratios. Four quantities, a, b, c, d are said to be in proportion if:
> $$\frac{a}{b} = \frac{c}{d} \text{ or } a:b = c:d$$

Demonstrating the product of the means as equal to the product of the extremes also can be accomplished through cross multiplication of the ratios expressed as fractions.

$$\frac{1}{2} \times \frac{4}{8}$$

$$2 \times 4 = 1 \times 8$$

$$8 = 8$$

The expression of ratios as fractions and the use of cross multiplication is the most useful way to express this mathematical relationship when using a proportion to solve a problem. What if one of the numbers in a proportion is unknown?

$$\frac{1}{2} = \frac{x}{8}$$

By applying the technique of cross multiplication, the equation then can be solved for x.

$$2 \times x = 1 \times 8$$

$$2x = 8$$

$$\frac{2x}{2} = \frac{8}{2}$$

$$x = 4$$

> **◆ Definition ◆ *Cross Multiplication***
> When using cross multiplication in a proportion, the product of the means equals the product of the extremes.
> $$\text{If } \frac{a}{b} = \frac{c}{d}, \text{ then } ad = bc$$

Using Ratios and Proportions to Make Unit Conversions

Ratios and proportions are useful in converting a value from one unit of measure to another. The process of conversion requires the use of known equivalents. Tables of equivalents can be found in reference books, dictionaries, and in the appendices of many technical books on horticulture, including this one.

Equivalents provide us with a known ratio. For example, it is known that 1 pound (lb) is equivalent to 453.59243 grams (g). This can be written as a ratio.

$$\frac{453.59243 \text{ g}}{1 \text{ lb}}$$

EXAMPLE 1-22

How many grams are equivalent to 3.75 lb? Using the ratio representing the equivalent and the value needed to convert to grams, a proportion can be set up to answer this question.

$$\frac{453.59243 \text{ g}}{1 \text{ lb}} = \frac{x \text{ g}}{3.75 \text{ lb}}$$

The proportion can be read as follows: If 453.59243 grams is equivalent to 1 lb, then how many grams are equivalent to 3.75 lb? Using the process of cross multiplication, solve for x.

$$\frac{453.59243 \text{ g}}{1 \text{ lb}} = \frac{x \text{ g}}{3.75 \text{ lb}}$$

$$1 \text{ lb} \times x \text{ g} = 453.59243 \text{ g} \times 3.75 \text{ lb}$$

$$\frac{1 \text{ lb} \times x \text{ g}}{1 \text{ lb}} = \frac{453.56243 \text{ g} \times 3.75 \text{ lb}}{1 \text{ lb}}$$

$$x \text{ g} = 1{,}700.9716 \text{ g}$$

This example is a simplified one. Many horticulturists can visualize that simply multiplying 453.59243 by 3.75 would provide the answer to the problem. Taking the problem back to a proportion can be a useful check. Notice that extra care was taken in the example to demonstrate the cancellation of units.

Take a moment to observe how the units are arranged in the ratio above. They are arranged in this way:

$$\frac{\text{g}}{\text{lb}} = \frac{\text{g}}{\text{lb}}$$

Consistency is important in arranging the units of a proportion; otherwise, the result will be an incorrect answer. Two methods are acceptable and they are the following:

$$\frac{x}{y} = \frac{x}{y} \quad \text{or} \quad \frac{x}{x} = \frac{y}{y}$$

Using Ratios and Proportions to Solve Problems

In addition to routine unit conversions, ratios and proportions can be used to solve common mathematical problems encountered in a horticultural operation.

EXAMPLE 1-23

A horticulturist needs to prepare a gasoline and oil mixture for a gas-powered string trimmer that is equipped with a 2-cycle engine. The recommended ratio is 32 parts gasoline to 1 part oil or a 32:1 ratio of gasoline to oil. How much oil will be needed to add to 4.50 gal of gasoline?

Step 1

Set up a proportion to represent the problem.

$$\frac{1 \text{ part oil}}{32 \text{ parts gas}} = \frac{x \text{ gallons oil}}{4.50 \text{ gallons gas}}$$

Notice that the units are arranged properly.

$$\frac{\text{parts}}{\text{parts}} = \frac{\text{gallons}}{\text{gallons}}$$

Step 2

Isolate and solve for x.

$$1 \times 4.5 = 32 \times x$$
$$4.5 = 32x$$
$$\frac{4.5}{32} = \frac{32x}{32}$$
$$x = \frac{4.50}{32}$$
$$x = 0.140625 \text{ gal}$$

Step 3

Convert gallons (gal) to fluid ounces (fl oz) by setting up a proportion and isolating and solving for x. Note that 1 gallon is equivalent to 128 fluid ounces.

$$\frac{1 \text{ gal}}{128 \text{ fl oz}} = \frac{0.140625 \text{ gal}}{x \text{ fl oz}}$$
$$x = 128 \times 0.140625$$
$$x = 18 \text{ fl oz}$$

SOLUTION

Eighteen fluid ounces of oil are needed to add to 4.5 gallons of gas.

EXAMPLE 1-24

A horticulturist has a 2 gal gas can that needs to be filled with a 32:1 ratio of gas to oil. The goal is to put the appropriate amount of oil in the can first and then fill up the can with gas to the 2 gal mark. How much oil is in 2 gal of a 32:1 mix of gas:oil?

Step 1

Set up the proportion.

It is important to note that in a 32:1 gas:oil mixture there are 33 parts of the mixture (gas and oil) to every 1 part of oil. The ratio is between the component oil to the mix of oil and gas. If 1 fluid ounce of oil is found in every 33 fl oz of gas and oil mix, how many ounces of oil are found in 2 gal (or 256 fl oz) of the gas and oil mixture?

$$\frac{1 \text{ fl oz oil}}{33 \text{ fl oz mix}} = \frac{x \text{ fl oz oil}}{256 \text{ fl oz mix}}$$

Step 2

Isolate and solve for x.

$$1 \times 256 = 33 \times x$$
$$256 = 33x$$
$$\frac{256}{33} = \frac{33x}{33}$$
$$x = \frac{256}{33}$$
$$x = 7.76 \text{ fl oz of oil}$$

SOLUTION

Place 7.76 fl oz of oil in a 2 gal gas can and then fill the can to 2 gal to achieve a 32:1 gas:oil mixture.

EXAMPLE 1-25

A greenhouse manager would like to prepare 35 ft³ of a root medium mix with a ratio of 3 parts peat moss to 2 parts perlite to 2 parts bark. If a total of 35 ft³ of root medium is desired, how many cubic feet of peat moss are needed to prepare this potting mix?

Step 1

Determine the ratio of peat moss to the root medium mix as a whole.

The ratio representing the amount of peat moss in the root medium mix is:

$$\frac{3 \text{ parts peat}}{7 \text{ parts root medium mix}}$$

The numerator (top number) represents the portion of the mix that is peat (3 parts). The denominator (bottom number) is the mix as a whole unit. The mix as a unit has 7 parts or 3+2+2=7. The mix and its components are represented generically in parts but, in this case, the components are measured in cubic feet.

Step 2

Set up the proportion.

$$\frac{3 \text{ parts peat}}{7 \text{ parts root medium mix}} = \frac{x \text{ cubic feet peat}}{35 \text{ cubic feet root medium mix}}$$

Step 3

Isolate and solve for x.

$$3 \times 35 = 7 \times x$$
$$\frac{3 \times 35}{7} = \frac{7x}{7}$$
$$x = 15$$

SOLUTION

Fifteen cubic feet of peat moss are needed to prepare 35 ft³ of a 3 parts peat moss: 2 parts perlite: 2 parts bark-root medium mix.

Other Common Ratios

Rate

Rate is a ratio of a variable quantity that occurs within the limits of a fixed quantity. Rate or ratios that describe rate are used in many horticulture operations. Using rate in a fraction form and then using it in a proportion allows the conversion of a standard rate to a rate that is useful to a horticulturist in specific situations. Examples of the use of rate in horticulture follow.

Speed or *rate of speed* is used when calibrating boom sprayers for pesticide applications or calibrating traveling boom-watering systems found in the greenhouse.

Rate of speed is described as follows:

$$\text{Rate of Speed} = \frac{\text{Distance}}{\text{Time}}$$

By manipulating the definition of rate ($r = \frac{d}{t}$), solutions can be found for distance ($d = r \times t$) and for time ($t = \frac{d}{r}$).

EXAMPLE 1-26

If a boom sprayer is to be operated at the speed of $\frac{3 \text{ ft}}{\text{second}}$, how far will the sprayer travel in 30 seconds?

$$\text{Rate of Speed} = \frac{\text{Distance}}{\text{Time}}$$

$$\frac{3 \text{ ft}}{\text{second}} = \frac{\text{Distance}}{30 \text{ seconds}}$$

$$\frac{3 \text{ ft}}{\text{second}} \times 30 \text{ seconds} = \frac{\text{Distance}}{30 \text{ seconds}} \times 30 \text{ seconds}$$

$$\text{Distance} = 90 \text{ ft}$$

Rate or *application rate* is used to describe fertilizer, pesticide, and growth regulator applications in horticulture. These rates may be expressed as pounds, ounces, fluid ounces of product, or active ingredient to be applied over a given area (square feet or acres). Occasionally, rate may be expressed as a volume of product or grams of active ingredient applied per pot in a greenhouse.

Examples of application rate:

$$\text{Fertilizer Product Application Rate} : \frac{x \text{ lb 5-10-5}}{100 \text{ ft}^2}$$

$$\text{Fertilizer Element Application Rate} : \frac{x \text{ lb N}}{1,000 \text{ ft}^2}$$

$$\text{Pesticide Active Ingredient Application Rate} : \frac{x \text{ grams active ingredient}}{\text{acre}}$$

$$\text{Pesticide Product Application Rate} : \frac{x \text{ fluid ounces of product}}{1,000 \text{ ft}^2}$$

EXAMPLE 1-27

A 5-10-5 fertilizer product is to be applied at the rate of 2 lb/100 ft^2 of garden space. A horticulturist needs to make this application to a perennial flower border that measures 250 ft^2. How many pounds of 5-10-5 are required for this application?

Step 1

Set up a proportion.

$$\frac{2 \text{ lb 5-10-5}}{100 \text{ ft}^2} = \frac{x \text{ lb 5-10-5}}{250 \text{ ft}^2}$$

Step 2

Solve for x.

$$2 \times 250 = 100 \times x$$

$$\frac{2 \times 250}{100} = \frac{100 \times x}{100}$$

$$x = \frac{2 \times 250}{100}$$

$$x = 5$$

SOLUTION

Five pounds of 5-10-5 are required to fertilize a 250-ft² perennial flower border.

Percent

Percent is a ratio expressed as a fraction with 100 as the denominator. It also may be described as a part as related to a whole. Numerous uses of percent are found in horticulture. A few are presented.

◀ Definition ▶ *Percent*

Percent comes from Latin *per centum* which translates to: per hundred or of each hundred. Percent is expressed as a fraction or a ratio with 100 as the denominator, as a ratio of a part to a whole, as a decimal, and as a whole number with the percent sign (%).

$$\text{Percent} = \frac{x}{100} = \frac{\text{part}}{\text{whole}} = \text{decimal} = \text{whole number }\%$$

For example:

$$\frac{50}{100} = \frac{15}{30} = 0.50 = 50\%$$

To convert a fraction or a ratio to a decimal, divide the numerator by the denominator.

For example:

$$\frac{15}{30} = 15 \div 30 = 0.50$$

To convert a decimal to a whole number with a percent sign, multiply the decimal times 100 and add the percent sign.

For example:

$$0.50 \times 100 = 50\%$$

A percent expressed as a fraction with 100 as the denominator is a common way to describe fertilizer analysis. Fertilizer analysis is the percent of N (nitrogen), P_2O_5 (phosphoric acid), and K_2O (potash) in a fertilizer product and can be found prominently displayed on a fertilizer product label. An example of an analysis is 5-10-5. This product has 5% N, 10% P_2O_5, and 5% K_2O in it. Every 100 lb of 5-10-5 has 5 lb of nitrogen in it.

EXAMPLE 1-28

If 1 lb of nitrogen is to be applied to 1,000 ft² of garden area, how much 5-10-5 is required to deliver that 1 lb of nitrogen?

Step 1

Set up a ratio to describe the amount of nitrogen found in 5-10-5.

$$\frac{5 \text{ lb N}}{100 \text{ lb of 5-10-5}}$$

There are 5 lb of nitrogen in every 100 lb of 5-10-5 product.

Step 2

Set up a proportion to determine how much 5-10-5 is needed to deliver 1 lb nitrogen.

$$\frac{5 \text{ lb N}}{100 \text{ lb of 5-10-5}} = \frac{1 \text{ lb N}}{x \text{ lb of 5-10-5}}$$

Step 3

Solve for x.

$$5 \times x = 100 \times 1$$
$$\frac{5 \times x}{5} = \frac{100}{5}$$
$$x = \frac{100}{5}$$
$$x = 20$$

SOLUTION

Twenty pounds of 5-10-5 is needed to deliver 1 lb N to 1,000 ft² or $\frac{20 \text{ lb 5-10-5}}{1,000 \text{ ft}^2}$ will deliver $\frac{1 \text{ lb N}}{1,000 \text{ ft}^2}$

EXAMPLE 1-29

A greenhouse manager begins a production cycle with 200 pots of mums. At the end of the production cycle, only 180 pots were suitable for sale. What percentage of loss (shrinkage) did the greenhouse manager experience with this crop?

Step 1

What portion of the crop was lost?

The grower started with 200 pots and has 180 pots remaining at the time of sale. What is the difference?

$$200 \text{ pots} - 180 \text{ pots} = 20 \text{ pots were lost}$$

Step 2

Twenty pots represent what percentage of shrinkage (loss) for the crop? Set up a ratio.

$$\frac{20 \text{ pots lost}}{200 \text{ pots total}}$$

Step 3

Calculate the percentage lost.

$$\frac{20 \text{ pots lost}}{200 \text{ pots total}} = 20 \div 200 = 0.1 \text{ percentage shrinkage}$$

Another way to look at this calculation is:

$$\frac{20}{200} = \frac{10}{100} = 10\% = 0.1$$

Notice that a percentage can be expressed as a decimal (e.g., 0.1) or expressed with a percent sign (%) by multiplying the decimal times 100.

$$0.1 \times 100 = 10\%$$

Practice Problem Set 1-8 *Ratios and Proportions*

1. Convert 1.75 lb to grams. (1 lb = 453.59243 g)
2. How many fluid ounces of oil need to be added to 1.5 gal of gas if the desired ratio of gas to oil is 32:1? (Hint: 1 gal = 128 fl oz)
3. If a nursery manager would like to prepare 2 yd^3 of potting mix with a ratio of 2 parts peat moss to 2 parts aged bark to 1 part sand, how many cubic feet of each component are needed to prepare the mix? (Hint: 1 yd^3 = 27 ft^3)
4. Rate of speed equals the ratio of distance to time $\left(\text{Rate of Speed} = \dfrac{\text{Distance}}{\text{Time}}\right)$. If a tractor travels at a rate of speed of 3 miles per hour, then how long will it take (in seconds) for the tractor to travel 100 ft? (Hint: 5,280 ft = 1 mile and 1 hour = 3,600 seconds)
5. If the recommended rate of fertilizer is $\dfrac{1.5 \text{ lb fertilizer}}{100 \text{ ft}^2 \text{ of garden}}$, then how many pounds of fertilizer are required for 750 ft^2 of garden space?
6. If 750 pots of a 25,000-pot chrysanthemum crop are not suitable for sale, what is the percentage of loss for this production cycle?

Chapter 2

Measurement and Calculations with Measured Values

Introduction

This chapter focuses on measurement and the use of measured values in calculations. A review of the concepts of significant digits and the rules of rounding are included with application to measured and calculated values found in horticultural operations. Units of measure from U.S. Customary and Metric systems are compared and techniques for effortless conversion between and within the systems are presented.

Measurement, Accuracy, and Precision

Measurement

To measure is to find the size or amount of an object or substance. The numerical value reflecting a measurement is reported in units of measure. For example, when something is weighed, the weight is reported in units such as ounces, pounds, grams, or kilograms. The value also is reported in reference to a standard. When reporting the weight (mass) of an object in grams, it is reported relative to a gram standard.

Measurements and calculations that involve measurements are merely estimates of the actual or true value. An estimate can be farther from or closer to the actual or true value, depending on the limits of the device used for the measurement. The measuring device, in this case a gram scale, is calibrated using standards so that the person performing the measurement is sure that the measurement is as close to the actual weight as possible—given the limits of the device. The person using the measuring device and their skill level also can impact how close a measurement is to the true value.

Defining Accuracy and Precision

Accuracy is how close a measured value comes to the actual or true value. Accuracy can be determined by one measurement. If a standard 10-gram weight is placed

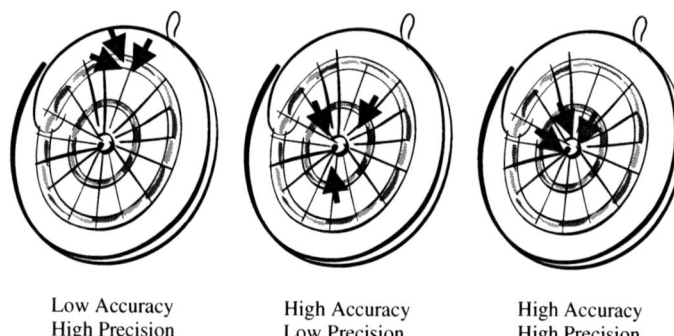

Figure 2-1
The dartboard analogy.

Low Accuracy High Accuracy High Accuracy
High Precision Low Precision High Precision

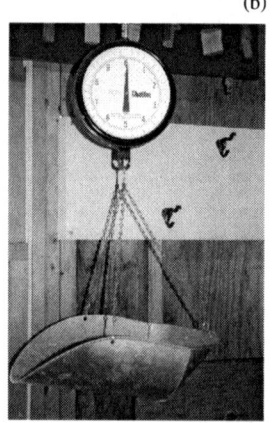

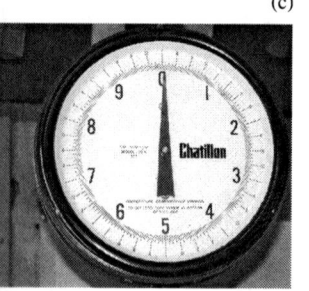

Figure 2-2
Scaled instruments: a ruler (a), a produce scale (b), and a close-up of the produce scale dial (c).

on a gram scale and measured 10.00 g, that gram scale is considered accurate. A measurement of 7.54 g is not considered accurate.

Precision is how close multiple measured values are to one another. Precision is defined via multiple measurements. If after five repetitive measures, the gram scale measured the weight of the 10-gram standard as: 9.99 g, 9.98 g, 10.00 g, 9.99 g, and 10.00 g, this gram scale is considered to be precise and accurate. It is important to note that if after five repetitive measures of the 10-gram standard, the weight measurements are: 7.55 g, 7.54 g, 7.53 g, 7.55 g, and 7.54 g, this gram scale is also precise. The accuracy of the scale in this situation would be low. The dartboard analogy illustrated in Figure 2-1 demonstrates the concepts of accuracy and precision.

Impact of the Measuring Instrument on Accuracy

The degree of accuracy of a measuring instrument can be demonstrated by the comparison of a scale that measures in grams to two decimal places, or to one hundredth of a gram, with an analytical balance that weighs to four decimal places, or to one ten-thousandth of a gram. The analytical balance offers a greater degree of accuracy. This would be similar to comparing a ruler with tic lines indicating one-fourth of an inch increments to a ruler with tic lines indicating one-sixteenth of an inch increments. The latter is more accurate.

Digital measuring devices often have published degrees of uncertainty. For example, a 200 g digital balance typically has an uncertainty of ±0.01 g. When reporting measurements from digital devices, report all the numbers displayed, including terminal zeros. All of the digits, including terminal zeros, indicate the level of accuracy of the device.

For scaled instruments or measuring devices (for example: scales, rulers, meter sticks), report one more digit than can be read on the scale (Figure 2-2). This involves reading the space between the smallest tics on the scale. For example, if an object measures exactly 11 mm long, report the result as 11.0 mm. If an object measures between 11 and 12 mm long, then report the result as 11.5 mm. Similarly, the length of an object measured with a ruler that is marked in U.S. Customary units may lie between the three-fourth-inch tic mark and the thirteen-sixteenth-inch tic mark on the scale (Figure 2-3). This is equivalent to measuring between $\frac{24}{32}$ and $\frac{26}{32}$ and would be reported as $\frac{25}{32}$.

Factors that Impact the Need for Accuracy and Precision in the Green Industry

Figure 2-3
Reading a measurement that lies between the tic marks 3/4 and 13/16.

It would be an unusual sight to find an analytical balance in a horticulture operation. Horticulturists simply do not need to measure quantities of materials that are that small. Rather, a simple scale that measures pounds and ounces is often found in a horticultural operation. Today, horticulturists find that a good quality 200 g capacity digital balance that measures to a degree of accuracy of ± 0.01 g is useful as well.

Pesticide and plant growth regulator active ingredients that have greater efficacy than their predecessors have resulted in the use of smaller quantities of product. This means that horticulturists have to use a greater degree of accuracy when measuring pesticide and plant growth regulator products. Greater accuracy and precision in the measurement and calculation of the area to receive a pesticide or plant growth regulator application is warranted as well.

In contrast, if a horticulturist is measuring an area for mulch application, accuracy may not be as important as the speed at which a professional can make the area measurements and calculation. A minor discrepancy in area calculation is not likely to affect a mulch application and the profitability of that application. However, a similar discrepancy in area calculation may affect an herbicide application and result in poor product performance (lack of weed control) or phytotoxicity of nontarget plants (damage to ornamental plants).

✐ Practice Problem Set 2-1   ✐ *Measurement, Accuracy, and Precision*

1. Which of the following represents the proper way to report a measurement of 25 g from a gram-scale with a degree of accuracy of ± 0.01?
   a. 25
   b. 25.0
   c. 25.00
   d. 25.000

2. Which of the following represents the proper way to report a measurement of 5 hundredths of a gram from an analytical balance with a degree of accuracy of ± 0.0001?
   a. 0.0500
   b. 0.5000
   c. 0.0005
   d. 0.0050

3. For each of the measurements indicated in Figure 2-4, report the result to the proper degree of accuracy.

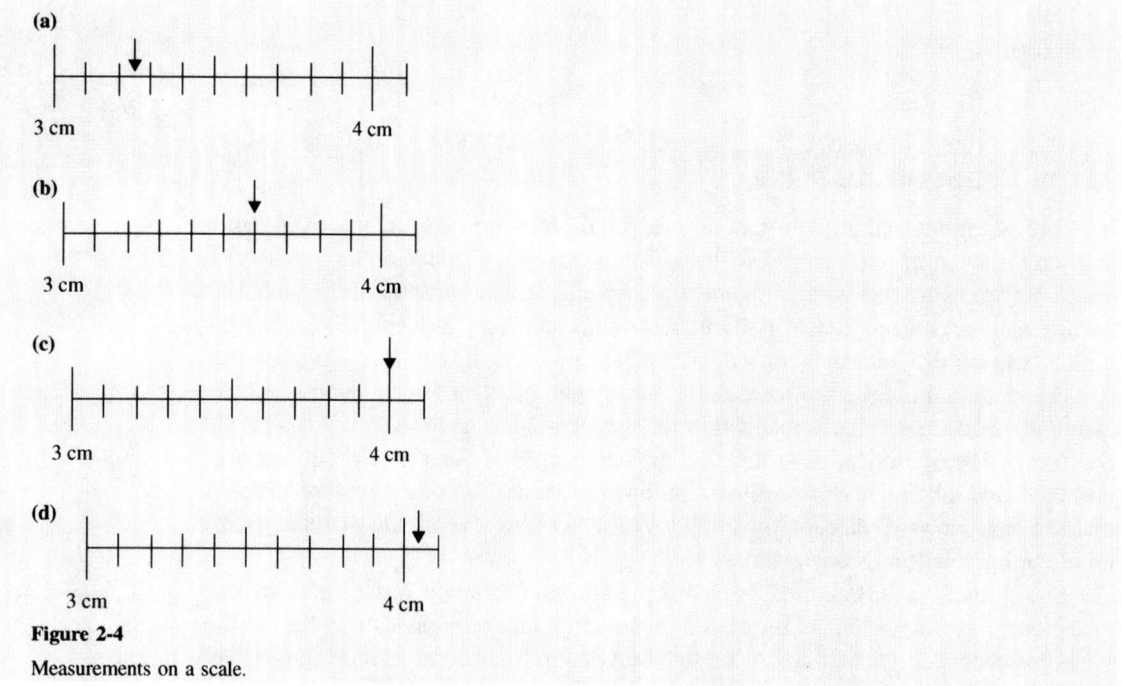

Figure 2-4
Measurements on a scale.

Significant Digits and Rules for Rounding Numbers

Application of the Concepts of Significant Digits and Rounding to Horticulture

What if a horticulturist determines through a calculation that exactly 45.67 marigolds are needed to complete a planting bed? The solution of 45.67 plants to fill a planting bed is ridiculous. One cannot plant 0.67 plants. It makes sense to round the number of plants to 46 plants or to the closest whole number. Is 46 a good estimate? Planting beds with lots of curves generally need more plants than calculated. A better estimate for a curvy bed design might be determined by rounding up to the nearest 10 or to an estimate of 50 plants. The use of good math and measurement skills is a great start, but bring along some common sense and horticulture knowledge when making plant number estimates.

What if it was determined that 1.87 pounds of fertilizer are required for a small planting bed? Can 1.87 pounds be measured with ease? Is there a need to measure fertilizer for a garden bed to this degree of accuracy? If the equipment used to measure quantities to that level of accuracy is available, use the more accurate equipment. Another option is to round the quantity up to two pounds or calculate how many ounces or grams are equivalent to 1.87 pounds. Then use the equipment appropriate for weighing those amounts of material. Most horticulturists agree that two pounds as an estimate of fertilizer application is just fine and to take extra time to measure to a greater degree of accuracy will not impact plant response. Combine your knowledge of plant response to fertilizer with your mathematical skills for the correct solution.

When is it critical to use the exact amount of product calculated? As stated previously, it becomes more critical when measuring pesticides and plant growth regulators. Even when measuring concentrates of these newer products by weight or by volume, a degree of accuracy of two significant digits is adequate. A good quality gram scale will measure in hundredths of grams and this would require rounding to two decimal places to the right of the decimal point.

Measuring small volumes of liquid materials might require the use of a 50 ml graduated cylinder or a 50 ml pipet. Each of these measuring devices would have a degree of accuracy of ± 0.2 ml and ± 0.02 ml, respectively. As a result, rounding to one decimal place to the right of the decimal point for the 50 ml graduated cylinder and two decimal places to the right of the decimal point for the 50 ml pipet is appropriate. In each case, rounding to the number of significant digits to the right of the decimal point that represent the degree of accuracy of the measuring device is the correct technique.

Measured numbers should include all the digits warranted by the measuring device. The weight (mass) of an object is measured on gram scale at 22.51 g. That means the object weighs between 22.50 g and 22.52 g. There are four significant digits in 22.51. If the same object was weighed on an analytical balance, it could measure at 22.5147 g. That means it weighs between 22.5146 g and 22.5148 g. There are 6 significant digits in 22.5147. Notice that in each of these numbers, the final digit is an estimate or is uncertain and all digits to the left of it are certain. Since the measurements from these devices are produced as numbers on an LCD panel, it does not require interpretation of a scale. The operator's manual indicates that the gram scale is reliable to ± 0.01 and the analytical balance is reliable to ± 0.0001. Use this information to determine the proper number of significant digits to report.

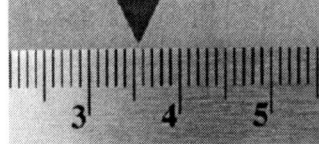

Figure 2-5
A measurement of 3.55 cm.

Measurement of length using a metric ruler is different than measuring with a device that has a digital numerical display. The increments in whole numbers are centimeters (cm). Each centimeter is divided up into tenths of centimeters (millimeters, or mm). The human eye can interpret a measurement between the millimeter tics on the ruler. If a measurement reads between the 3.5 and 3.6 tic marks on the ruler and you estimate that it is half way between the two, then one would report the measurement as 3.55 (Figure 2-5). The last digit is an estimate but is reliable enough to report as a significant digit. If a measurement reads exactly 3.5, then it would be reported as 3.50 (Figure 2-6). The last digit is still an estimate and is needed to indicate the degree of accuracy of the instrument.

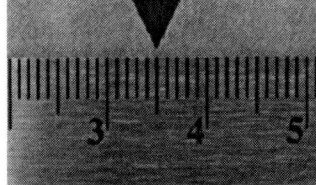

Figure 2-6
A measurement of 3.5 cm should be reported as 3.50 cm to indicate the degree of accuracy of the instrument.

Whenever a number is expressed, make sure that all of the digits are meaningful and with the understanding that the last significant digit is an estimate. This is important especially when expressing the results of a computation. The result of a mathematical operation is only as reliable as the least accurate number used in the computation. Perform the mathematical operation with all the digits that are significant for each measurement or number, and then round to the appropriate number of significant digits found to the right of the decimal point (decimal places). Use all of the digits found in mathematical constants and mathematical equivalents because they are exact numbers. If the mathematical operation has intermediate then final results, round the intermediate result to one additional decimal place after the first computations; perform the final computations; and report the final result in the appropriate number of decimal places. It also is appropriate to keep all the digits through the intermediate calculations and round only when the final calculation is complete. Remember rounding to the appropriate number of decimal places is based on the least accurate measurement in the computation.

Significant Digits: Definition and Rules

> **◄ Definition ►** *Significant digits*
>
> ***Significant digits*** are the digits of the decimal form of a number beginning with the leftmost digit, as long as it is not zero, and continuing to the right to include all digits as determined by the accuracy of the measuring device. Significant digits are also called significant figures.

Exact numbers are numbers for which all digits are significant. Counting numbers, mathematical constants, and equivalents are exact numbers. When using exact numbers in computations, always use all the digits provided by your source. For the mathematical constant pi (π), it is best to use the pi function on your calculator.

> **◄ Definition ►** *Exact numbers, counting numbers, mathematical constants, and equivalents*
>
> ***Exact numbers*** are numbers in which all digits are significant. Exact numbers include the following: ***counting numbers***, ***mathematical constants***, and ***equivalents***.
>
> ***Counting numbers***, also known as natural numbers, are the positive integers 1, 2, 3, 4, ... and so on and are the numbers used to count things. Decimal numbers, negative integers, and fractions are not counting numbers.
>
> ***Mathematical constants*** are mathematical quantities with fixed values. The number pi (π), which describes the ratio of the circumference of a circle to the diameter, is a commonly used mathematical constant.
>
> ***Equivalents*** are numbers with values that describe a one-to-one relationship. For example, 453.59237 grams are equivalent to 1.0 pound.

Significant Digit Rules

Rule 1: Digits one through nine (non-zero digits) are significant

When measuring an object, the use of a measuring device will allow you to obtain a number. All the digits you can obtain from the measuring device are significant. For example, the number 333 has three significant digits, and the number 33.33 has four significant digits.

Rule 2: Zero is significant when it is found between two non-zero digits

For example, the number 608 has three significant digits. The number 6,008 has four significant digits.

Rule 3: Zero is significant if it is the terminal zero (or zeros) to the right of the decimal point and if it is also to the right of a non-zero digit

For example, the number 7.040 has four significant digits. The number 0.7040 has four significant digits. The number 704.0 has four significant digits. The number 7,040 has three significant digits. The number 0.70400 has five significant digits.

Rule 4: Zeros to the left of the decimal points are not significant

For example, the number 0.67 has two significant digits. Zeros placed to the left of the decimal point are placed there out of convention. It prevents confusion between a period in punctuation versus a decimal point in mathematical notation.

Rule 5: A zero used to fix a decimal point or used as a place holder for numbers less than one is not significant

For example, the number 0.003040 has four significant digits. The number 0.000563245 has six significant digits. Writing these numbers in scientific notation can help clarify the number of significant digits. The number 0.003040 may be written as 3.040×10^{-3}. The number 0.000563245 may be written as 56.3245×10^{-5}.

Rule 6: Zeros to the right of a whole number may or may not be significant

In normal convention, the number 500 has one significant digit and the number 2,500 has two significant digits. If 500 were written as 5.00×10^2, then it would have three significant digits. If 2,500 were written as 2.5×10^3, then it would have two significant digits. If it were written as 2.50×10^3, then it would have three significant digits. Only the individual reporting the number would have knowledge of the measuring instrument and the degree of its accuracy. That individual would then choose the appropriate number of significant digits to report.

✐ Practice Problem Set 2-2 ✐ *Significant Digits*

For each of the numbers below, report the number of significant digits.

1. 4.36
2. 300
3. 310
4. 312
5. 7.3×10^{-3}
6. 0.00063
7. 00.2000
8. 5,673

9. 10,213.789
10. 3.75×10^{12}

Rounding Numbers

As previously discussed, measurements and calculations that involve measurements are merely estimates of the real number or value. When a professional horticulturist understands the concepts of instrument accuracy and precision and human influences on accuracy and precision, they realize that, in the real world, we can only estimate a value. After performing calculations that use measured numbers alone or in combination with exact numbers (counted numbers, mathematical constants, and conversion factors), our calculations produce numbers with as many digits as the calculator will allow. Reducing these numbers to a reliable or realistic number of significant digits is based on what we know about the measuring instrument and the rules of math with significant digits. Once the number of significant digits has been determined, reducing a number to the appropriate number of significant digits involves the use of the rules of rounding.

Rules for Rounding Numbers

Rule 1: If the number to the right of the last significant digit is less than five, drop this digit and all the digits to the right of it

A horticulturist has, through calculation, determined that 11.5437 g of a growth-regulator product needs to be used to mix a spray solution. The gram scale that the horticulturist will be using is accurate to ± 0.01 g. Round 11.5437 g down to two significant digits to the right of the decimal point as warranted by the measuring instrument. Since the value needs to be rounded down to two digits to the right of the decimal point, look at the third digit to the right of the decimal point (3). Since three is less than five, the result after rounding is 11.54 g.

Rule 2: If the number to the right of the last significant digit is greater than or equal to five, increase the last significant digit by one and drop all the digits to the right of it

Using the situation described above, a horticulturist determined that 25.969 g of a product are needed to mix a spray solution. Round this value to two significant digits to the right of the decimal point. Look at the third digit to the right of the decimal point (9). Since nine is greater than five, the result after rounding is 25.97 g.

Rule 3: Do not round exact numbers such as counted numbers, mathematical constants (i.e., pi), equivalents, or unit conversion factors

A horticulturist needs to convert a value measured in pounds to grams. The equivalent for one pound is 453.59237 g. Use all the digits in this equivalent; *do not* round to

Significant Digits and Rules for Rounding Numbers 39

454 g. Once an exact number like this equivalent is used in a calculation, reduction to the appropriate number of significant digits can be performed as a last step.

EXAMPLE 2-1

Round 656.7439 to five significant digits. Examine the sixth digit. It is three and it is less than five. Drop every digit after the fifth significant digit. The resulting number is: 656.74.

EXAMPLE 2-2

Round 2.4732 to two significant digits. Examine the third digit. It is seven and it is more than five. Increase the last significant digit (four) to five and drop every digit to the right of it. The resulting number is: 2.5.

EXAMPLE 2-3

Round 7.07973 to four significant digits. Examine the fifth digit. It is seven and it is more than five. Increase the last significant digit (9) to 10 and drop every digit to the right of it. The resulting number is: 7.080. Keep the terminal zero (four significant digits) to indicate the precision of the instrument.

EXAMPLE 2-4

Round 4.3625 to four significant digits. Examine the fifth digit. It is exactly five. Increase the last significant digit (two) to three and drop every digit to the right of it. The resulting number is: 4.363.

Practice Problem Set 2-3 Rounding Numbers

Round the following numbers to four significant digits:

1. 3.75
2. 5.3
3. 4.03
4. 4.5

Round the following numbers to three significant digits:

5. 365.6
6. 0.004573
7. 0.4207
8. 7.655×10^3

Round the following numbers to two significant digits:

9. 0.253
10. 75.4
11. 0.00729
12. 0.301

Units of Measure

There are two systems of measurement commonly used by horticulturists: the U.S. Customary System and the metric system. In the United States, the U.S. Customary System is the most popular. The metric system is used in science and by most of the rest of the world. It is important to know both systems and to be able to convert values between the two measurement systems. A third system, the Systéme Internationale, is used for expressing units of measure in scientific publications. The Systéme Internationale, commonly known as SI units, is primarily a metric system and allows for uniform expression of values in the international scientific community.

The metric system is a base-10 system. A uniform system of prefixes are used with metric units of measure to describe multiples or submultiples of that unit of measure. Table A-1, found in Appendix A, shows this system of prefixes.

Units of Length

The fundamental unit of length in the U.S. System is the yard and the fundamental unit of length in the metric system is the meter. One yard is equivalent to 0.9144 meter. A yard is subdivided into feet and one foot equals one-third of a yard. One foot is subdivided into inches and one foot is equivalent to 12 inches. A longest measure of length in the U.S. System is a statute mile that is equal to 5,280 feet or 1,760 yards, and the nautical mile is equal to 1.151 statute miles.

A meter is subdivided into a decimeter (1/10 of a meter or 0.1 meter), a centimeter (1/100 of a meter or 0.01 meter), a millimeter (1/1,000 of a meter or 0.001 meter), a micrometer (1/10,000 of a meter or 0.0001 meter), and a nanometer (1/1,000,000 of a meter or 0.00001 meter). Units larger than one meter are dekameters (10 meters), hectometers (100 meters), and kilometers (1,000 meters). Tables B-1 and B-2, found in Appendix B, compare units of length.

Units of Area

Area is a unit of length taken to the power of two or a unit of square measure. Units of area in the U.S. Customary System are square inch, square foot, square yard, acre, and square mile. Units of area in the metric system are square centimeters, square meters, are, hectare, and square kilometers. Tables B-3 and B-4, found in Appendix B, compare units of area.

Units of Volume

Volume is a unit of length taken to the power of three or a unit of cubic measure. Units of volume in the U.S. Customary System are cubic inch, cubic foot, and cubic yard. Metric system units of volume are cubic meter, cubic decimeter, and cubic centimeter. Tables B-5 and B-6, found in Appendix B, compare units of volume. Table B-7, found in Appendix B, compares units of dry volume. Units of dry volume are U.S. Customary units and include dry pint, dry quart, peck, and bushel.

Units of Capacity

A liter (L) is the primary unit of capacity (liquid measure) for the metric system. Smaller units of capacity for the metric system include deciliter (0.1 liter), centiliter (0.01 liter), milliliter (0.001 liter), and microliter (0.000001 liter). Larger units of capacity for the metric system include dekaliter (10 liters), hectoliter (100 liters), and kiloliter (1,000 liters). The gallon is the primary unit of liquid measure for the U.S. Customary System and it is equivalent to 128 fluid ounces, 8 pints, 4 quarts, and 16 cups. Smaller units of liquid measure that are sometimes used in horticulture are the tablespoon and teaspoon. There are 16 tablespoons in a cup and 3 teaspoons in 1 tablespoon. These units of capacity are commonly used in cooking. Tables B-8 and B-9, found in Appendix B, compare units of capacity.

Units of Weight (Mass)

The fundamental unit of weight for the metric system is the gram. Smaller units than a gram are decigram (0.1 gram), centigram (0.01 gram), milligram (0.001 gram), and microgram (0.000001 gram). Larger units than a gram are dekagram (10 grams), hectogram (100 grams), kilogram (1,000 grams), and metric ton (1,000,000 grams).

The fundamental unit of weight for the U.S. Customary System is the pound. An ounce is a smaller unit than a pound, and there are 16 ounces in a pound. Note that units of capacity include fluid ounces and units of weight include ounces. Mistakes and confusion occur when these units are interchanged incorrectly. Always use the two words *fluid ounces* to describe the volume of a liquid and the word *ounces* to describe the weight of a solid. Larger units than a pound include a short ton, which is 2,000 pounds, and a long ton, which is equivalent to 2,240 pounds. Tables B-10 and B-11, found in Appendix B, compare units of weight.

Units of Concentration

Concentration is commonly expressed in units of parts per million and as percents in horticulture applications. Concentration is the ratio of a solute present in a solution relative to a standard amount of solvent.

> **Definition** *Solute*
>
> A *solute* is a substance that is dissolved or dispersed in a solvent. In a solution, the solute is the substance found in smaller quantities.

In a liquid fertilizer application, the solute is the soluble fertilizer product. Pesticides and growth regulator products, both liquid and solid forms, also are solutes when prepared with a solvent prior to application.

> **◄ Definition ► Solvent**
> A *solvent* is a substance capable of dissolving or dispersing a solute. In a solution, the solvent is the substance found in the largest quantity.

In a liquid fertilizer application, the solvent is water. Pesticides and plant growth regulators are either commonly dissolved or dispersed in water (the solvent) prior to application.

> **◄ Definition ► Solution**
> A *solution* is a homogenous mixture of solute and solvent.

Concentration can be expressed three different ways. It can be expressed as a ratio on a weight-to-weight basis, a weight-to-volume basis, and a volume-to-volume basis. Table B-12, found in Appendix B, compares these three methods of expression.

Units of Temperature

The primary unit of temperature in the United States is degrees Fahrenheit. The Fahrenheit scale defines the freezing point of water at 32°F and the boiling point of water at 212°F. The Celsius or centigrade scale is commonly used in science and by most of the rest of the world. It defines the freezing point of water at 0°C and the boiling point of water at 100°C. The relationship of the Celsius and Fahrenheit scales is expressed in the following equations:

$$°C = \frac{5}{9}(°F - 32)$$

$$°F = \frac{9}{5}(°C) + 32$$

Converting temperature between the Celsius and Fahrenheit scales involves inserting the known value into one of the equations above and solving for the unknown value.

EXAMPLE 2-5

Convert 72°F to the Celsius equivalent.

$$°C = \frac{5}{9}(°F - 32)$$

$$°C = \frac{5}{9}(72 - 32)$$

$$°C = \frac{5}{9}(40)$$

$$°C = 22.2\overline{2}$$

SOLUTION
72°F is equivalent to 22.2$\overline{2}$°C.

EXAMPLE 2-6
Convert 34°C to the Fahrenheit equivalent.

$$°F = \frac{9}{5}(°C) + 32$$

$$°F = \frac{9}{5}(34) + 32$$

$$°F = 61.2 + 32$$

$$°F = 93.2$$

SOLUTION
34°C is equivalent to 93.2°F.

Practice Problem Set 2-4 — Converting Units of Temperature

Convert the following temperatures in Celsius to Fahrenheit:

1. 30°C
2. 27°C
3. 5°C

Convert the following temperatures in Fahrenheit to Celsius:

4. 65°F
5. 78°F
6. 29°F

Units of Time

The fundamental unit of time is the second. There are sixty seconds in one minute, sixty minutes in one hour, twenty-four hours in a day, seven days in a week, and fifty-two weeks in a year. In production floriculture, a year is described in weeks and crops are shipped, potted, and scheduled based on crop weeks numbered one to fifty-two. Table B-13, found in Appendix B, compares units of time.

Unit Conversion Techniques

Effortless conversion of one unit of measure to another unit of measure is an essential skill for a horticulturist. Two methods are commonly used to convert units between measurement systems: conversion using equivalents and conversions using conversion factors. Tables of equivalents and conversion factors can be found in Appendices B and C of this text.

Performing Unit Conversions Using Equivalents

An equivalent is expressed as a ratio in the form of a fraction. For example, the equivalent $\frac{2.54 \text{ centimeter}}{1 \text{ inch}}$ reads: 2.54 centimeters is equivalent to 1 inch. By using an equivalent expressed as a ratio and setting up a proportion, the conversion of a value from inches to centimeters or vice versa is possible. Remember that a ratio is a mathematical statement that two ratios are equal.

EXAMPLE 2-7

Convert 10.5 in. to centimeters.

Step 1

Set up a proportion that reads: If 2.54 cm are equivalent to 1 in., then how many centimeters are equivalent to 10.5 in.?

$$\frac{2.54 \text{ cm}}{1 \text{ in.}} = \frac{x \text{ cm}}{10.5 \text{ in.}}$$

Step 2

Perform cross multiplication.

$$\frac{2.54 \text{ cm}}{1 \text{ in.}} \times \frac{x \text{ cm}}{10.5 \text{ in.}}$$

$$1 \text{ in.} \times x \text{ cm} = 2.54 \text{ cm} \times 10.5 \text{ in.}$$

Step 3

Isolate and solve for x.

$$\frac{1 \text{ in.} \times x \text{ cm}}{1 \text{ in.}} = \frac{2.54 \text{ cm} \times 10.5 \text{ in.}}{1 \text{ in.}}$$

$$\frac{1 \cancel{\text{in.}} \times x \text{ cm}}{1 \cancel{\text{in.}}} = \frac{2.54 \text{ cm} \times 10.5 \cancel{\text{in.}}}{1 \cancel{\text{in.}}}$$

$$x \text{ cm} = 26.67 \text{ cm}$$

SOLUTION

10.5 in. are equivalent to 26.67 cm.

Note that the units cancel properly and that the desired unit (centimeters) is found in the numerator.

Performing Unit Conversions Using Conversion Factors

Conversion factors are multipliers used to convert one unit to another and are found in tables of conversion factors (Table C-1, found in Appendix C). Unit conversion is performed by multiplying the number that will be converted by the appropriate conversion factor.

EXAMPLE 2-8

Convert 30 L to gallons.

Step 1

Locate the conversion factor for liters to gallons in a conversion factor table.
The multiplier is 0.2642 to convert liters to gallons.

Step 2

Multiply the conversion factor by the value that is to be converted.

$$30 \text{ liters} \times 0.2642 \frac{\text{gallons}}{\text{liter}} = 7.926 \text{ gal}$$

SOLUTION

Thirty liters are equivalent to 7.926 gal.
Use care in selecting the conversion factor from the tables. Reversing the units in the conversion will result in an incorrect result.

Practice Problem Set 2-5 Unit Conversions

Perform the following unit conversions using equivalents:

1. 4 inches to centimeters
2. 7,200 square inches to square feet
3. 100 gallons to cubic feet
4. 3,484,800 square feet to acres
5. 2.5 gallons to fluid ounces

Perform the following unit conversions using conversion factors:

6. 212.5 centimeters to inches
7. 7 liquid pints to liters
8. 45.72 meters to yards
9. 27 acres to square meters
10. 91 liters to gallons

Chapter 3 Geometry

Introduction

Practical geometry is used extensively throughout the green industry. This chapter will review the formulae for calculating the area and volume of common geometric figures. In addition, there is a presentation of mathematical techniques used to calculate the area of irregular figures.

Area of Geometric Figures

Area is reported in units of square measure. Common units of square measure are square feet (ft^2), square inches (in.^{2}), and acres (ac). There are 144 in.^{2} in 1 ft^2 and 43,560 ft^2 in 1 ac. One important rule to remember when calculating area is to always use like units of measurement. When moving between units, particularly when converting between in.^{2} and ft^2, remember that one square foot is equivalent to 12 in. × 12 in. or 144 square inches. Figure 3-1 illustrates this principle.

Area of Regular Figures

Regular figures in geometry are those that mathematicians have been able to define using mathematical derivations. These figures are familiar to all of us and include classic shapes such as rectangles and circles. Area calculations for each geometric figure involve the use of simple formulas unique to each figure type. In the sections that follow, each class of figures is identified, and then the formula for determining the area of each figure type is presented with example calculations.

Quadrilaterals

Quadrilaterals are four-sided figures with four angles. The most common quadrilaterals used in the green industry are squares, rectangles, parallelograms, and trapezoids.

Square

A square is a figure with four right angles (90°), two pairs of parallel sides, and all four sides are of equal length.

Figure 3-1

Comparison of square feet to square inches

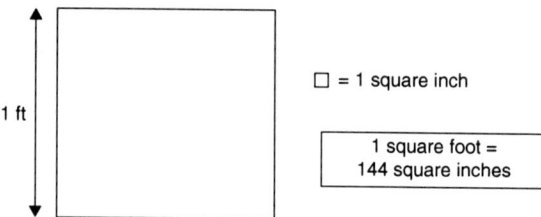

❧ Definition ❧ Area of a Square

The *area of a square (A)*, shown in Figure 3-2, is calculated by squaring the length of one side (*s*).

$$\text{Area} = \text{side} \times \text{side}$$

$$A = s^2$$

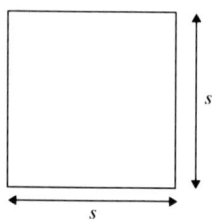

Figure 3-2

Calculating the area of a square

EXAMPLE 3-1

Calculate the area of a square that measures 3 ft on each side.

$$\text{Area} = \text{Side} \times \text{Side}$$

$$A = s^2$$

$$A = (3 \text{ ft})^2$$

$$\text{Area} = 9 \text{ ft}^2$$

Rectangle

A rectangle is a figure with four right angles (90°) and two pairs of opposite sides that are equal in length.

❧ Definition ❧ Area of a Rectangle

The *area of a rectangle (A)*, shown in Figure 3-3, is calculated by multiplying length (*l*) by width (*w*).

$$\text{Area} = \text{Length} \times \text{Width}$$

$$A = l \times w$$

Area of Geometric Figures **49**

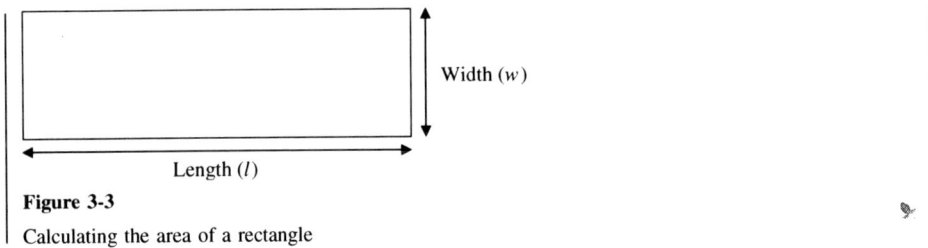

Figure 3-3
Calculating the area of a rectangle

EXAMPLE 3-2

Calculate the area of a rectangle that measures 50 ft long by 15 ft wide.

$$\text{Area} = \text{Length} \times \text{Width}$$
$$A = l \times w$$
$$A = 50 \text{ ft} \times 15 \text{ ft}$$
$$\text{Area} = 750 \text{ ft}^2$$

EXAMPLE 3-3

Calculate the area of a rectangle (in square feet) that measures 4 ft long by 6 in. wide.

Step 1

In this situation, all units need to be converted to the same unit to calculate area. Six inches is easily converted to 0.50 ft by setting up the following proportion:

$$\frac{1 \text{ ft}}{12 \text{ in.}} = \frac{x \text{ ft}}{6 \text{ in.}}$$

Isolate and solve for x.

$$12x = 6$$
$$\frac{12x}{12} = \frac{6}{12}$$
$$x = \frac{6}{12}$$
$$x = 0.50 \text{ ft}$$

Step 2

Apply the formula to the dimensions with the same units.

$$\text{Area} = \text{Length} \times \text{Width}$$
$$A = l \times w$$
$$A = 4 \text{ ft} \times 0.50 \text{ ft}$$
$$\text{Area} = 2 \text{ ft}^2$$

Parallelogram

A parallelogram has two pairs of parallel opposite sides.

> ◄ **Definition** ► **Area of a Parallelogram**
>
> *Area of a parallelogram (A)*, shown in Figure 3-4, is calculated by multiplying the base (b) by the height (h).
>
> $$\text{Area} = \text{base} \times \text{height}$$
> $$A = b \times h$$

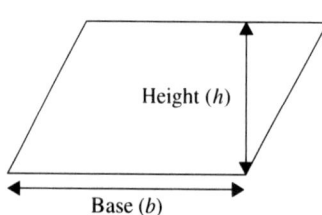

Height (h)

Base (b)

Figure 3-4
Calculating the area of a parallelogram

EXAMPLE 3-4

Calculate the area of a parallelogram with a base measuring 3 yd and a height measuring 1.5 yd.

$$\text{Area} = \text{base} \times \text{height}$$
$$A = b \times h$$
$$A = 3 \text{ yd} \times 1.50 \text{ yd}$$
$$\text{Area} = 4.50 \text{ yd}^2$$

Trapezoid

A trapezoid is a quadrilateral that has one pair of parallel opposite sides.

> ◄ **Definition** ► **Area of a Trapezoid**
>
> The *area of a trapezoid* (A), shown in Figure 3-5, is found by multiplying the average length of the parallel sides, base$_1$ and base$_2$ (b_1 and b_2), by the height (h).
>
> $$\text{Area} = [(\text{base}_1 + \text{base}_2) \div 2] \times \text{height}$$
> $$A = [(b_1 + b_2) \div 2] \times h$$

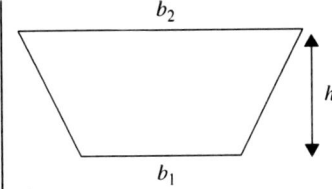

Figure 3-5
Calculating the area of a trapezoid

EXAMPLE 3-5

Calculate the area of a trapezoid with a base₁ length measuring 5 ft, a base₂ length measuring 9 ft, and a height measuring 4 ft.

$$\text{Area} = [(\text{base}_1 + \text{base}_2) \div 2] \times \text{height}$$
$$A = [(b_1 + b_2) \div 2] \times h$$
$$A = [(5 \text{ ft} + 9 \text{ ft}) \div 2] \times 4 \text{ ft}$$
$$A = [14 \text{ ft} \div 2] \times 4 \text{ ft}$$
$$A = 7 \text{ ft} \times 4 \text{ ft}$$
$$\text{Area} = 28 \text{ ft}^2$$

✎ Practice Problem Set 3-1 ✎ *Area of Quadrilaterals*

Calculate the area of each rectangle or square.

1. $l = 25$ ft, $w = 15$ ft (unit = ft²)
2. $l = 5$ ft, $w = 27$ in. (unit = ft²)
3. $s = 11$ ft (unit = ft²)

Calculate the area of each trapezoid.

4. $b_1 = 8$ ft, $b_2 = 13$ ft, $h = 5$ ft (unit = ft²)
5. $b_1 = 18$ in., $b_2 = 16$ in., $h = 0.75$ ft (unit = in.²)
6. $b_1 = 8,000$ ft, $b_2 = 10,000$ ft, $h = 250$ ft (unit = ac)

Triangles

A triangle is a polygon with three sides. The sum of all angles in a triangle is equal to 180°. Triangles are classified by both the length of their sides and by their angles.

Identifying Triangles by the Length of the Sides (Figure 3-6)

Equilateral An equilateral triangle has three equal sides. Each angle measures 60°.

Isosceles An isosceles triangle has two sides of equal length. The angles that are opposite the equal sides are equal.

Scalene A scalene triangle has three sides of differing length. All angles will be different.

Figure 3-6

Triangles that are identified by the length of the sides

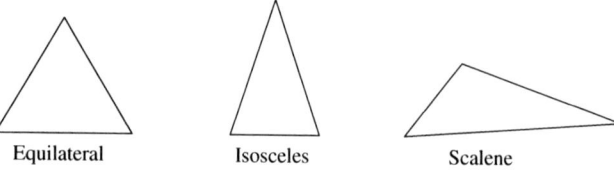

Equilateral   Isosceles   Scalene

Identifying Triangles by Angles (Figure 3-7)

Acute Triangle An acute triangle has three angles that each measure less than 90 degrees.

Obtuse Triangle An obtuse triangle has one angle that is greater than 90 degrees.

Right Triangle A right triangle has one angle that is exactly 90 degrees.

Figure 3-7

Triangles that are identified by the size of the angles

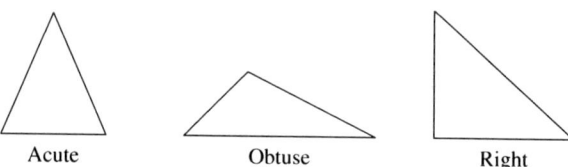

Acute   Obtuse   Right

Calculating the Area of a Triangle

There are two formulas used to calculate the area of a triangle. The first formula requires the measurement of the triangle base and height. The second formula, known as Heron's Formula, is based on measurements of the length each side of the triangle.

Calculating the Area of a Triangle Using Height and Base Measurements

When measuring a triangle, the base and height must be the line segments that meet forming a right angle. When two line segments form a right angle, they are described as perpendicular lines. The small square at the intersection of these two lines indicates that the lines are at right angles (90°) or that they are perpendicular (Figure 3-8).

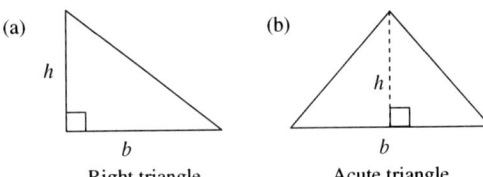

Figure 3-8
Determining height of a triangle

The height is sometimes indicated as one of the sides of a triangle (Figure 3-8a). This situation involves a right triangle or a triangle with one of the angles being 90 degrees.

Sometimes the height of a triangle is measured from a line that is perpendicular to the base and extends from the base to the opposite corner of the triangle (Figure 3-8b). It is possible to determine the height of most triangles, even though the triangle itself does not have a right angle.

> **◆ Definition ◆** *Calculating the Area of a Triangle Using Height and Base Measurements*
>
> The basic formula for the *area of a triangle (A)* is base (*b*) multiplied by the height (*h*) and then divided by two.
>
> $$\text{Area} = (\text{base} \times \text{height}) \div 2$$
>
> or
>
> $$A = \tfrac{1}{2}(b \times h) \text{ or } A = (b \times h) \div 2$$

EXAMPLE 3-6

Calculate the area of a triangle with a base measuring 7 ft and a height measuring 4 ft.

$$\text{Area} = (\text{base} \times \text{height}) \div 2$$
$$A = (b \times h) \div 2$$
$$A = (7 \text{ ft} \times 4 \text{ ft}) \div 2$$
$$A = 28 \text{ ft}^2 \div 2$$
$$\text{Area} = 14 \text{ ft}^2$$

Calculating Area of a Triangle Using Heron's Formula

Heron's formula is an alternate way to calculate the area of a triangle when it is difficult to determine the height of the triangle. Remember, height is measured at a right angle to the base and sometimes this is difficult to accomplish in the field. If the lengths of the three sides can be measured, Heron's Formula will help ensure greater accuracy when calculating the area of a triangle if the height cannot be easily measured.

> **◄ Definition ►** *Calculating the Area of a Triangle Using Heron's Formula*
> Heron's formula is as follows:
> $$\text{Area } (A) = \sqrt{s(s-a)(s-b)(s-c)}$$
> $$\text{Where } s = \tfrac{1}{2}(a+b+c)$$
> and a, b, and c represent the length of each of the three sides of the triangle.

EXAMPLE 3-7

Determine the area of the triangle using Heron's formula (Figure 3-9).

$$s = \tfrac{1}{2}(12 + 17 + 25) = \tfrac{1}{2}(54) = 27$$
$$\text{Area } (A) = \sqrt{s(s-a)(s-b)(s-c)}$$
$$A = \sqrt{27(27-12)(27-17)(27-25)}$$
$$A = \sqrt{27(15)(10)(2)}$$
$$A = \sqrt{8,100}$$
$$\text{Area} = 90 \text{ ft}^2$$

Figure 3-9

Calculate the area of the triangle using Heron's formula

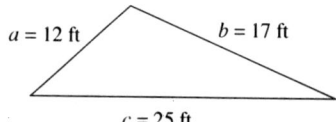

The Pythagorean Theorem

The Pythagorean Theorem states that the sum of the squares of the two shortest sides of a triangle is equal to the square of the longest side (hypotenuse) of a triangle. This relationship, proven by the Pythagorean Theorem, is valid only for a right triangle.

> **◄ Definition ►** *Pythagorean Theorem*
> The **Pythagorean Theorem** proves that the sum of the squares of the two shortest sides of a right triangle equal the square of the hypotenuse (or longest side) of a right triangle, and it is represented by the following equation:
> $$a^2 + b^2 = c^2$$
> where a and b represent the two short sides of a right triangle and c represents the hypotenuse. (See Figure 3-10).

Area of Geometric Figures **55**

Figure 3-10
A Triangle with the Elements of the Pythagorean Theorem Identified

If the equation (Pythagorean Theorem) holds true for a triangle, then the triangle is determined to be a right triangle. There are some practical uses for the Pythagorean Theorem. If the length of one of the sides of a right triangle is unknown, then the Pythagorean Theorem can be used to mathematically determine the length of that side. This is useful if the hypotenuse is represented by the rafter of a greenhouse or garden structure and the correct length of that rafter is unknown.

Since the Pythagorean Theorem always holds true for a right triangle, then a horticulturist can make use of that information to make sure the corner of a garden or garden structure is square. See Appendix D for a description of how to use the Pythagorean Theorem to ensure right angles in a garden design or structure or to "square-up" a garden plot.

EXAMPLE 3-8

Using the Pythagorean Theorem, the two triangles in Figure 3-11 are compared to determine if they are right triangles.

(a)

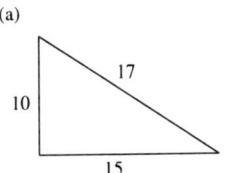

(b)

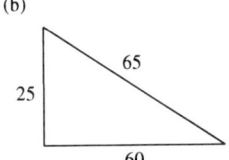

Figure 3-11
A comparison of two triangles: Which one is a right triangle?

TRIANGLE A

$$a^2 + b^2 = c^2, \text{ where } a = 10, b = 15, \text{ and } c = 17$$
$$a^2 = 10^2 = 100$$
$$b^2 = 15^2 = 225$$
$$c^2 = 17^2 = 289$$
$$a^2 + b^2 = 100 + 225 = 325$$

The sum of $a^2 + b^2 (325)$ does not equal $c^2 (289)$; thus, triangle A is not a right triangle.

TRIANGLE B

$$a^2 + b^2 = c^2, \text{ where } a = 25, b = 60, \text{ and } c = 65$$
$$a^2 = 25^2 = 625$$
$$b^2 = 60^2 = 3{,}600$$
$$c^2 = 65^2 = 4{,}225$$
$$a^2 + b^2 = 625 + 3{,}600 = 4{,}225$$

The sum of $a^2 + b^2$ (4,225) does equal c^2 (4,225); thus, triangle B is a right triangle.

Practice Problem Set 3-2 Area of Triangles

Calculate the area of each triangle.

1. $h = 10$ ft, $b = 15$ ft
2. $h = 3$ ft, $b = 21$ ft
3. $a = 5$ ft, $b = 10$ ft, $c = 12$ ft
4. Use the Pythagorean Theorem to determine the missing length of a right angle in Figure 3-12 that has a base of 13 ft and a height of 4 ft.

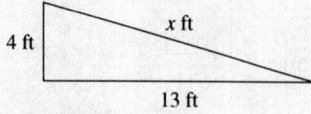

Figure 3-12
Calculate the length of the hypotenuse using the Pythagorean Theorem

Circle and Ellipse

Circle

A circle, shown in Figure 3-13, is a closed curve. Every point on the curve is equidistant from a center point.

Figure 3-13
Circle with center point identified

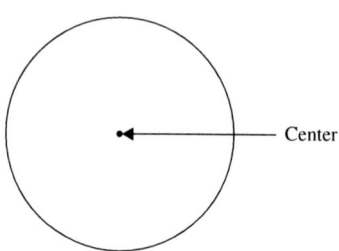

Area of Geometric Figures **57**

> ◄ Definition ► **Definitions Related to a Circle**

AREA

The area (A) of a circle, shown in Figure 3-14, is the area enclosed by a circle since a circle is a line. It is derived mathematically by using one of the following formulae:

$$A = \pi r^2 \text{ or } A = \frac{\pi d^2}{4} \text{ or } A = \frac{c^2}{4\pi}$$

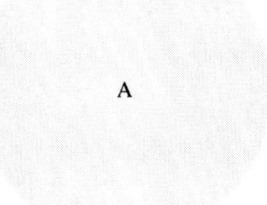

Figure 3-14
Area of a circle

CIRCUMFERENCE

The circumference (c) of a circle, shown in Figure 3-15, is the distance from one point on the circle around and back to that point. It is sometimes called the perimeter of a circle, but the word *perimeter* is usually reserved for figures composed of straight lines. It is derived mathematically by using one of the following formulae:

$$c = 2\pi r \text{ or } c = \pi d \text{ or } c = \sqrt{\frac{4\pi}{A}}$$

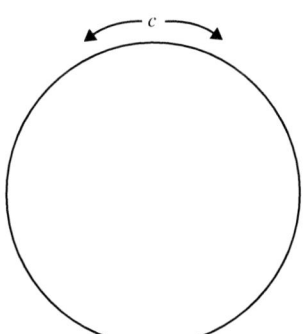

Figure 3-15
Circle with circumference identified

DIAMETER

The diameter (d) of a circle, shown in Figure 3-16, is the length of a straight line that extends from any point on the circle, through the center point, and on to the opposite side of the circle. It is derived mathematically by using one of the following formulae:

$$d = 2r \text{ or } d = \frac{c}{\pi} \text{ or } d = \sqrt{\frac{4A}{\pi}}$$

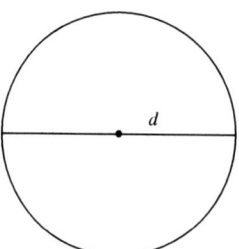

Figure 3-16
Circle with diameter identified

PI (π)

Pi is a mathematical constant and represents the ratio of the circumference of a circle to the diameter of a circle ($\frac{c}{d}$). The number pi written to three significant digits is 3.142, but pi has an infinite number of significant digits. We recommend that you use the pi button on your calculator to enter it into mathematical calculations.

$$\pi = \frac{c}{d}$$

RADIUS

The radius (r) of a circle, shown in Figure 3-17, is the length of a line that extends from the center point to any point on the circle. It is derived mathematically by using one of the following formulae:

$$r = \frac{d}{2} \text{ or } r = \frac{c}{2\pi} \text{ or } r = \sqrt{\frac{A}{\pi}}$$

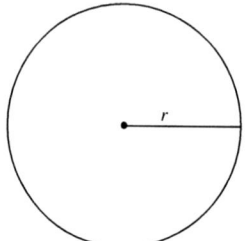

Figure 3-17
Circle with radius identified

EXAMPLE 3-9

Determine the area of the circle that has a radius of 10 ft.

$$\text{Area} = \pi(r^2)$$
$$A = \pi(10 \text{ ft})^2$$
$$A = \pi(100 \text{ ft}^2)$$
$$A = 314.2 \text{ ft}^2$$

The area of a circle can be calculated by knowing the circumference of a circle. The circumference is the distance around the perimeter of a circle. There is a direct relationship between the area of a circle and the circumference of a circle. The formula used to determine area is:

$$\text{Area} = \frac{\text{circumference}^2}{4\pi}$$
$$A = \frac{c^2}{4\pi}$$

EXAMPLE 3-10

Determine the area of the circle that has a circumference of 100 ft.

$$\text{Area} = \frac{\text{circumference}^2}{4\pi}$$
$$A = \frac{c^2}{4\pi}$$
$$A = \frac{100^2}{4\pi}$$
$$A = \frac{10{,}000 \text{ ft}^2}{12.56}$$
$$A = 796.2 \text{ ft}$$

There is also a direct relationship of the circumference of a circle and the diameter of the circle. Circumference (c) equals the product of pi (π) and the diameter (d) of the circle. The formula is as follows:

$$c = \pi d$$

This relationship is useful if a horticulturist cannot directly measure the diameter of a tree. The circumference of a tree can be measured using a measuring tape, and then the diameter (caliper) of the tree can be determined mathematically.

EXAMPLE 3-11

Determine the diameter of the circle that has a circumference of 100 in.

$$c = \pi d$$
$$100 \text{ in.} = \pi d$$

Isolate and solve for d.

$$\frac{100 \text{ in.}}{\pi} = \frac{\pi d}{\pi}$$

$$\frac{100 \text{ in.}}{\pi} = d$$

$$d = 31.83 \text{ in.}$$

Ellipse

An ellipse is a curved line that has an oval shape where the sum of the distance from two points (foci) to any place on the curve is equal. Figure 3-18 demonstrates this principle.

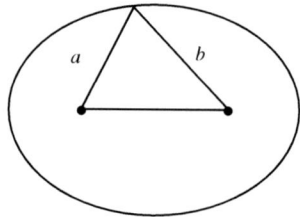

Figure 3-18

Ellipse with foci and lengths a and b identified

> **Definition** — **Definitions Related to an Ellipse**
>
> ### AREA
>
> The area (A) of an ellipse, shown in Figure 3-19, is calculated by multiplying pi (π) times the product of the length of the major radius (r_{major}) and the length of the minor radius (r_{minor}).
>
> $$\text{Area} = \pi \times \text{major radius} \times \text{minor radius}$$
>
> $$A = \pi(r_{major} \times r_{minor})$$
>
>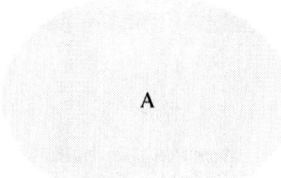
>
> **Figure 3-19**
>
> Area of an ellipse
>
> ### CENTER
>
> The center of an ellipse, shown in Figure 3-20, is the point at which the center of the major axis crosses the center of the minor axis at a 90-degree angle.

Area of Geometric Figures **61**

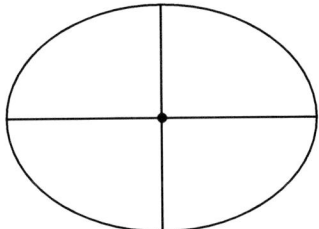

Figure 3-20
Center of an ellipse

FOCI OR FOCUS POINTS

Two points within an ellipse that define the shape of the ellipse, the foci are located on the major axis equidistant from the center (see Figure 3-21). If lines are drawn from each focus point to a single point on the ellipse, then the sum of the length of the two lines is always equal regardless of the location of the point on the ellipse.

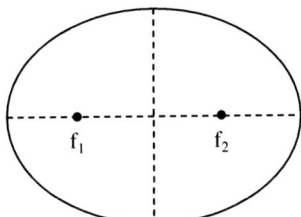

Figure 3-21
Location of the foci on an ellipse

MAJOR AXIS

The major axis (a_{major}), shown in Figure 3-22, is the longest diameter of an ellipse. It is equivalent to the major radius (r_{major}) multiplied by two.

$$\text{major axis} = \text{major radius} \times 2$$

$$a_{major} = r_{major} \times 2$$

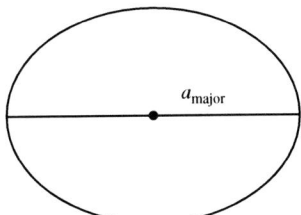

Figure 3-22
Location of the major axis

MINOR AXIS

The minor axis (a_{minor}), shown in Figure 3-23, is the shortest diameter of an ellipse. It is equivalent to the minor radius (r_{minor}) multiplied by two.

$$\text{minor axis} = \text{minor radius} \times 2$$

$$a_{minor} = r_{minor} \times 2$$

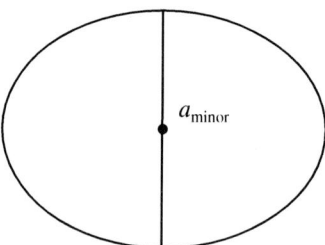

Figure 3-23
Location of the minor axis

MAJOR RADIUS

The major radius (r_{major}), shown in Figure 3-24, is the longest radius of an ellipse. It is equivalent to the major axis (a_{major}) divided by two.

$$\text{major radius} = \text{major axis} \div 2$$

$$r_{major} = a_{major} \div 2$$

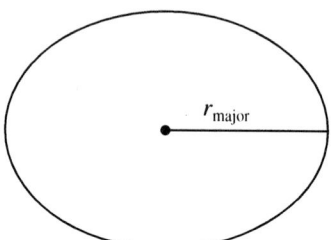

Figure 3-24
Location of the major radius

MINOR RADIUS

The minor radius (r_{minor}), shown in Figure 3-25, is the shortest radius of an ellipse. It is equivalent to the minor axis (a_{minor}) divided by two.

$$\text{minor radius} = \text{minor axis} \div 2$$

$$r_{minor} = a_{minor} \div 2$$

Area of Geometric Figures **63**

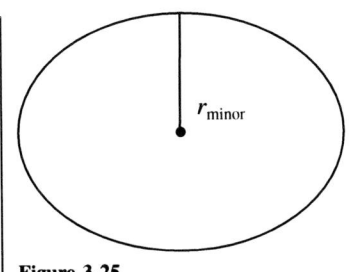

Figure 3-25
Location of the minor radius

EXAMPLE 3-12

Calculate the area of an ellipse that has a major radius length of 5 ft and a minor radius length of 2.5 ft, as shown in Figure 3-26.

$$\text{Area} = \pi \times \text{major radius} \times \text{minor radius}$$
$$A = \pi(r_{major} \times r_{minor})$$
$$A = \pi(2.5 \text{ ft})(5 \text{ ft})$$
$$A = \pi(12.5 \text{ ft}^2)$$
$$A = 39.25 \text{ ft}$$

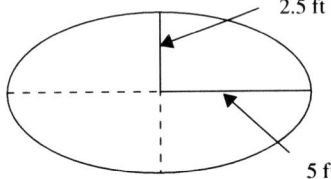

Figure 3-26
Determine the area of the ellipse

⚙ Practice Problem Set 3-3 ⚙ *Area of Circles and Ellipses*

Calculate the area of each circle.

1. $r = 12$ ft (unit = ft^2)
2. $d = 52$ ft (unit = ft^2)
3. $c = 27$ yd (unit = yd^2)

Calculate the area of each ellipse.

4. major axis = 29 ft, minor axis = 11 ft (unit = ft^2)
5. major axis = 6 ft, minor axis = 2 yd (unit = ft^2)

Area of Irregular Figures

Irregular figures are shapes that do not conform to the shapes of classic geometric forms; therefore, there is no standard mathematical formula that can be applied to irregular figures. We see many irregular forms in the green industry. Garden beds, patios, and lawn areas all will commonly be irregular or nonconforming in shape. Three methods are offered for calculating the area of irregular figures. The Offset Method is useful for calculating the area of spaces that can be easily traversed such as a newly established perennial garden. The Modified Offset Method is an indirect way of calculating area when a space, such as a pond, cannot be easily traversed. Finally, Simpson's Rule operates similarly to the Offset Method for measurement but employs an algebraic derivation (formula) of Simpson's Rule that is based in calculus. Simpson's Rule is valuable because it offers greater accuracy than the Offset Method.

The Offset Method

The Offset Method is a way of estimating the area of an irregularly shaped figure. It divides an irregularly shaped area into a series of trapezoids that are equally spaced along a measured line. The Offset Method has four steps, as listed in Example 3-13.

> **◄ Definition ►** *Formula for the Offset Method*
> The formula for the **Offset Method** for calculating area is: the area of an irregular figure is equivalent to the product of the sum (Σ) of the lengths of the offsets ($L_1, L_2, L_3\ldots$) and the offset interval length (h).
> $$\text{Area} = (\Sigma L_1, L_2, L_3 \ldots) \times h$$

EXAMPLE 3-13

Calculate the area of an irregular shaped figure using the Offset Method.

Step 1

Establish a line along the longest axis of the irregularly shaped figure. Label the end points of the line A and B (Figure 3-27).

Figure 3-27
Establish the long axis

A ——————————— B

Step 2

Establish offset lines perpendicular to line AB. The offset lines should be equally spaced. For example, if line AB was 50 ft long, it could be divided into five 10 ft sections (Figure 3-28). If more precision is desired, or if the area is very irregular, it could be divided into ten 5 ft sections. Precision is increased with a greater number of offset lines. The distance between offsets is called h, and $h = 10$ ft in this example.

Area of Geometric Figures **65**

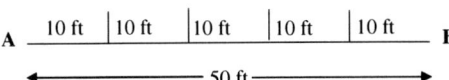

Figure 3-28

Establish offsets perpendicular to the long axis

Step 3
Extend each offset line to the perimeter of the irregularly shaped figure. Measure each offset line from end to end (Figure 3-29).

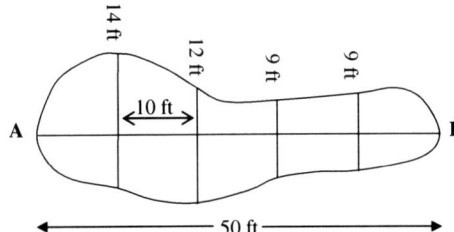

Figure 3-29

Measure the length of each of the offset lines

Step 4
Sum the lengths of all the offset lines.

$$(14 \text{ ft} + 12 \text{ ft} + 9 \text{ ft} + 9 \text{ ft}) = 44 \text{ ft}^2$$

Step 5
Multiply the sum by the distance between the offset lines (h).

$$(44 \text{ ft})(10 \text{ ft}) = 440 \text{ ft}^2$$

SOLUTION

The estimated area of this irregular figure is: 440 ft^2.

The Modified Offset Method

The Modified Offset Method is used to measure areas that are irregular in shape but cannot easily be traversed to set the offset lines. It divides an irregularly shaped area into a series of trapezoids that are equally spaced along a measured line just like the Offset Method, but the method for measuring the length of the offsets is indirect.

> **◆ Definition ◆** *Formula for the Modified Offset Method*
> The formula for the *Modified Offset Method* for calculating area is: the area of the irregular figure is equivalent to the product of the sum (Σ) of the length of the

offsets (L_1, L_2, L_3...) and the offset interval length (h).

$$\text{Area} = (\Sigma\ L_1, L_2, L_3 \ldots) \times h$$

where L_1, L_2, L_3, etc. are determined indirectly by calculating the difference between the sum (Σ) of the offsets E_1, E_2 or F_1, F_2 and the length of the line AC or BD, for example:

$$L_1 = AC - \Sigma E$$

where $\Sigma E = E_1 + E_2$

EXAMPLE 3-14

Calculate the area of of an irregular-shaped figure using the Modified Offset Method, as follows.

Step 1

Create a rectangle around the area that is to be measured. Measure the length of line segments AB or CD. These are equivalent to the length of the rectangle. Measure the length of line segments AC or BD. These are equivalent to the width of the rectangle. In Figure 3-30, the length (l) is 30 ft and the width (w) is 14 ft.

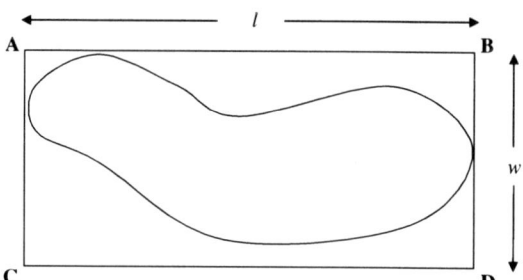

Figure 3-30

Establish a rectangle around the area to be measured

Step 2

Establish offset lines perpendicular to lines AB and CD as shown in Figure 3-31. The offset lines should be equally spaced. For this example line AB is 30 ft long and the offsets are spaced at 5 ft intervals. There will be two measurements for each offset line (for example, E1 and E2).

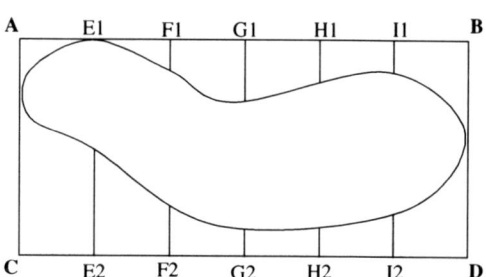

Figure 3-31

Establish offset lines

Step 3

Measure the lengths of each of the offset line segments. This is done by measuring from the point at which the line touches the figure to the point at which the line touches the rectangle surrounding the figure. There will be a pair of measurements (i.e., E1 and E2) for each offset. Add up each pair of offset measurements. The sigma sign (Σ) is used to designate that the pair has been added together.

The measurements for the offset lines found in Figure 3-31 are: (E1 = 0 and E2 = 7), (F1 = 2 and F2 = 3), (G1 = 4 and G2 = 5), (H1 = 3 H2 = 2) and (I1 = 2 and I2 = 3).

$$\Sigma E = E1 + E2 = 0 + 7 = 7$$
$$\Sigma F = F1 + F2 = 2 + 3 = 5$$
$$\Sigma G = G1 + G2 = 4 + 5 = 9$$
$$\Sigma H = H1 + H2 = 3 + 2 = 5$$
$$\Sigma I = I1 + I2 = 2 + 3 = 5$$

Step 4

Subtract each of the sums of the line segments (E through I) from the width of the rectangle. Each of these results is equal to the actual width of the figure at each of the offset locations.

$$E = (w - \Sigma E) \text{ or } 14 \text{ ft} - 7 \text{ ft} = 7 \text{ ft}$$
$$F = (w - \Sigma F) \text{ or } 14 \text{ ft} - 5 \text{ ft} = 9 \text{ ft}$$
$$G = (w - \Sigma G) \text{ or } 14 \text{ ft} - 9 \text{ ft} = 5 \text{ ft}$$
$$H = (w - \Sigma H) \text{ or } 14 \text{ ft} - 5 \text{ ft} = 9 \text{ ft}$$
$$I = (w - \Sigma I) \text{ or } 14 \text{ ft} - 5 \text{ ft} = 9 \text{ ft}$$

Step 5

Sum the widths of the figure at each offset as determined in Step 4.

$$7 \text{ ft} + 9 \text{ ft} + 5 \text{ ft} + 9 \text{ ft} + 9 \text{ ft} = 39 \text{ ft}$$

Step 6

Multiply the summed value found in Step 5 by the distance between offsets (5 ft). This is an estimate of the area of the figure.

$$(39 \text{ ft})(5 \text{ ft}) = 195 \text{ ft}^2$$

SOLUTION

The estimated area of this irregular figure is: 195 ft^2.

Simpson's Rule

Simpson's Rule provides a more accurate method of estimating the area of an irregular figure than the Offset or Modified Offset Methods. The Offset and Modified Offset Methods segment the area into trapezoids, while Simpson's Rule approximates a quadratic (or curvature) rather than a straight line within each subinterval. The steps for this procedure follow in Example 2-15.

> **◄ Definition ►** *Formula for Simpson's Rule*
> The formula for **Simpson's Rule** for calculating area (A) is:
>
> $$A = \frac{h}{3}[(L_1 + L_9) + 2(L_3 + L_5 + L_7) + 4(L_2 + L_4 + L_6 + L_8)]$$
>
> where h is the distance between offsets and L_1, L_2, L_3, are the lengths of the offsets.
> Note that a greater number of offsets is appropriate, but there must be an odd number of offsets. The use of a greater number of offsets results in greater accuracy of the estimate. Also note that the first and last offsets are entered into the first set of parentheses, the odd-numbered offsets are entered into the second set of parentheses, and the even-numbered offsets are entered into the third set of parentheses. ►

EXAMPLE 3-15

Calculate the area for the irregularly shaped illustration in Figure 3-32 using Simpson's Rule.

Step 1

Measure the distance between the two points at the farthest ends of the figure. This is the length of the line between points A and B. The length of the line in Figure 3-32 is 50 ft.

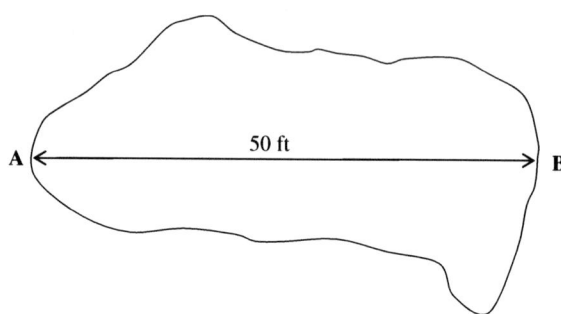

Figure 3-32

Establish the long axis

Step 2

Divide the line AB into an even number of equal subsegments (Figure 3-33). Increasing the number of subsegments will increase the accuracy. Use as many subsegments as is practical. This example will use 10 subsegments. The length of the subsegments is called h. For our example $h = 5$ ft.

Area of Geometric Figures **69**

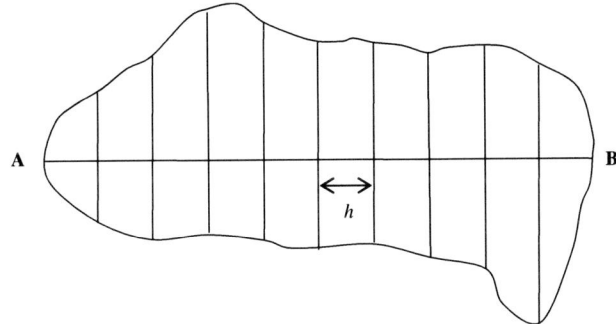

Figure 3-33
Establish the distance between offset lines at equally spaced intervals of length 'h'

Step 3
Draw lines perpendicular from the line that connects A to B at each mark. The lines will extend to the edges of the figure. Label each line L_1 to L_9 (Figure 3-34).

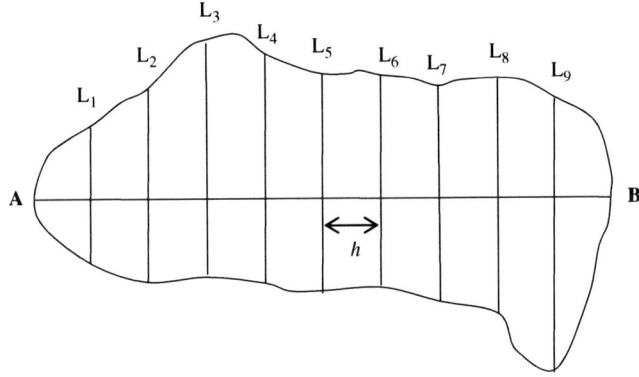

Figure 3-34
Establish offset lines that extend to the edge of the figure

Step 4
Measure the length of each line.

$L_1 = 10$ ft $L_4 = 7$ ft $L_7 = 6$ ft
$L_2 = 12$ ft $L_5 = 5$ ft $L_8 = 5$ ft
$L_3 = 10$ ft $L_6 = 6$ ft $L_9 = 4$ ft

Step 5
Insert your measurements into the following formula:

$$A = \frac{h}{3}[(L_1 + L_9) + 2(L_3 + L_5 + L_7) + 4(L_2 + L_4 + L_6 + L_8)]$$

Note that the first and last line lengths go into the first group of numbers in parentheses. Even-numbered line lengths go into the last group of numbers in parentheses. The remaining odd-numbered line lengths go into the middle group of numbers in parentheses.

$$A = \frac{5}{3}[(10 + 4) + 2(10 + 6 + 6) + 4(12 + 7 + 5 + 5)]$$

$$A = \frac{5}{3}[(14) + 2(22) + 4(29)]$$

$$A = \frac{5}{3}[14 + 44 + 116]$$

$$A = \frac{5}{3}(174)$$

$$A = 290 \text{ ft}^2$$

SOLUTION

The estimated area of this irregular figure is 290 ft².

Practice Problem Set 3-4 Area of Irregular Figures

1. Calculate the area of Figure 3-35 using the Offset Method.

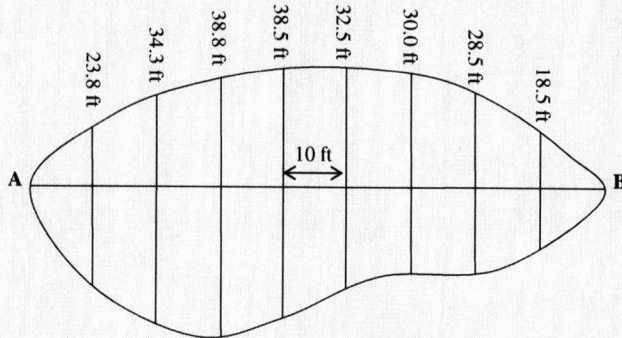

Figure 3-35
Calculate the area of this figure using the offset method

2. Calculate the area of Figure 3-36 using the Modified Offset Method.

The offset lines are spaced every 3 ft, and each pair of offset lines measures as follows: (E1 = 3 ft and E2 = 1.25 ft), (F1 = 1 ft and F2 = 0.75 ft), (G1 = 0.75 ft and G2 = 0.75 ft), (H1 = 1.25 ft and H2 = 1 ft), (I1 = 2.25 ft and I2 = 1.75 ft), (J1 = 3.25 ft and J2 = 1.25 ft),

(K1 = 3 ft and K2 = 0.75 ft), (L1 = 2.25 ft and L2 = 1.5 ft), (M1 = 2 ft and M2 = 3 ft), and (N1 = 2.75 ft and N2 = 4 ft).

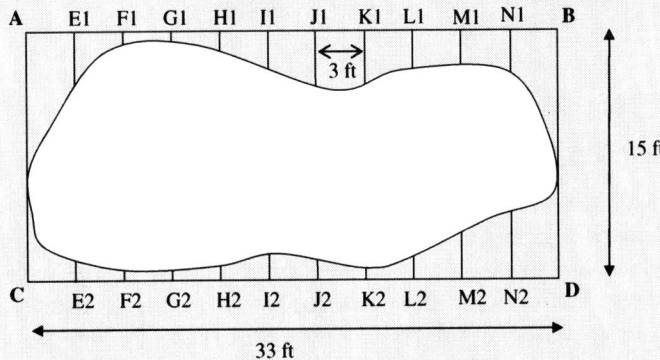

Figure 3-36
Calculate the area of this figure using the Modified Offset Method

1. Calculate the area of Figure 3-37 using Simpson's Rule.

$$L_1 = 26 \text{ ft} \quad L_4 = 36 \text{ ft} \quad L_7 = 36 \text{ ft}$$
$$L_2 = 34 \text{ ft} \quad L_5 = 33 \text{ ft} \quad L_8 = 34 \text{ ft}$$
$$L_3 = 38 \text{ ft} \quad L_6 = 35 \text{ ft} \quad L_9 = 30 \text{ ft}$$

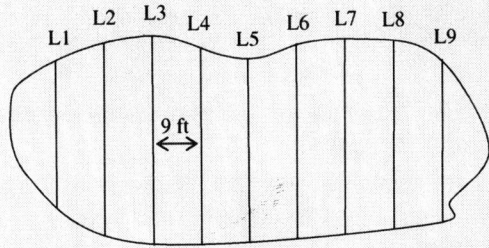

Figure 3-37
Calculate the area of this figure using Simpson's rule

Volume of Geometric Figures

Volume is described in terms of cubic measure. Common cubic measures are: cubic feet (ft^3), cubic yards (yd^3), and gallons (gal).

Watch unit conversions in U.S. measures of feet and yards. There are 27 ft^3 in every 1 yd^3 (Figure 3-38).

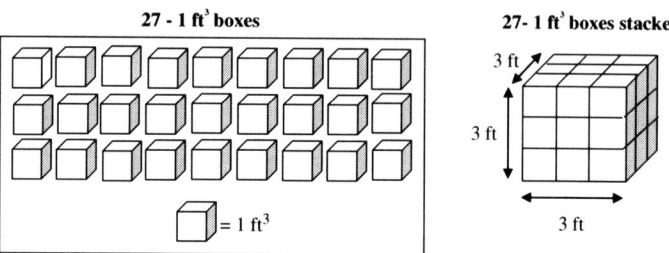

Figure 3-38

One cubic yard (yd³)

Figures with Parallel Bases and Equal Cross-Sections

The phrase "figures with parallel bases and equal cross-sections" sounds complex, but what it means is that if you take a slice or a cross-section of this type of solid geometric form, the slices are all the same and they are parallel or in line with one another. Think of a sliced loaf of bread that is perfectly square, round, rectangular, or triangular as a way to describe this type of figure. These figures are very common and familiar to most horticulturists. Cubes and cylinders are probably the most familiar. In geometry, a cube also is known as a square prism with all sides having the same length. Extend the length of one of the sides, and you still have a square prism because the cross-section or base is square. Change the cross-section to a triangle, and a triangular prism results. A triangular prism is the same shape as the gable portion of an even-span greenhouse. If the cross-section or base is a rectangle, it is called a rectangular prism. The generalized formula for calculating the volume of this type of figure is:

$$\text{Volume } (V) = Bh$$

Where B equals the area of one of the parallel bases or cross-sections and h equals the length of the side perpendicular to the base. Notice that the formula to calculate the area of the base (B) will vary depending on the shape of the base (or cross-section).

> **◄ Definition ►** **Formula for Calculating the Volume of Figures with Parallel Bases and Equal Cross-Sections**
>
> The formula for calculating the volume of figures with parallel bases and equal cross-sections, shown in Figure 3-39, is volume (V) is equal to the product of the area of the base (B) and the height (h).
>
> $$\text{Volume} = \text{Area of base} \times \text{height}$$
> $$V = Bh$$

Figure 3-39

Figures with parallel bases

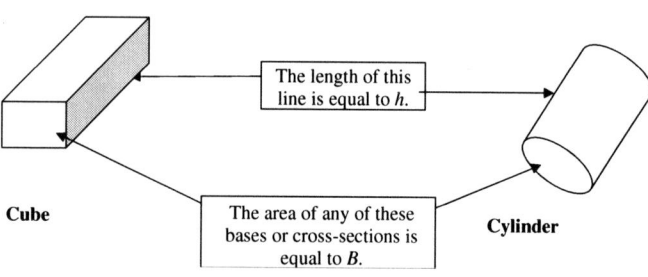

Volume of Geometric Figures 73

EXAMPLE 3-16

Calculate the volume (V) of a rectangular box that has a base (B) that measures 5 in. by 7 in. and a height (h) of 27 in. (Figure 3-40).

$$\text{Area of Base} = \text{Length} \times \text{Width}$$
$$B = l \times w$$
$$B = 7 \text{ in.} \times 5 \text{ in.}$$
$$B = 35 \text{ in.}^2$$
$$\text{Volume} = \text{base} \times \text{height}$$
$$V = B \times h$$
$$V = 35 \text{ in.}^2 \times 27 \text{ in.}$$
$$V = 945 \text{ in.}^3$$

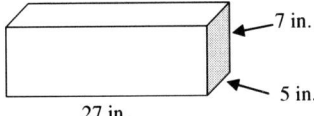

Figure 3-40
Calculating the volume of a rectangular box

Figures with Pointed Tops

Figures with pointed tops are commonly called cones or pyramids. Upon close examination, a cone has a round base, and pyramids commonly have square or triangular bases. The formula to determine the volume of these shapes with pointed tops is:

$$V = \tfrac{1}{3} B h$$

where B equals the area of the base of the figure and h represents the vertical height of the figure measured from the center of the base to the top of the point. Figure 3-41 illustrates the formula for each type of figure.

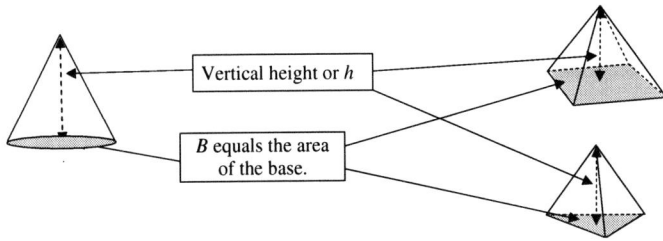

Figure 3-41
Figures with pointed tops

Definition ❧ Formula for Calculating the Volume of Figures with Pointed Tops

The formula for calculating the volume of figures with pointed tops is volume (V) is equal to one-third the product of the area of the base (B) and the height (h).

$$\text{Volume} = \frac{1}{3}(\text{Area of base} \times \text{height})$$

$$V = \frac{1}{3}Bh$$

EXAMPLE 3-17

Calculate the volume of a cone-shaped figure in Figure 3-42 that has a base (B) with a diameter of 22 in. and a height (h) of 18 in. (radius = $\frac{1}{2}$ diameter, thus the radius of the base = $\frac{1}{2}$ of 22 in. or 11 in.).

$$B = \pi r^2$$
$$B = \pi(11^2)$$
$$B = \pi \times 121 \text{ in.}^2$$
$$B = 380.1 \text{ in.}^2$$
$$V = \frac{1}{3}(B \times h)$$
$$V = \frac{1}{3}(380.1 \text{ in.}^2 \times 18 \text{ in.})$$
$$V = \frac{1}{3}(6842.3 \text{ in.}^3)$$
$$V = 2280.8 \text{ in.}^3$$

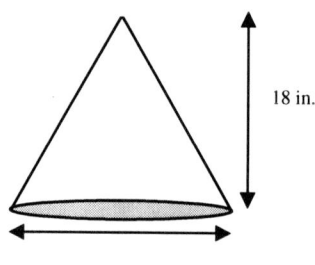

Figure 3-42
Calculating the volume of a cone

Figures with Tapered Sides

Figures with tapered sides (Figure 3-43) are common among plant containers used in the green industry. This would include pots used for growing plants in greenhouses or nurseries and containers (some of them quite large) that can be used in the interior landscape industry.

Volume of Geometric Figures **75**

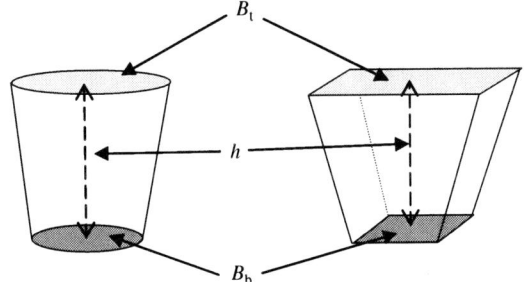

Figure 3-43
Figures with tapered sides

The formula to determine the volume of figures with tapered sides is:

$$V = [(B_t + B_b) \div 2](h)$$

where: B_t = Area of the top of the figure; B_b = Area of the bottom of the figure; and h = height measured from the center of the base to the center of the top.

> **◆ Definition ◆** *Formula for Calculating the Volume of Figures with Tapered Sides*
>
> The formula for calculating the volume of figures with tapered side is volume (V) is equal to the product of the average of area of the bottom (B_b) and the area of the top (B_t) and the height (h).
>
> Volume = [(Area of top + Area of bottom) ÷ 2] × height)
>
> $$V = [(B_t + B_b) \div 2] \times h$$

EXAMPLE 3-18

Calculate the volume of the figure illustrated in Figure 3-44 that has tapered sides. Dimensions are the Base top (B_t) with a radius of 2.0 in., a Base bottom (B_b) of 1.5 in. and a height (h) of 4 in.

$$B_b = \pi r^2$$
$$B_t = \pi r^2$$
$$B_b = \pi(1.5^2)$$
$$B_t = \pi(2.0^2)$$
$$B_b = \pi \times 2.25 \text{ in.}^2$$
$$B_t = \pi \times 4.0 \text{ in.}^2$$
$$B_b = 7.069 \text{ in.}^2$$

$B_t = 12.57 \text{ in.}^2$

$V = [(B_t + B_b) \div 2](h)$

$V = [(12.57 \text{ in.}^2 + 7.069 \text{ in.}^2) \div 2](4)$

$V = [(19.639 \text{ in.}^2) \div 2](4)$

$V = (9.8195 \text{ in.}^2)(4)$

$V = 39.28 \text{ in.}^3$

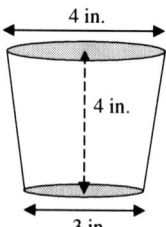

Figure 3-44

Calculate volume of this figure with tapered sides

Spherical and Hemispherical Figures

This type of figure is not as common in horticulture. Perfectly rounded bowl-shaped containers like hanging moss baskets are hemispherical. Some garden pools may be semi-spherical. Round topiary standards are spherical. The formula for the area of a sphere is:

$$V = \frac{4}{3}\pi r^3$$

The formula for hemisphere or one-half a sphere is:

$$V = \left(\frac{4}{3}\pi r^3\right) \div 2$$

Figure 3-45 illustrates spheres and hemispheres.

Figure 3-45

Figures with spherical and hemispherical shapes

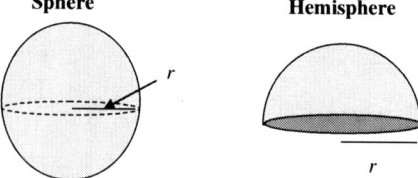

Volume of Geometric Figures 77

❖ Definition ❖ Formulae for Calculating the Volume of Spheres and Hemispheres

The formula for calculating the volume of a sphere is volume (V) is equal to four-thirds the product of pi and the radius cubed.

$$\text{Volume} = \frac{4}{3}(\text{pi})(\text{radius}^3)$$

$$V = \frac{4}{3}\pi r^3$$

The formula for calculating the volume of a hemisphere is volume (V) is equal to four-thirds the product of pi and the radius cubed divided by two.

$$\text{Volume} = \left[\frac{4}{3}(\text{pi})(\text{radius}^3)\right] \div 2$$

$$V = \left[\frac{4}{3}\pi r^3\right] \div 2$$

EXAMPLE 3-19

Calculate the volume of a sphere (Figure 3-46) that has a radius (r) of 6 in.

$$V = \frac{4}{3}\pi r^3$$

$$V = \frac{4}{3}\left[\pi\left(6^3\right)\right]$$

$$V = \frac{4}{3}\left[\pi\left(216 \text{ in.}^3\right)\right]$$

$$V = \frac{4}{3} \times 678.5 \text{ in.}^3$$

$$V = 904.8 \text{ in.}$$

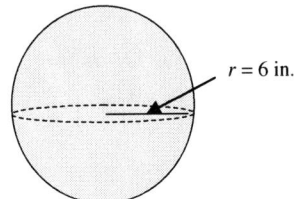

Figure 3-46

Calculate the volume of this spherical figure

EXAMPLE 3-20

Calculate the volume of a hemisphere (Figure 3-47) that has a radius (r) of 5 in.

$$V = \left(\frac{4}{3}\pi r^3\right) \div 2$$

$$V = \left[\frac{4}{3}\left[\pi\left(5^3\right)\right]\right] \div 2$$

$$V = \left[\frac{4}{3}\left[\pi\left(125 \text{ in.}^3\right)\right]\right] \div 2$$

$$V = \left[\frac{4}{3} \times 392.7 \text{ in.}^3\right] \div 2$$

$$V = 523.6 \div 2$$

$$V = 261.8 \text{ in.}$$

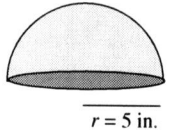

Figure 3-47
Calculating the volume of a hemisphere

$r = 5$ in.

Practice Problem Set 3-5 *Volume of Geometric Figures*

1. Calculate the volume of a rectangular-shaped prism that has a base that measures 6 ft by 10 ft and a height of 14 ft.
2. Calculate the volume of a cylinder that has a base with a diameter of 12 in. and a height of 3 ft.
3. Calculate the volume of a cone that has a base with a radius of 1 ft 3 in. and a height of 2.5 ft.
4. Calculate the volume of a sphere that has a diameter of 18 in.

PART TWO: GREEN INDUSTRY APPLICATIONS

Chapter 4: Calculating the Area of Landscape Features

Introduction

Accurately calculating the area of different landscape features is an important skill for many horticulture professionals. For example, calculating the correct area for a landscape bed or turfgrass area is the first step in making appropriate and accurate fertilizer and pesticide applications. The amount of mulch or soil amendment a landscape contractor will purchase for a project depends on an accurate calculation of the area. An accurate area calculation also is essential for landscape professionals who design and install either annual bedding plant displays or groundcover plantings. The number of plants used in these types of plantings depends on the total area to be covered and the plant spacing. Finally, landscape professionals involved in designing and building hardscape projects (for example, patios and walkways) must calculate the hardscape area to estimate the number of paving units required for a project.

There are several strategies for determining the area of landscape features. The strategy chosen by a horticulturist is dependent on the shape of the space, the ease by which an area can be measured, and the importance of the accuracy of the calculation. The area of landscape features can be determined using the following techniques or strategies: the Geometric Method, the Offset Method, the Modified Offset Method, or through the use of Simpson's Rule. Chapter 3 discusses how to use the Offset Method, the Modified Offset Method, and Simpson's Rule for equations with simple geometric forms.

Geometric Method

The Geometric Method is used to calculate the area of landscape spaces that have simple geometric shapes or combinations of simple geometric shapes (composite figures). Any time a horticulturist can superimpose a simple geometric shape or a combination of simple geometric shapes over a landscape space, the Geometric Method can be used. The accuracy of this method depends on how perfectly the shape fits the space.

Simple Geometric Forms

In many situations, a single, simple geometric shape can be superimposed over a landscape feature and used to determine the area of the feature. One of the most recognizable geometric shapes is the rectangle, and many landscape spaces take that form. A good example of a rectangular landscape space is a football field.

EXAMPLE 4-1

Calculate the area of a football field (Figure 4-1).

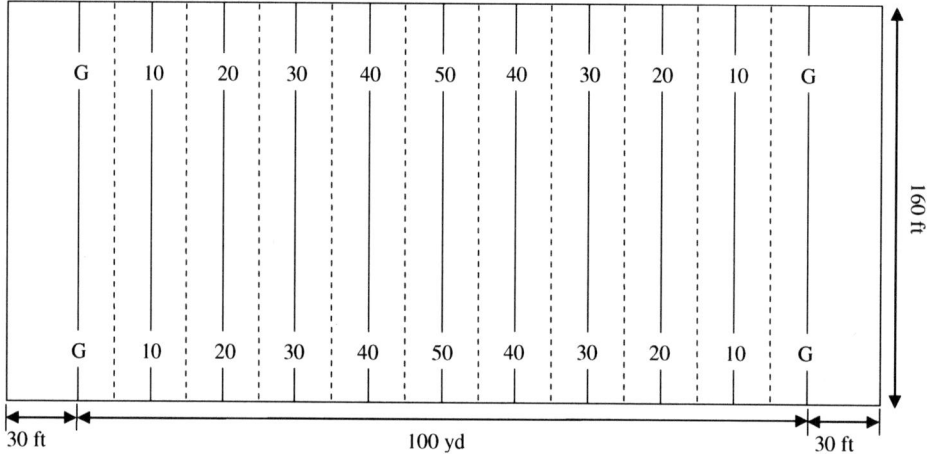

Figure 4-1
A regulation football field

Step 1

Measure the space. A football field measures 100 yd long with 30 ft end zones beyond each goal line. The width of a football field is 160 ft.

Step 2

The measurements have mixed units; therefore, convert all of the measurements to feet. Use a proportion and isolate and solve for x to determine how many feet are equivalent to 100 yd. Table B-1 in Appendix B determines that 3 ft is equivalent to 1 yd.

$$\frac{x \text{ ft}}{100 \text{ yd}} = \frac{3 \text{ ft}}{1 \text{ yd}}$$

Cross multiply:

$$x = \frac{(3 \text{ ft})(100 \text{ yd})}{1 \text{ yd}}$$

$$x = 300 \text{ ft}$$

One hundred yards are equivalent to 300 ft.

Step 3
Determine the length of the football field in feet.

$$30 \text{ ft end zone} + 300 \text{ ft field} + 30 \text{ ft end zone} = 360 \text{ ft}$$

A football field is 360 ft long.

Step 4
Calculate the area of the football field by using the formula for the area of a rectangle.

$$\text{Area} = \text{Length} \times \text{Width}$$
$$\text{Area} = 360 \text{ ft} \times 160 \text{ ft}$$
$$\text{Area} = 57{,}600 \text{ ft}^2$$

SOLUTION

The area of a football field is 57,600 ft^2.

Composite Geometric Forms

Landscape features can be more complex and calculating the area can require that several similar or dissimilar forms be superimposed onto the space. For example, driveways can be rectangular, L-shaped, or a combination of rectangles and squares. Sidewalks are commonly composed of several square- or rectangle-shaped subunits. However, occasionally paved areas are round or oval in shape. Lawn spaces and planting beds can be combinations of round, oval, trapezoidal, rectangular, or square shapes.

EXAMPLE 4-2

Calculate the area of a sidewalk found in Figure 4-2.

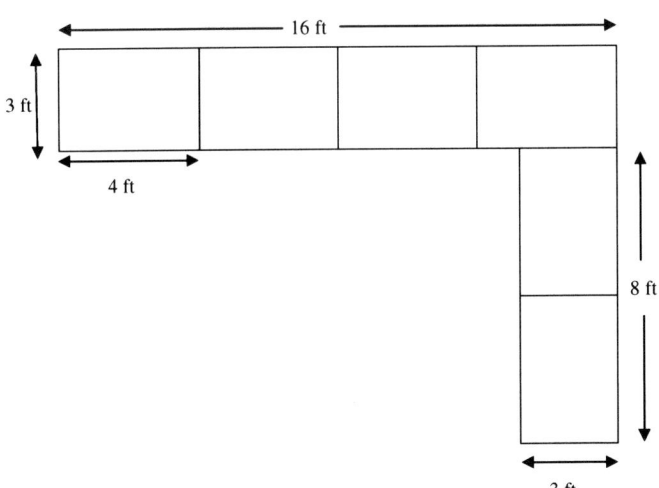

Figure 4-2
Dimensions of a sidewalk

There are three ways that the Geometric Method can be used to determine the area of this sidewalk, as follows:

1. Calculate the area of the sidewalk by dividing the space into six 3 ft by 4 ft rectangles.
2. Calculate the area of the sidewalk by dividing it into two rectangles with one measuring 16 ft by 3 ft (Area 1) and the other measuring 8 ft by 3 ft (Area 2).
3. Calculate the area by determining the total length in linear feet and multiplying by the width.

METHOD 1

Calculate the area of the sidewalk by dividing the space into six 3 ft by 4 ft rectangles.

Step 1

Calculate the area of one of the six rectangles.

$$Area = Length \times Width$$
$$Area = 3\,ft \times 4\,ft$$
$$Area = 12\,ft^2$$

Step 2

Multiply the number of rectangles by the area of each.

$$6 \times 12\,ft^2 = 72\,ft^2$$

SOLUTION

The area of this sidewalk is 72 ft².

METHOD 2

Calculate the area of the sidewalk by dividing it up into two rectangles with one measuring 16 ft by 3 ft (Area 1) and the other measuring 8 ft by 3 ft (Area 2).

Step 1

Calculate the area of each rectangle.

$$Area = Length \times Width$$
$$Area\ 1 = 16\,ft \times 3\,ft$$
$$Area\ 1 = 48\,ft^2$$
$$Area\ 2 = 8\,ft \times 3\,ft$$
$$Area\ 2 = 24\,ft^2$$

Step 2
Add the two areas to find the sum.

$$\text{Total Area} = 48\,\text{ft}^2 + 24\,\text{ft}^2$$
$$\text{Total Area} = 72\,\text{ft}^2$$

SOLUTION

The area of this sidewalk is 72 ft^2.

METHOD 3
Calculate the area by determining the total length in linear feet and multiplying by the width.

Step 1
Add the length of the sidewalk to determine the total linear feet of sidewalk.

$$16\,\text{ft} + 8\,\text{ft} = 24\,\text{ft}$$

Step 2
Multiply the total linear feet of sidewalk times the width of the sidewalk.

$$\text{Area} = \text{Length} \times \text{Width}$$
$$\text{Area} = 24\,\text{ft} \times 3\,\text{ft}$$
$$\text{Area} = 72\,\text{ft}^2$$

SOLUTION

The area of this sidewalk is 72 ft^2.

Most lawns can be divided into multiple geometric figures like rectangles, trapezoids, triangles, and ellipses.

EXAMPLE 4-3

Using the landscape example in Figure 4-3, determine the area of the lawn.

Step 1
Superimpose simple geometric shapes over the lawn area. Figure 4-4 illustrates the lawn area segmented into six geometric shapes. There is one ellipse, two triangles, one trapezoid, and two rectangles.

Step 2
Calculate the area of each of the geometric shapes.

  Area A—Rectangle

$$\text{Area A} = \text{Length} \times \text{Width}$$
$$\text{Area A} = 57.5\,\text{ft} \times 46.5\,\text{ft}$$
$$\text{Area A} = 2{,}674\,\text{ft}^2$$

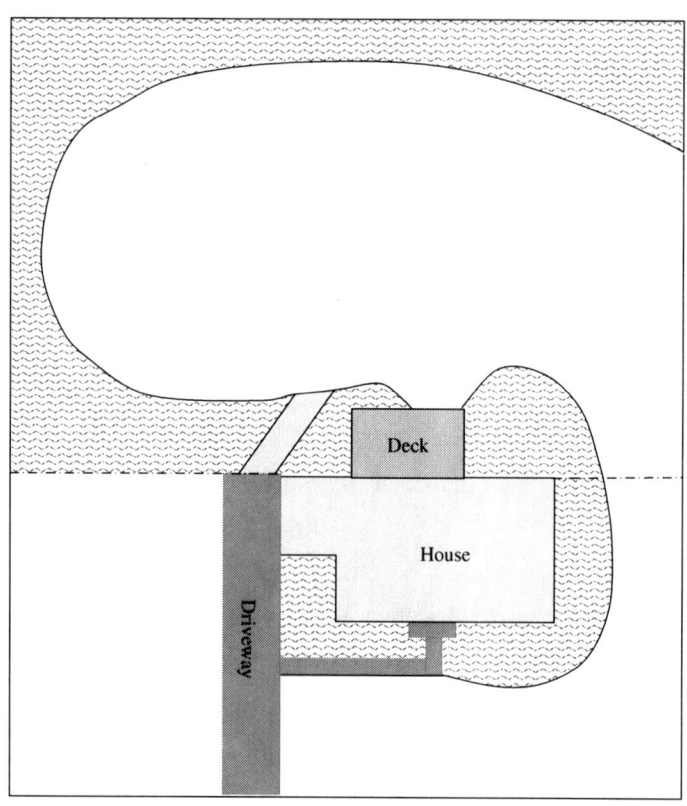

Figure 4-3
Smith property

Area B—Rectangle

$$\text{Area B} = \text{Length} \times \text{Width}$$
$$\text{Area B} = 85 \text{ ft} \times 22 \text{ ft}$$
$$\text{Area B} = 1{,}870 \text{ ft}^2$$

Area C—Trapezoid

$$\text{Area C} = [(b_1 + b_2) \div 2] \times \text{height}$$
$$\text{Area C} = [(120.5 \text{ ft} + 109.5 \text{ ft}) \div 2] \times 21 \text{ ft}$$
$$\text{Area C} = [230 \text{ ft} \div 2] \times 21 \text{ ft}$$
$$\text{Area C} = 115 \text{ ft} \times 21 \text{ ft}$$
$$\text{Area C} = 2{,}415 \text{ ft}^2$$

Area D—Triangle

$$\text{Area D} = (b \times h) \div 2$$
$$\text{Area D} = (27 \text{ ft} \times 9 \text{ ft}) \div 2$$

Geometric Method **85**

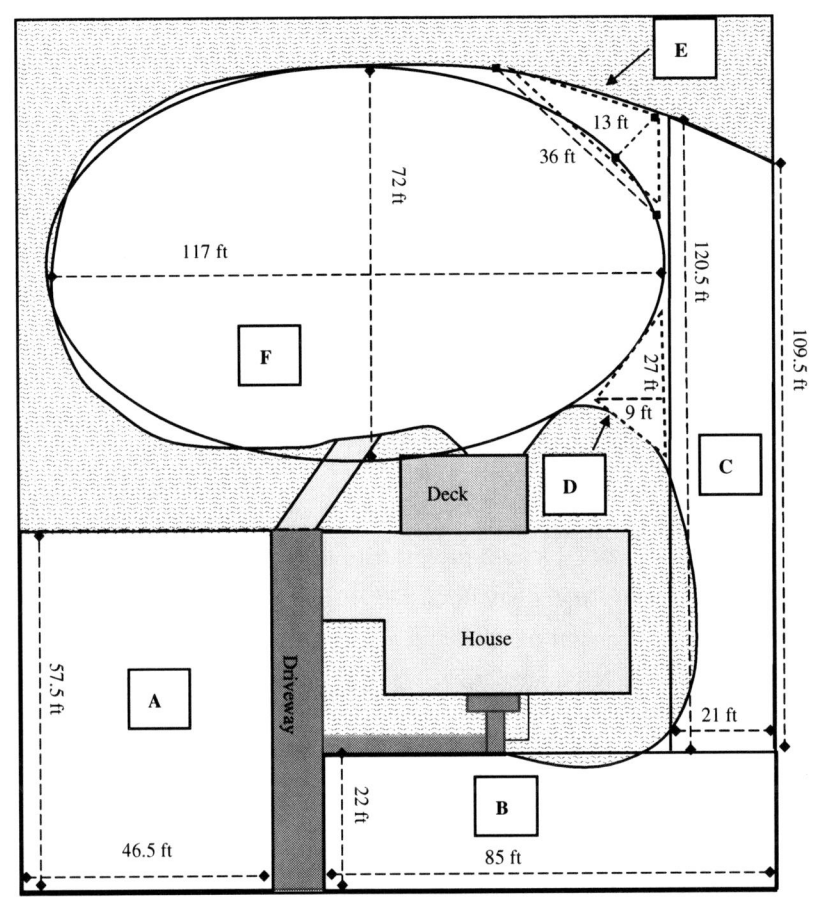

Figure 4-4

Smith property and geometric figures

$$\text{Area D} = 243\,\text{ft}^2 \div 2$$
$$\text{Area D} = 122\,\text{ft}^2$$

Area E—Triangle

$$\text{Area E} = (b \times h) \div 2$$
$$\text{Area E} = (36\,\text{ft} \times 13\,\text{ft}) \div 2$$
$$\text{Area E} = 468\,\text{ft}^2 \div 2$$
$$\text{Area E} = 234\,\text{ft}^2$$

Area F—Ellipse

$$\text{Area} = \pi\,[(wr)(lr)]$$
$$wr = 72\,\text{ft} \div 2 = 36\,\text{ft}$$
$$lr = 117\,\text{ft} \div 2 = 58.5\,\text{ft}$$

$$\text{Area F} = \pi\,[(36\text{ ft})(58.5\text{ ft})]$$
$$\text{Area F} = \pi\,(2{,}106\text{ ft}^2)$$
$$\text{Area F} = 6{,}616\text{ ft}^2$$

Step 3

Add the areas of each of the geometric shapes to determine the total lawn area.

$$2{,}674\text{ ft}^2 + 1{,}870\text{ ft}^2 + 2{,}415\text{ ft}^2 + 122\text{ ft}^2 + 234\text{ ft}^2 + 6{,}616\text{ ft}^2 = 13{,}931\text{ ft}^2$$

SOLUTION

The lawn area is 13,931 ft^2.

Practice Problem Set 4-1 — Area of Geometric Figures

Calculate the area of each of the following:

1. A rugby field that has a width of 225 ft and a length of 330 ft (unit = ft^2).
2. A lacrosse field that has a width of 180 ft and a length of 330 ft (unit = ft^2).
3. An oval-shaped flower bed that has a diameter of 25 ft and a length of 65 ft (unit = ft^2).
4. There are 12 trees in a landscape that have circular tree rings that measure 8 ft in diameter.
   a. What is the area of one tree ring?
   b. What is the total area of all the tree rings? (unit = ft^2)

Area of Irregular Landscape Features

Many landscape features are irregular in shape and do not conform to common geometric shapes or composites of geometric shapes. Perennial garden beds, shrub borders, foundation planting beds, ponds, and pools are examples of landscape features that often have irregular shapes. The Offset Method and Simpson's Rule are used in landscapes when direct measurements can be made across an area. The Modified Offset Method is used for either water features or when obstacles prevent direct measurements. The Offset Method and Modified Offset Method reduce an irregular shape into smaller trapezoids. While this provides a good approximation of an area, it does not account for the curvature of a landscape feature. When the landscape feature has a lot of curvature, Simpson's Rule should be used. Simpson's Rule accounts for the curvature of a landscape feature, thereby providing a more accurate estimate of a curved area.

Method 1: Offset Method

The Offset Method is a way of estimating the area of an irregularly shaped landscape feature. It divides an irregularly shaped area into a series of trapezoids that are equally spaced along a measured line.

Area of Irregular Landscape Features **87**

EXAMPLE 4-4

Using the Offset Method, determine the area of the flower bed in Figure 4-5.

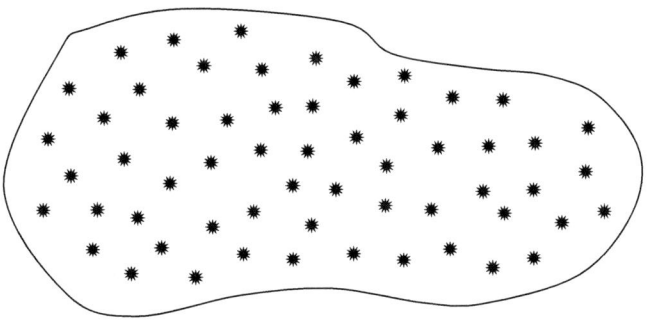

Figure 4-5
Flower bed

Step 1
Establish and measure the length line and label it AB (Figure 4-6).

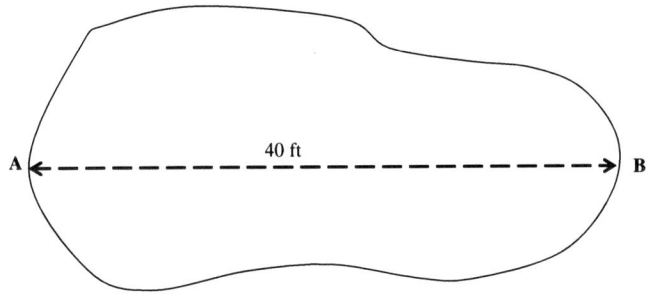

Figure 4-6
Length line

Step 2
Establish offset lines at 4 ft spacing along the length line (Figure 4-7).

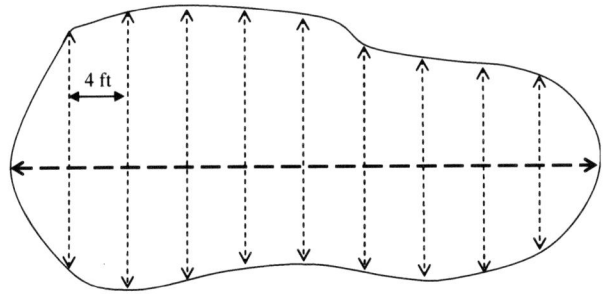

Figure 4-7
Establish offset lines

Step 3
Measure each offset line from end to end (Figure 4-8).

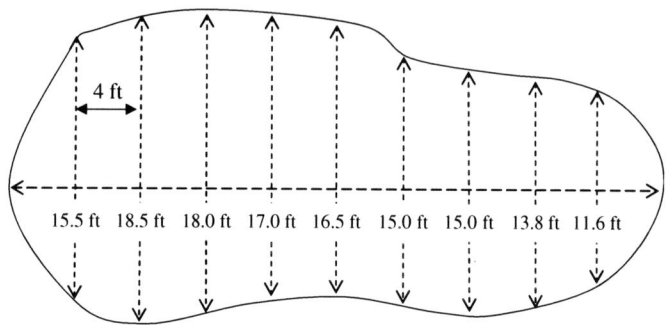

Figure 4-8

Measure the length of the offset lines

Step 4

Add the lengths to find the sum of the offset lines.

$15.5\,\text{ft} + 18.5\,\text{ft} + 18\,\text{ft} + 17\,\text{ft} + 16.5\,\text{ft} + 15\,\text{ft} + 15\,\text{ft} + 13.8\,\text{ft} + 11.6\,\text{ft} = 140.9\,\text{ft}$

Step 5

Multiply the sum of the offset lines by the distance between the offset lines.

$$140.9\,\text{ft} \times 4\,\text{ft} = 563.6\,\text{ft}^2$$

SOLUTION

The surface area of the flower bed is $563.6\,\text{ft}^2$.

Method 2: Modified Offset Method

The Modified Offset Method is used to measure areas that are irregular in shape but cannot easily be traversed to set the offset lines. It divides an irregularly shaped area into a series of trapezoids that are equally spaced along a measured line.

EXAMPLE 4-5

Using the Offset Method, determine the area of the pond in Figure 4-9.

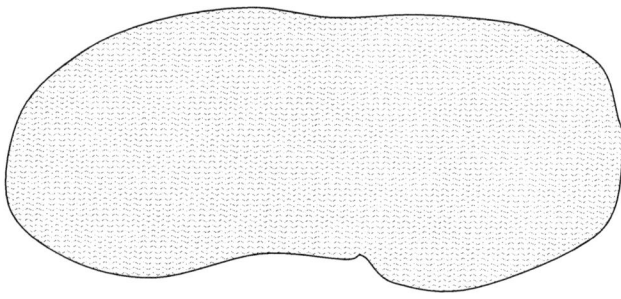

Figure 4-9

Surface area of the pond

Step 1

Create a rectangle around the area that is to be measured (Figure 4-10). Measure the length of line segments AB or CD. These are equivalent to the length of the rectangle. Measure the length of line segments AC or BD. These are equivalent to the width of the rectangle.

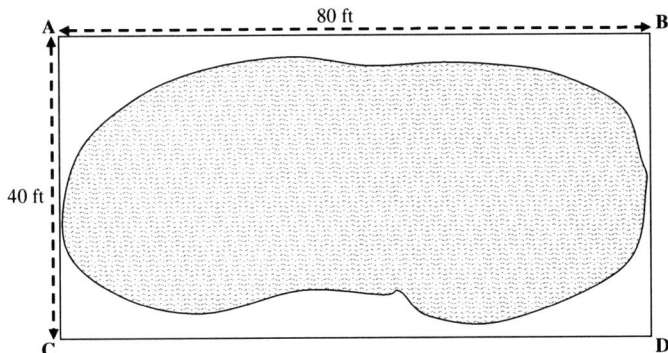

Figure 4-10
Determine length and width lines

Step 2

Establish offset lines every 4 ft perpendicular to lines AB and CD. Label each offset line as shown in Figure 4-11.

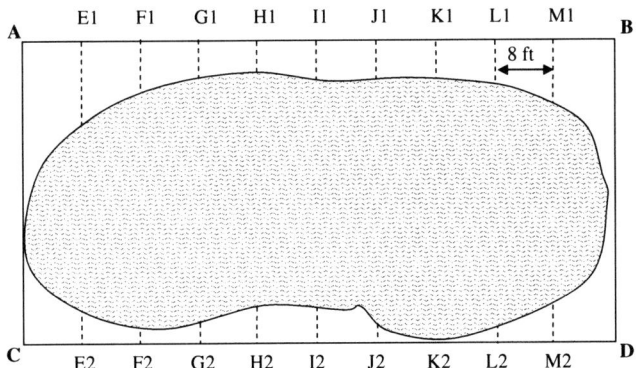

Figure 4-11
Establish offset lines

Step 3

Measure each offset line from the edge of the rectangle to the perimeter of the landscape feature.

E1 = 10.5 ft: F1 = 7 ft: G1 = 5 ft: H1 = 4 ft: I1 = 5 ft: J1 = 5 ft:

K1 = 5 ft: L1 = 5.5 ft: M1 = 8 ft

$$E2 = 5.5 \text{ ft}: F2 = 1 \text{ ft}: G2 = 2.5 \text{ ft}: H2 = 5.5 \text{ ft}: I2 = 5.5 \text{ ft}: J2 = 2 \text{ ft}:$$
$$K2 = 0.50 \text{ ft}: L2 = 2 \text{ ft}: M2 = 5.5 \text{ ft}$$

Step 4

Add the paired offset lines.

$$\Sigma E = E1 + E2 = 16 \text{ ft}:$$
$$\Sigma F = F1 + F2 = 8 \text{ ft}:$$
$$\Sigma G = G1 + G2 = 7.5 \text{ ft}:$$
$$\Sigma H = H1 + H2 = 9.5 \text{ ft}:$$
$$\Sigma I = I1 + I2 = 10.5 \text{ ft}:$$
$$\Sigma J = J1 + J2 = 7 \text{ ft}:$$
$$\Sigma K = K1 + K2 = 5.5 \text{ ft}:$$
$$\Sigma L = L1 + L2 = 7.5 \text{ ft}:$$
$$\Sigma M = M1 + M2 = 13.5 \text{ ft}:$$

Step 5

Subtract the sum of each offset line from the width to obtain the distance across the landscape feature.

$$E: 40 \text{ ft} - 16 \text{ ft} = 24 \text{ ft}$$
$$F: 40 \text{ ft} - 8 \text{ ft} = 32 \text{ ft}$$
$$G: 40 \text{ ft} - 7.5 \text{ ft} = 32.5 \text{ ft}$$
$$H: 40 \text{ ft} - 9.5 \text{ ft} = 30.5 \text{ ft}$$
$$I: 40 \text{ ft} - 10.5 \text{ ft} = 29.5 \text{ ft}$$
$$J: 40 \text{ ft} - 7 \text{ ft} = 33 \text{ ft}$$
$$K: 40 \text{ ft} - 5.5 \text{ ft} = 34.5 \text{ ft}$$
$$L: 40 \text{ ft} - 7.5 \text{ ft} = 32.5 \text{ ft}$$
$$M: 40 \text{ ft} - 13.5 \text{ ft} = 26.5 \text{ ft}$$

Step 6

Add the lengths to find the sum of the offset lines.

$$24 \text{ ft} + 32 \text{ ft} + 32.5 \text{ ft} + 30.5 \text{ ft} + 29.5 \text{ ft} + 33 \text{ ft} + 34.5 \text{ ft} + 32.5 \text{ ft} + 26.5 \text{ ft} = 275 \text{ ft}$$

Step 7

Multiply the sum of the offset lines by the distance between the offset lines.

$$275 \text{ ft} \times 8 \text{ ft} = 2{,}200 \text{ ft}^2$$

Area of Irregular Landscape Features **91**

SOLUTION

The surface area of the pond is 2,220 ft².

Method 3: Simpson's Rule

Simpson's Rule provides a more accurate method of estimating the area of an irregular figure than the Offset or Modified Offset Methods. The mathematics involved is a little more complicated, but worth the effort if the area being calculated has many curves.

EXAMPLE 4-6

Using Simpson's Rule, determine the area of the garden bed in Figure 4-12.

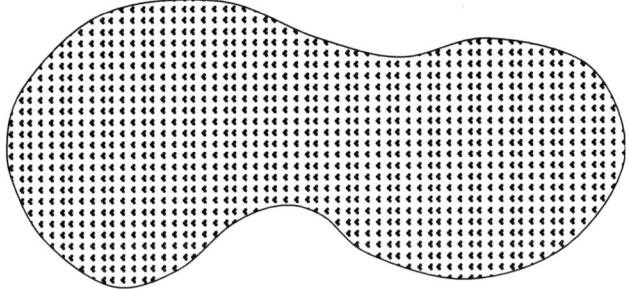

Figure 4-12

Surface area of a garden bed

Step 1

Measure the distance between the two points at the farthest ends of the figure. This is the length of the line between points A and B (Figure 4-13).

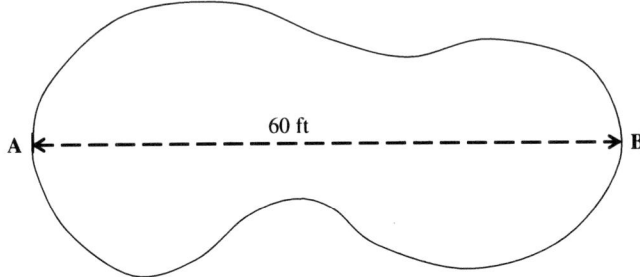

Figure 4-13

Determine the length line

Step 2

Divide the line into an even number of equally spaced subsegments (Figure 4-14). This example will use 10 subsegments. The width of each subsegment is called h. For this example $h = 6$ ft.

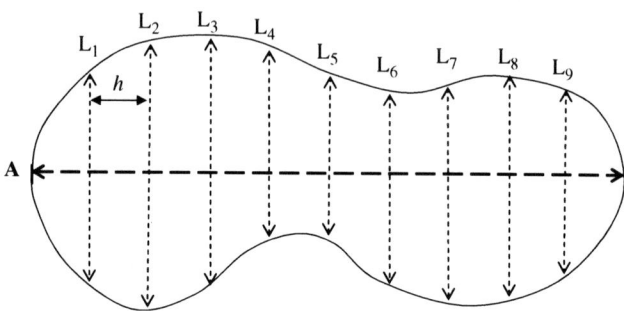

Figure 4-14

Establish the distance between offset lines at equally spaced intervals of length h.

Step 3

Draw lines perpendicular from the line that connects A to B at each mark. The lines will extend to the edges of the figure. Label each line L_1 to L_9 (Figure 4-14).

Step 4

Measure the length of each line.

$L_1 = 20$ ft $L_4 = 19.5$ ft $L_7 = 21$ ft
$L_2 = 26$ ft $L_5 = 16$ ft $L_8 = 22$ ft
$L_3 = 24.5$ ft $L_6 = 19$ ft $L_9 = 18.5$ ft

Step 5

Insert your measurements into the following formula:

$$\text{Area} = \frac{h}{3}[(L_1 + L_9) + 2(L_3 + L_5 + L_7) + 4(L_2 + L_4 + L_6 + L_8)]$$

$$\text{Area} = \frac{6}{3}[(20 + 18.5) + 2(24.5 + 16 + 21) + 4(26 + 19.5 + 19 + 22)]$$

$$\text{Area} = \frac{6}{3}[(38.5) + 2(61.5) + 4(86.5)]$$

$$\text{Area} = \frac{6}{3}[38.5 + 123 + 346]$$

$$\text{Area} = \frac{6}{3}(507.5)$$

$$\text{Area} = 1{,}015 \text{ ft}^2$$

SOLUTION

The surface area of the garden bed is $1{,}015$ ft^2.

Practice Problem Set 4-2 — *Area of Irregular Landscape Features*

1. Calculate the area of the wildflower planting in Figure 4-15. The total length of the area is 110 ft^2.
   a. Using the Offset Method
   b. Using Simpson's Rule

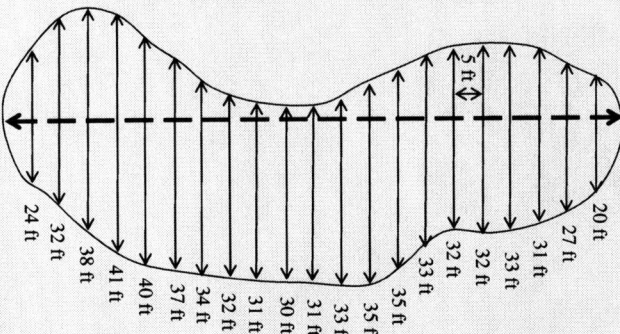

Figure 4-15
Surface area of a wildflower bed

2. Calculate the area of a small garden pond in Figure 4-16 using the Modified Offset method.

The offset lines measure as follows:

$$E1 = 2.5\,\text{ft} \quad F1 = 1.5\,\text{ft} \quad G1 = 1.0\,\text{ft} \quad H1 = 1.0\,\text{ft} \quad I1 = 1.3\,\text{ft}$$
$$E2 = 2.7\,\text{ft} \quad F2 = 2.5\,\text{ft} \quad G2 = 2.0\,\text{ft} \quad H2 = 1.5\,\text{ft} \quad I2 = 1.3\,\text{ft}$$

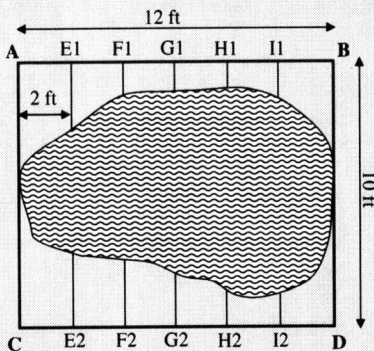

Figure 4-16
Surface area of a small garden pond

Chapter 5: Fertilizer Calculations

Introduction

Professionals in the green industry use many different types of fertilizer products during the course of their work. These products are essential to maintaining good plant health. Inadequate fertilizer applications result in poor plant growth and poor plant performance, and excessive fertilizer applications result in poor plant health and risk environmental damage. Knowledge of fertilizer products and application rates, understanding fertilizer labels, and mastering the mathematical skills to calculate the correct amount of product for an application are essential skills for the green industry professional. This chapter will focus on the skills necessary to read a fertilizer label, interpret a rate of application, and calculate the proper amount of fertilizer for a given application.

Fertilizer Terminology

Working with fertilizer calculations requires an understanding of the terminology used in the fertilizer industry. Terms that are most important in understanding how to read a label and perform fertilizer calculations are included in this chapter.

Fertilizer Analysis

Fertilizer analysis is the key to performing fertilizer calculations. An analysis of a fertilizer product provides information about the percentage of the nutrients (by weight) in the bag, box, or container of liquid that holds the fertilizer. It is usually listed in large letters in a prominent place on the container; however, it may be found in a small, detailed label. It is presented as three numbers separated by a hyphen (-). A common garden fertilizer analysis is 13-13-13.

These three numbers provide important information on which to base calculations. The first number is nitrogen (N). Nitrogen is expressed on an elemental basis, and this material contains 13 percent by weight elemental N. The second number is the percentage by weight of phosphoric acid (P_2O_5). It is important to note that this is not the same as elemental phosphorus (P). The third number represents the percentage by weight of potash (K_2O). Again, this is not the same as elemental potassium (K). The use of P_2O_5 or phosphoric acid and K_2O or potash does not mean the fertilizer contains these specific molecules, but these oxide forms of the elements result from elemental analysis of a product and allows comparisons between products. A 13-13-13 fertilizer contains by weight 13% N, 13% P_2O_5, and 13% K_2O. The manufacturer guarantees this analysis, and the analysis always reads in the above order.

> **◄ Definition ►** *Fertilizer Analysis*
>
> **Fertilizer analysis** or guaranteed analysis is the amount of N, P_2O_5, and K_2O expressed as a percent of the total weight of a fertilizer product. This information is expressed in the format:
>
> $$N\text{-}P_2O_5\text{-}K_2O$$
>
> For example, a 13-13-13 product contains 13% N, 13% P_2O_5, and 13% K_2O.

Form or Formulation

Fertilizer products are available in many forms or formulations. Dry forms of fertilizer products include granules, prills, and water-soluble granules. Granules are spread over a lawn or garden area in dry form, or they may be incorporated into root media (potting soil) during formulation. Prills include sulfur-coated products that are popular in lawn care and polymer-encapsulated products that are popular in the greenhouse and nursery industries. Sulfur-coated products are spread over turf areas and polymer-encapsulated products are usually applied to the surface of root media of containerized crops (top-dressed). Water-soluble products are easy to dissolve in water and are applied as a periodic liquid treatment or are applied with irrigation water. Water-soluble granule products are used often in conjunction with fertilizer injector or proportioner equipment, and those calculations can be found in Chapter 10. Liquid fertilizer products are sold as liquids and are diluted in water prior to application. Commercial liquid product use is more common in the lawn-care industry, and those calculations can be found later in this chapter. The advantage of using commercial liquid products is that pesticide and growth regulator products can be added to the application.

Rate

Rate or rate of application of a fertilizer can be expressed in several formats. The format that is used depends on the industry and the source of the recommendation. Textbook resources often recommend rates based on the nutritional elements. Commercial resources, such as fertilizer manufacturers, often recommend rates based on the amount of fertilizer product to be used. The primary methods for reporting fertilizer rate follow.

Fertilizer Terminology

Amount of Fertilizer Product per Unit Area

Commercial fertilizer products used to fertilize annual display beds, perennial beds, and shrub beds commonly list rate in this format on the label. All consumer products labels in the lawn and garden industry list rate in this format as well. These fertilizer rates are expressed in this way:

$$\frac{\text{Amount of Product by Weight}}{\text{Square Feet of Area}}$$

For example:

$$\frac{2 \text{ lb } 5\text{-}10\text{-}5}{100 \text{ ft}^2}$$

Amount of a Nutritional Element (N, P, or K) per Unit Area

Technical publications, textbooks, and soil test results report the rate of application in the following way:

$$\frac{\text{Amount of a Nutritional Element by Weight}}{\text{Square Feet of Area}}$$

For example:

$$\frac{1 \text{ lb N}}{1{,}000 \text{ ft}^2}$$

These resources are objective and will not recommend a specific product. The focus is on plant nutritional needs, not on a fertilizer product recommendation.

Amount of a Nutritional Element Expressed in Oxide Form (P_2O_5 or K_2O) per Unit Area

Technical publications, textbooks, and soil test results report the rate of application in the following way:

$$\frac{\text{Amount of a Nutritional Element in the Oxide Form by Weight}}{\text{Square Feet of Area}}$$

For example:

$$\frac{1.5 \text{ lb } K_2O}{1{,}000 \text{ ft}^2}$$

These resources are objective and will not recommend a specific product. The focus is on plant nutritional needs, not on a fertilizer product recommendation.

Amount of a Nutritional Element Expressed in Parts per Million (ppm)

Technical and textbook resources, particularly those in the greenhouse and nursery industries, report fertilizer rate in the following way:

Parts per Million of a Nutritional Element

For example:

250 ppm N

All the products used are water-soluble granules, and they are dissolved in water prior to application to the plants. Most of the time, fertilizer proportioners or injectors are used to make these applications. These mathematical calculations are found in Chapter 10.

> **◄ Fertilizer Fact ►**
> Fertilizer Rate can be expressed as:
>
> $$\frac{\text{Amount of Product by Weight}}{\text{Square Feet of Area}}$$
>
> or
>
> $$\frac{\text{Amount of a Nutritional Element by Weight}}{\text{Square Feet of Area}}$$
>
> or
>
> $$\frac{\text{Amount of a Nutritional Element in the Oxide Form by Weight}}{\text{Square Feet of Area}}$$
>
> or
>
> $$\text{Parts per Million of a Nutritional Element}$$

How to Read a Fertilizer Label

Fertilizer labels contain more information than just the guaranteed analysis (N-P_2O_5-K_2O). Oftentimes the nitrogen sources (listed as percent by weight), percent by weight of other nutritional elements (including minor elements), and fertilizer salts contained in the product are included. Figure 5-1 illustrates a typical fertilizer label.

How Much N, P, or K Is in the Bag?

Understanding how much N, P_2O_5 or P, and K_2O or K is found in a particular fertilizer product can help the green industry professional in a couple of ways. First, comparing two fertilizer products in this way can help in determining which product is more economical to use. In other words, how much N, P, and/or K am I getting for my money? Second, when applications over large areas are needed, knowing the amount of N in the bag can help the green industry professional decide how many bags to order or how many bags to bring out of storage for an application.

How Much N, P_2O_5 and K_2O Are in the Bag?

To determine the weight of each nutritional element, begin by multiplying the total weight of the container by the percentage of each material in the analysis written in

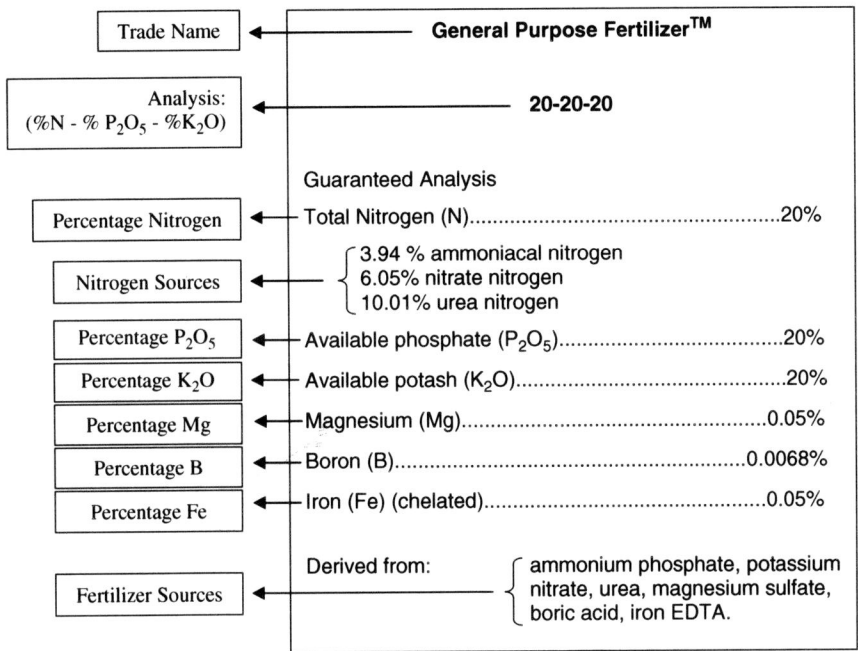

Figure 5-1
Illustration of a Fertilizer Product Label.

decimal format. Assume that a 13-13-13 product is sold in a 50 lb bag. The amount of N-P$_2$O$_5$-K$_2$O in this bag would be the following:

$$(50\,\text{lb})(0.13) = 6.5\,\text{lb N}$$

$$(50\,\text{lb})(0.13) = 6.5\,\text{lb P}_2\text{O}_5$$

$$(50\,\text{lb})(0.13) = 6.5\,\text{lb K}_2\text{O}$$

This 50 lb bag would contain 6.5 lb N, 6.5 lb P$_2$O$_5$, and 6.5 lb K$_2$O.

> **≼ Fertilizer Fact ≽**
> When using the fertilizer analysis (%N-%P$_2$O$_5$-%K$_2$O) in a calculation, you can convert the percentage to a decimal form by dividing by 100.
>   For example, 13% N converts to a decimal as follows:
>
> $$13\% \div 100 = 0.13$$
>
> Another example, 5% K$_2$O converts to a decimal as follows:
>
> $$5\% \div 100 = 0.05$$

An alternate method for performing this calculation is to use a proportion. For the N calculation, the proportion reads 13 pounds N is to 100 pounds of 13-13-13 as x pounds N is to 50 pounds of 13-13-13. Mathematically, it is presented the following way:

$$\frac{13 \text{ lb N}}{100 \text{ lb } 13\text{-}13\text{-}13} = \frac{x \text{ lb N}}{50 \text{ lb } 13\text{-}13\text{-}13}$$

Cross-multiply.

$$13 \times 50 = 100x$$
$$650 = 100x$$

Isolate and solve for x.

$$\frac{650}{100} = \frac{100x}{100}$$

$$x = \frac{650}{100}$$

$$x = 6.5 \text{ lb N}$$

> **Fertilizer Fact**
>
> When using the fertilizer analysis (%N-%P$_2$O$_5$-%K$_2$O) in a calculation, you can convert the percentage to a ratio format that can be used in a proportion by using 100 in the denominator.
>
> For example, 13% N converts to a ratio format by using 100 as the denominator.
>
> $$13\% = \frac{13 \text{ lb N}}{100 \text{ lb } 13\text{-}13\text{-}13}$$
>
> Then it can be used in a proportion.
>
> $$\frac{13 \text{ lb N}}{100 \text{ lb } 13\text{-}13\text{-}13} = \frac{x \text{ lb N}}{50 \text{ lb } 13\text{-}13\text{-}13}$$
>
> Notice that the units in the proportion match.
>
> $$\frac{\text{lb N}}{\text{lb } 13\text{-}13\text{-}13} = \frac{\text{lb N}}{\text{lb } 13\text{-}13\text{-}13}$$
>
> If the units in a proportion do not match, an incorrect answer will result.

How Much P and K Are in the bag?

A frequent mistake is to interpret the analysis as though the numbers are representative of N-P-K, rather than the amount of N-P$_2$O$_5$-K$_2$O. An understanding of how an analysis is presented is very important because a misinterpretation can lead to some serious errors in the calculations involving P and K.

When dealing with calculations involving P and K, we need to be reminded that only a portion of P$_2$O$_5$ is P and only a portion of K$_2$O is K. To calculate how much P is in P$_2$O$_5$ and how much K is on K$_2$O, a high school chemistry refresher is needed. Begin by calculating the molecular weight of each molecule.

By referring to the periodic table of elements, we know that P has an atomic weight of 31 and O has an atomic weight of 16. Therefore, the molecular weight of P_2O_5 is:

$$[2 \times 31(P)] + [5 \times 16(O)] = 62 + 80 = 142$$

The molecular weight of P_2O_5 is 142.

The portion of the molecule that is P is 62, so the percentage of P in P_2O_5 by weight is:

$$\frac{62}{142} = 0.44 \text{ or } 44\%$$

By referring to the periodic table of elements, we know that K has an atomic weight of 39 and O has an atomic weight of 16. Therefore, the molecular weight of K_2O is:

$$[2 \times 39(K)] + [1 \times 16(O)] = 78 + 16 = 94$$

The molecular weight of K_2O is 94.

The portion of the molecule that is K is 78 so the percentage of K in K_2O by weight is:

$$\frac{78}{94} = 0.83 \text{ or } 83\%$$

❈ Fertilizer Fact ❈
P_2O_5 CONTAINS 44% P.
K_2O CONTAINS 83% K.

In the case of the 13-13-13 fertilizer, the amount of actual P and K in the material would be determined by multiplying the pounds of the oxide form of P or K by the percentage of P or K in the oxide form written as a decimal. Mathematically, this is expressed as follows:

$$(6.5 \text{ lb } P_2O_5)(0.44) = 2.86 \text{ lb P}$$

$$(6.5 \text{ lb } K_2O)(0.83) = 5.4 \text{ lb K}$$

This 50 lb bag of 13-13-13 fertilizer contains 6.5 lb N, 2.86 lb P, and 5.4 lb K.

EXAMPLE 5-1

A 50 lb bag of fertilizer has an analysis of 20-5-10. How much N, P, and K does this bag contain?

Step 1

Multiply the weight of the bag by the percent of N, P_2O_5 and K_2O listed in the analysis written in the decimal form.

$$(50\,\text{lb})(0.20) = 10\,\text{lb N}$$
$$(50\,\text{lb})(0.05) = 2.5\,\text{lb P}_2\text{O}_5$$
$$(50\,\text{lb})(0.10) = 5\,\text{lb K}_2\text{O}$$

Step 2

For P and K, multiply the pounds of P_2O_5 and K_2O calculated above by the percent of P or K in each of the respective molecules.

$$(2.5\,\text{lb P}_2\text{O}_5)(0.44) = 1.1\,\text{lb P}$$
$$(5\,\text{lb K}_2\text{O})(0.83) = 4.15\,\text{lb K}$$

SOLUTION

A 50 lb bag of 20-5-10 contains:

1. 10 lb of N
2. 1.1 lb of P
3. 4.15 lb of K

Practice Problem Set 5-1   How Much N, P, or K Is in the Bag?

1. How many pounds of N, P, and K are there in a 20 lb box of a 10-10-10 garden fertilizer?
2. How many pounds of N, P, and K are there in a 40 lb bag of a 20-3-9 fertilizer?

Calculating the Proper Amount of Fertilizer for an Application

The following section reviews calculations necessary for determining the proper amount of fertilizer to use based on a rate recommendation. As previously discussed, rates can be expressed in several different formats. Calculations are categorized by the way in which the rate is expressed, and we have chosen to present them in order of increasing difficulty. Rates expressed as ppm of a nutritional element are specific to the greenhouse and nursery industries and are presented in Chapter 10. Calculations specific to liquid fertilizer products are presented here. Finally, some examples of calculations based on rate expressed in metric units are presented.

Calculations Based on Rate Expressed as the Amount of Fertilizer Product per Unit Area

When fertilizer rate is expressed as the amount of fertilizer product per unit area, the calculation is a one-step process. If the area receiving the application is known and the rate of product per square foot area is known, the process is straightforward.

Calculating the Proper Amount of Fertilizer for an Application **103**

EXAMPLE 5-2

A 375 ft² perennial border is to receive a fertilizer application at the rate of 2 lb 5-10-5/100 ft². How many pounds of 5-10-5 are required for this application?

Step 1

Set-up a proportion that answers the following question:

If 2 lb of 5-10-5 are required for 100 ft² of garden, then how many pounds (x) are required for 375 ft² of garden?

$$\frac{2 \text{ lb } 5\text{-}10\text{-}5}{100 \text{ ft}^2} = \frac{x \text{ lb } 5\text{-}10\text{-}5}{375 \text{ ft}^2}$$

Cross-multiply.

$$100x = 2 \times 375$$

Isolate and solve for x.

$$\frac{100x}{100} = \frac{750}{100}$$
$$x = 7.5 \text{ lb } 5\text{-}10\text{-}5$$

SOLUTION

A total of 7.5 pounds of 5-10-5 are needed for an application to a 375 ft² perennial border.

Practice Problem Set 5-2   Determining How Much Fertilizer to Apply When Rate Is Expressed as the Amount of Fertilizer Product per Unit Area

1. A 3,200 ft² garden bed is being prepared for a tulip bulb planting. If a 5-20-20 fertilizer product is to be applied at the rate of 3 lb/100 ft² of garden area, then how many pounds of 5-20-20 are required for this garden?
2. A shrub border measuring 15 feet by 125 feet requires a starter fertilization using 10-10-10 at the rate of 1 lb product/100 ft² of garden area. How many pounds of 10-10-10 are needed for this application?

Calculations Based on Rate Expressed as the Amount of a Nutritional Element (N, P, or K) per Unit Area

When fertilizer rate is expressed as the amount of nutritional element per unit area, the calculation becomes a two-step process for N and a three-step process for P and K. For N, the first step is to adjust the rate from pounds of N per unit area to pounds of fertilizer product per unit area, and the second step is to calculate pounds of product required for the specific amount of surface area called for in the application.

For P and K, since the analysis is expressed in terms of P_2O_5 and K_2O, the first step is to adjust the rate from pounds of P or K per unit area to pounds of P_2O_5 or K_2O per unit area. The second step is to adjust the rate to pounds of product per unit area, and the third step is to calculate pounds of product required for the specific amount of surface area called for in the application.

When Rate Is Expressed as the Amount of N Required per Unit Area

Nitrogen is the material of greatest importance in most fertilizer applications. Many fertilizer rate recommendations are based on the amount of nitrogen required for a given number of square feet of lawn or garden area. For instance, a standard recommendation for lawns is to apply 1 lb of nitrogen/1,000 ft^2 of surface area. This leads to the question of how much fertilizer product is needed to achieve that amount of nitrogen? If a 13-13-13 fertilizer product is used, how many pounds of 13-13-13 contain a total of one pound of N?

Method 1—Expressing Analysis as a Ratio

One method that can be used to answer this question is to express the problem in terms of a proportion: if 13 pounds of N are found in 100 pounds of 13-13-13 fertilizer product, then one pound of N is found in how many pounds of 13-13-13 fertilizer product? This expression written in mathematical terms is as follows:

$$\frac{13 \text{ lb N}}{100 \text{ lb } 13\text{-}13\text{-}13} = \frac{1 \text{ lb N}}{x \text{ lb } 13\text{-}13\text{-}13}$$

Cross-multiply.

$$13x = 100$$

Isolate and solve for x.

$$\frac{13x}{13} = \frac{100}{13}$$

$$x = \frac{100}{13}$$

$$x = 7.7 \text{ lb } 13\text{-}13\text{-}13$$

One pound of N is found in 7.7 pounds of 13-13-13. Therefore:

$$\frac{1 \text{ lb N}}{1,000 \text{ ft}^2} \text{ is achieved with } \frac{7.7 \text{ lb } 13\text{-}13\text{-}13}{1,000 \text{ ft}^2}$$

The rate originally expressed in terms of pounds of N/1,000 ft^2 has been adjusted or converted to pounds of 13-13-13 fertilizer product/1,000 ft^2.

To complete the calculation process for a specific application, assume that the 13-13-13 product is to be applied to 26,250 ft^2 of lawn area at the above rate.

Use a proportion to describe the problem: If 7.7 pounds of 13-13-13 are needed for 1,000 ft^2 of lawn, then how many pounds of 13-13-13 are needed for 26,250 ft^2

Calculating the Proper Amount of Fertilizer for an Application 105

of lawn? This expression written in mathematical terms is as follows:

$$\frac{7.7 \text{ lb } 13\text{-}13\text{-}13}{1,000 \text{ ft}^2} = \frac{x \text{ lb } 13\text{-}13\text{-}13}{26,250 \text{ ft}^2}$$

Cross-multiply.

$$1,000x = 7.7 \times 26,250$$

Isolate and solve for x.

$$\frac{1,000x}{1,000} = \frac{202,125}{1,000}$$

$$x = \frac{202,125}{1,000}$$

$$x = 202 \text{ lb } 13\text{-}13\text{-}13$$

For a lawn area measuring 26,250 ft^2, an application of 202 pounds of 13-13-13 is necessary to achieve an application rate of 1 lb of N/1,000 ft^2.

EXAMPLE 5-3

METHOD 1—EXPRESSING ANALYSIS AS A RATIO

If nitrogen needs to be applied to a perennial bed at the rate of 1 lb N/1,000 ft^2, how many pounds of 5-10-5 are needed for a garden measuring 100 ft^2?

Step 1

Determine how many pounds of 5-10-5 contain one pound of N by setting up a proportion.

$$\frac{5 \text{ lb N}}{100 \text{ lb } 5\text{-}10\text{-}5} = \frac{1 \text{ lb N}}{x \text{ lb } 5\text{-}10\text{-}5}$$

Cross-multiply.

$$5x = 100$$

Isolate and solve for x.

$$\frac{5x}{5} = \frac{100}{5}$$

$$x = \frac{100}{5}$$

$$x = 20 \text{ lb } 5\text{-}10\text{-}5$$

Twenty pounds of 5-10-5 contain 1 lb of N; therefore:

$$\frac{1 \text{ lb N}}{1,000 \text{ ft}^2} \text{ is achieved with } \frac{20 \text{ lb } 5\text{-}10\text{-}5}{1,000 \text{ ft}^2}$$

Step 2

Determine how many pounds of 5-10-5 are needed for 100 ft² of garden area by setting up a proportion.

$$\frac{20 \text{ lb } 5\text{-}10\text{-}5}{1{,}000 \text{ ft}^2} = \frac{x \text{ lb } 5\text{-}10\text{-}5}{100 \text{ ft}^2}$$

Cross-multiply.

$$1{,}000x = 20 \times 100$$

Isolate and solve for x.

$$\frac{1{,}000x}{1{,}000} = \frac{2{,}000}{1{,}000}$$

$$x = \frac{2{,}000}{1{,}000}$$

$$x = 2 \text{ lb } 5\text{-}10\text{-}5$$

SOLUTION

Two pounds of 5-10-5 will provide N at the rate of 1 lb/1,000 ft² to a garden area measuring 100 ft².

Method 2—Expressing Analysis as a Decimal Number

It may be useful to draw pictures to help understand the concepts involved in solving fertilizer problems. In this case, focus on the 1 lb of nitrogen. This can be represented by rectangular box:

```
┌─────────────┐
│             │
│    1 lb N   │
└─────────────┘
```

The box represents the 1 lb of nitrogen that is to be applied to each 1,000 ft² of the area.

Remember that the fertilizer to be used will always be less than 100% of nitrogen; therefore, more than 1 lb of fertilizer will need to be applied. The question remains: How much more?

To visualize this unknown amount of fertilizer that needs to be applied to achieve the rate of 1 lb of nitrogen, draw a box with a dotted line above the 1 lb N (Figure 5-2).

This represents the amount of fertilizer that must be removed from the bag and placed on 1,000 ft² of the area to achieve the desired rate of 1 lb N. The x is used to indicate this unknown amount of fertilizer.

Before trying to solve this problem, go back to the earlier problem with the 13-13-13 garden fertilizer.

In this case, it is known that there are 50 lb of fertilizer that is 13% N. The problem is solved in the following way:

$$(50)(0.13) = 6.5 \text{ lb N}$$

Calculating the Proper Amount of Fertilizer for an Application 107

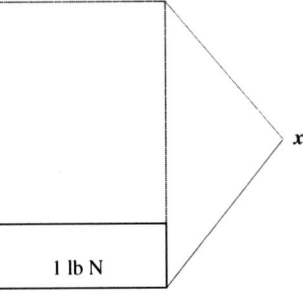

Figure 5-2
The amount of fertilizer (x) required to provide 1 lb N

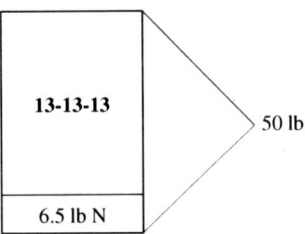

Figure 5-3
There are 6.5 lb N in 50 lb of 13-13-13 fertilizer

There are 6.5 lb N in the 50 lb bag (Figure 5-3).

Next, turn the problem around and ask the following question: How much 13-13-13 should be applied to 1,000 ft² to achieve the application rate of 1 lb N?

$$(x)(0.13) = 1 \text{ lb N}$$

Notice that this is exactly the same problem, with the exception that the amount of N that is desired is known (1 lb N) and the amount of fertilizer needed (x) is unknown. In solving this problem, it is important to think about it in the right way. The following words will help with the thought process: *13% of some weight of fertilizer (x) will need to be taken out of the bag and applied to 1,000 ft² to apply 1 lb of N.* The question is, how much fertilizer?

To solve this problem, both sides of the equation need to be divided by 0.13.

$$\frac{(x)(0.13)}{0.13} = \frac{1}{0.13}$$

The 0.13 on the left side of the equation cancels out and the equation becomes the following:

$$x = \frac{1}{0.13}$$

Enter the number 1 in the calculator and divide by 0.13. In one step the problem is solved.

$$x = 7.7 \text{ lb of 13-13-13 fertilizer}$$

It is now known that 7.7 lb of 13-13-13 needs to be applied to 1,000 ft² to apply 1 lb N.

The answer can be checked easily by answering this question: If 7.7 lb of 13-13-13 fertilizer were to be applied to 1,000 ft² of lawn, how much N would be applied?

$$(7.7 \text{ lb})(0.13) = 1 \text{ lb N}$$

The answer checks out correctly and the proper amount of fertilizer (7.7 lb) has been determined.

Generally, the area to be treated will be larger than 1,000 ft². Another step is needed to determine how much fertilizer will be used to treat the entire area.

EXAMPLE 5-4

METHOD 2—EXPRESSING ANALYSIS AS A DECIMAL NUMBER

How much 12-10-6 fertilizer would have to be purchased to apply 0.75 lb of N/1,000 ft² to a 12,000 ft² flower bed?

Step 1

Determine how much fertilizer product will deliver 0.75 lb of N. Always begin by first setting up the problem properly, as follows:

$$(x)(0.12) = 0.75 \text{ lb N}$$

Think of the following words to solve the problem: *12% of some amount of 12-10-6 fertilizer (x) will provide 0.75 lb of N.* Next, divide 0.75 by 0.12. (Enter 0.75 in the calculator and divide it by 0.12.)

$$x = \frac{0.75}{0.12}$$

$$x = 6.3 \text{ lb 12-10-6 fertilizer product}$$

The answer is 6.25, which can be rounded to 6.3 lb of 12-10-6.

A total of 6.3 lb 12-10-6 fertilizer will be needed for each 1,000 ft² of bed to apply a rate of 0.75 lb N/1,000 ft².

Step 2

The next part of the question is how much 12-10-6 will be needed to treat 12,000 ft²?

This is most easily calculated with a mathematical proportion, which is arranged as follows:

$$\frac{6.3 \text{ lb}}{1,000 \text{ ft}^2} = \frac{x \text{ lb}}{12,000 \text{ ft}^2}$$

In solving this problem, recite the following words: If 6.3 lb of 12-10-6 are required per 1,000 ft², how much (x) 12-10-6 fertilizer will be required to treat 12,000 ft²?

This type of mathematical relationship is solved by cross-multiplying and dividing:

$$(x)(1{,}000) = (6.3)(12{,}000)$$

$$(x)(1{,}000) = 75{,}600$$

$$x = \frac{75{,}600}{1{,}000}$$

$$x = 75.6 \text{ lb of 12-10-6 fertilizer}$$

SOLUTION

It has now been determined that 75.6 lb of 12-10-6 fertilizer will be needed to treat a 12,000 ft² flower bed at a N rate of 0.75 lb N/1,000 ft².

Practice Problem 5-3 — Determining How Much Fertilizer to Apply When Rate Is Expressed as the Amount of N Required per Unit Area

How much 15-3-10 fertilizer would be needed to apply 1 lb N/1,000 ft² to 57,000 ft² of a soccer field and surrounding area?

When Rate Is Expressed as the Amount of P or K Required per Unit Area

Remember that the fertilizer analysis is expressed as N-P₂O₅-K₂O and not N-P-K. This will have an impact on the calculations involving P and K. The principles involved in these calculations are the same as those used for N, but an additional step is required because of the way that the analysis is expressed.

A common P-containing fertilizer is triple super phosphate that has an analysis of 0-48-0. (It may also be purchased as 0-46-0 in some locations). This means that the fertilizer contains 48% by weight P₂O₅. The P₂O₅ is then 44% elemental P. The relationship of P to P₂O₅ within 0-48-0 is shown in Figure 5-4.

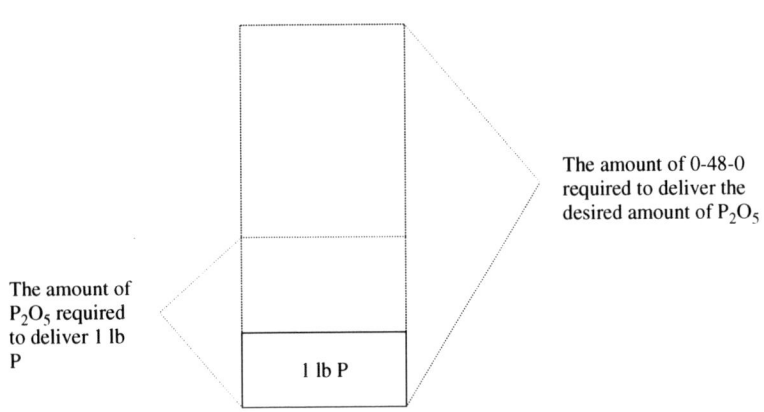

Figure 5-4
The amount of P₂O₅ and P in a phosphoric fertilizer

EXAMPLE 5-5

METHOD 1—EXPRESSING ANALYSIS AS A RATIO

How much 12-25-10 starter fertilizer is needed to apply 1 lb P/1,000 ft^2 to an 80,000 ft^2 seed bed?

Step 1

Determine how much P_2O_5 contains one pound P by setting up a proportion. Remember that P_2O_5 contains 44% P.

$$\frac{44 \text{ lb P}}{100 \text{ lb } P_2O_5} = \frac{1 \text{ lb P}}{x \text{ lb } P_2O_5}$$

Cross-multiply.

$$44x = 100$$

Isolate and solve for x.

$$\frac{44x}{44} = \frac{100}{44}$$

$$x = \frac{100}{44}$$

$$x = 2.3 \text{ lb } P_2O_5$$

One pound of P is provided by 2.3 lb P_2O_5; therefore:

$$\frac{1 \text{ lb P}}{1,000 \text{ ft}^2} \text{ is achieved with } \frac{2.3 \text{ lb } P_2O_5}{1,000 \text{ ft}^2}$$

Step 2

Determine how many pounds of 12-25-10 fertilizer product will provide 2.3 pounds of P_2O_5 by setting up a proportion.

$$\frac{25 \text{ lb } P_2O_5}{100 \text{ lb } 12\text{-}25\text{-}10} = \frac{2.3 \text{ lb } P_2O_5}{x \text{ lb } 12\text{-}25\text{-}10}$$

Cross-multiply.

$$25x = 100 \times 2.3$$

Isolate and solve for x.

$$\frac{25x}{25} = \frac{230}{25}$$

$$x = \frac{230}{25}$$

$$x = 9.2 \text{ lb } 12\text{-}25\text{-}10$$

Nine and two-tenths pounds of 12-25-10 will provide 2.3 pounds of P_2O_5, therefore:

$$\frac{1 \text{ lb P}}{1,000 \text{ ft}^2} \text{ is achieved with } \frac{2.3 \text{ lb } P_2O_5}{1,000 \text{ ft}^2} \text{ OR } \frac{9.2 \text{ lb } 12\text{-}25\text{-}10}{1,000 \text{ ft}^2}$$

Step 3

Determine how many pounds of 12-25-10 are required to treat an area measuring 80,000 ft² by setting up a proportion.

$$\frac{9.2 \text{ lb } 12\text{-}25\text{-}10}{1,000 \text{ ft}^2} = \frac{x \text{ lb } 12\text{-}25\text{-}10}{80,000 \text{ ft}^2}$$

Cross-multiply.

$$1,000x = 9.2 \times 80,000$$

Isolate and solve for x.

$$\frac{1,000x}{1,000} = \frac{736,000}{1,000}$$

$$x = \frac{736,000}{1,000}$$

$$x = 736 \text{ lb } 12\text{-}25\text{-}10$$

SOLUTION

A total of 736 lb of 12-25-10 will deliver 1 lb of P/1,000 ft² to 80,000 ft² of seed bed.

METHOD 2—EXPRESSING ANALYSIS AS A DECIMAL NUMBER

How much 12-25-10 starter fertilizer is needed to apply 1 lb P/1,000 ft² to an 80,000 ft² seed bed?

Notice that the problem asks how much P, not P_2O_5. This changes how the problem is worked. While the basic procedure and math is the same, this will require an additional step. Look at Figure 5-4. Before the amount of fertilizer can be determined, the amount of P_2O_5 required to achieve a rate of 1 lb P must be calculated. Once the amount of P_2O_5 is determined, the amount of 12-25-10 can then be calculated.

Step 1

First, determine the amount of P_2O_5 that would be required to deliver 1 lb of P. The P_2O_5 is 44% by weight P, therefore, *44% of some amount of P_2O_5 (x) will provide 1 lb P*. That amount of P_2O_5 can be determined as follows:

$$(x)(0.44) = 1 \text{ lb P}$$

$$x = \frac{1}{0.44}$$

$$x = 2.3 \text{ lb } P_2O_5$$

It has been calculated that 2.3 lb of P_2O_5 must be applied to each 1,000 ft² to achieve a rate of 1 lb P. The problem is now exactly like the N problems.

Step 2

The 12-25-10 fertilizer contains 25% P_2O_5 by weight. It already has been calculated that 2.3 lb P_2O_5 are required. Some unknown quantity (x) of 12-25-10 will be required to achieve the 2.3 lb of P_2O_5, therefore:

$$(x)(.25) = 2.3 \text{ lb } P_2O_5$$

$$x = \frac{2.3}{0.25}$$

$$x = 9.2 \text{ lb } 12\text{-}25\text{-}10$$

If 9.2 lb of 12-25-10 are applied to 1,000 ft^2, an application rate of 2.3 lb P_2O_5 and 1 lb P will be achieved.

Step 3

If 9.2 lb of 12-25-10 are applied to 1,000 ft^2, how much would be needed to treat 80,000 ft^2?

$$\frac{9.2 \text{ lb}}{1,000 \text{ ft}^2} = \frac{x}{80,000 \text{ ft}^2}$$

$$(1,000)(x) = (9.2)(80,000)$$

$$(1,000)(x) = 736,000$$

$$x = \frac{736,000}{1,000}$$

$$x = 736 \text{ lb of } 12\text{-}25\text{-}10 \text{ fertilizer}$$

SOLUTION

A total of 736 lb 12-25-10 fertilizer applied to 80,000 ft^2 will provide the 1 lb P/1,000 ft^2.

The concept and procedure used to work potassium problems are the same as that used in the phosphorus problems. The only difference is that K_2O contains 83% K.

EXAMPLE 5-6

METHOD 1—EXPRESSING ANALYSIS AS A RATIO

How much 0-0-50 (potassium sulfate) will be needed to apply 1 lb of K/1,000 ft^2 to a 36,000 ft^2 sports field?

Step 1

Determine how much K_2O contains 1 pound K by setting up a proportion. Remember that K_2O contains 83% K.

$$\frac{83 \text{ lb K}}{100 \text{ lb } K_2O} = \frac{1 \text{ lb K}}{x \text{ lb } K_2O}$$

Cross-multiply.
$$83x = 100$$

Isolate and solve for x.
$$\frac{83x}{83} = \frac{100}{83}$$
$$x = \frac{100}{83}$$
$$x = 1.2$$

One pound of K is provided by 1.2 lb K_2O; therefore:

$$\frac{1 \text{ lb K}}{1,000 \text{ ft}^2} \text{ is achieved with } \frac{1.2 \text{ lb } K_2O}{1,000 \text{ ft}^2}$$

Step 2
Determine how many pounds of 0-0-50 fertilizer product will provide 1.2 pounds of K_2O by setting up a proportion.

$$\frac{50 \text{ lb } K_2O}{100 \text{ lb 0-0-50}} = \frac{1.2 \text{ lb } K_2O}{x \text{ lb 0-0-50}}$$

Cross-multiply
$$50x = 100 \times 1.2$$

Isolate and solve for x
$$\frac{50x}{50} = \frac{120}{50}$$
$$x = \frac{120}{50}$$
$$x = 2.4 \text{ lb 0-0-50}$$

Two and four-tenths pounds of 0-0-50 will provide 1.2 pounds of K_2O, therefore:

$$\frac{1 \text{ lb K}}{1,000 \text{ ft}^2} \text{ is achieved with } \frac{1.2 \text{ lb } K_2O}{1,000 \text{ ft}^2} \text{ or } \frac{2.4 \text{ lb 0-0-50}}{1,000 \text{ ft}^2}$$

Step 3
Determine how many pounds of 0-0-50 are required to treat an area measuring 36,000 ft^2 by setting up a proportion.

$$\frac{2.4 \text{ lb 0-0-50}}{1,000 \text{ ft}^2} = \frac{x \text{ lb 0-0-50}}{36,000 \text{ ft}^2}$$

Cross-multiply.
$$1,000x = 2.4 \times 36,000$$

Isolate and solve for x.

$$\frac{1,000x}{1,000} = \frac{86,400}{1,000}$$

$$x = \frac{86,400}{1,000}$$

$$x = 86.4 \text{ lb } 0\text{-}0\text{-}50$$

SOLUTION

A total of 86.4 lb of 0-0-50 will deliver one pound of K/1000 ft^2 to 36,000 ft^2 of sports field.

METHOD 2—EXPRESSING ANALYSIS AS A DECIMAL NUMBER

How much 0-0-50 (potassium sulfate) will be needed to apply 1 lb of K/1,000 ft^2 to a 36,000 ft^2 sports field?

This problem calls for K and not K$_2$O. As with the phosphorus problems, this is an important difference, and large errors can occur if this fact is not taken into account. This again will be a two-step problem.

Step 1

The first step is to determine the amount of K$_2$O to attain a rate of 1 lb K/1,000 ft^2. Remember that K$_2$O contains 83% K by weight.

Some unknown quantity (x) of K$_2$O is required to provide this 1 lb K.

$$(x)(0.83) = 1 \text{ lb K}$$

$$x = \frac{1}{0.83}$$

$$x = 1.2 \text{ lb K}_2\text{O}$$

It has been calculated that 1.2 lb of K$_2$O must be applied to each 1,000 ft^2 of sports field to achieve a rate of 1 lb K. The problem is now exactly like the N problems.

Step 2

The second step is to determine how much 0-0-50 is required to achieve that amount of K$_2$O.

The 0-0-50 fertilizer contains 50% K$_2$O by weight. It already has been calculated that 1.2 lb K$_2$O are required. Some unknown quantity (x) of 0-0-50 is required to achieve the 1.2 lb of K$_2$O, therefore:

$$(x)(0.50) = 1.2 \text{ lb K}_2\text{O}$$

$$x = \frac{1.2}{0.50}$$

$$x = 2.4 \text{ lb } 0\text{-}0\text{-}50$$

Calculating the Proper Amount of Fertilizer for an Application **115**

If 2.4 lb of 0-0-50 are applied to 1,000 ft², an application rate of 1.2 lb K₂O and 1 lb K will be achieved. See Figure 5-5.

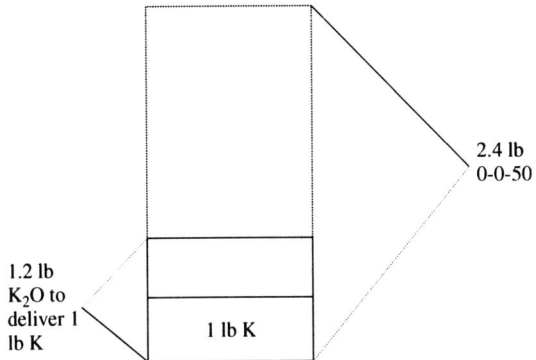

Figure 5-5
The amount of K₂O and K in a 0-0-50 fertilizer

If 2.4 lb of 0-0-50 are applied to 1,000 ft², how much would be needed to treat 36,000 ft²?

$$\frac{2.4}{1,000} = \frac{x}{36,000}$$

$$(1,000)(x) = (2.4)(36,000)$$

$$(1,000)(x) = 86,400$$

$$x = \frac{86,400}{1,000}$$

$$x = 86.4 \text{ lb } 0\text{-}0\text{-}50$$

SOLUTION

A total of 86.4 lb of 0-0-50 fertilizer is needed to treat 36,000 ft² of a sports field at a rate of 1 lb K/1,000 ft².

✏ Practice Problem Set 5-4 ✏ *Determining How Much Fertilizer to Apply When Rate Is Expressed as the Amount of P or K Required per Unit Area*

1. How much 18-46-0 fertilizer would be needed to apply 1.5 lb P/1,000 ft² to a 70,000 ft² seed bed?
2. How much 0-0-50 fertilizer would be needed to apply 1 lb of K/1,000 ft² to 80,000 ft² of sports fields?
3. A sports complex has 200,000 ft² of turf. A total of 3 lb N/1,000 ft² is to be applied during the season using a 22-2-8 fertilizer. The soil test indicates that a total of 3 lb of K should be applied to 1,000 ft²/year. How much *additional* potassium sulfate (0-0-50) will be needed to achieve the 3 lb K level for this sports complex?

Calculations Based on Rate Expressed as the Amount of the Oxide Form (P_2O_5 or K_2O) of a Nutritional Element per Unit Area

When fertilizer rate is expressed as the amount of P_2O_5 or K_2O/unit area, the calculation is a two-step process. First the rate is adjusted from P_2O_5 or K_2O/unit area to the pounds of fertilizer product per unit area. The second step is to calculate pounds of product required for the specific amount of surface area called for in the application.

EXAMPLE 5-7

METHOD 1—EXPRESSING ANALYSIS AS A RATIO

How much triple super phosphate (0-48-0) fertilizer would be needed to apply 2 lb P_2O_5/1,000 ft^2 to a newly seeded 10,000 ft^2 lawn?

Step 1

Determine how many pounds 0-48-0 contain 2 lb of P_2O_5 by setting up a proportion.

$$\frac{48 \text{ lb } P_2O_5}{100 \text{ lb } 0\text{-}48\text{-}0} = \frac{2 \text{ lb } P_2O_5}{x \text{ lb } 0\text{-}48\text{-}0}$$

Cross-multiply.

$$48x = 100 \times 2$$

Isolate and solve for x.

$$\frac{48x}{48} = \frac{200}{48}$$

$$x = \frac{200}{48}$$

$$x = 4.2 \text{ lb } 0\text{-}48\text{-}0$$

Four and two-tenths pounds of 0-48-0 contain two pounds of P_2O_5; therefore:

$$\frac{2 \text{ lb } P_2O_5}{1,000 \text{ ft}^2} = \frac{4.2 \text{ lb } 0\text{-}48\text{-}0}{1,000 \text{ ft}^2}$$

Step 2

Determine how many pounds of 0-48-0 are needed for 10,000 ft^2 of lawn area by setting up a proportion.

$$\frac{4.2 \text{ lb } 0\text{-}48\text{-}0}{1,000 \text{ ft}^2} = \frac{x \text{ lb } 0\text{-}48\text{-}0}{10,000 \text{ ft}^2}$$

Cross multiply.

$$1,000x = 4.2 \times 10,000$$

Isolate and solve for x.

$$\frac{1,000x}{1,000} = \frac{42,000}{1,000}$$

$$x = \frac{42,000}{1,000}$$

$$x = 42 \text{ lb } 0\text{-}48\text{-}0$$

SOLUTION

Forty-two pounds of 0-48-0 will provide P_2O_5 at the rate of 2 lb/1,000 ft^2 to a lawn area measuring 10,000 ft^2.

METHOD 2—EXPRESSING ANALYSIS AS A DECIMAL NUMBER

How much triple super phosphate (0-48-0) fertilizer would be needed to apply 2 lb P_2O_5/1,000 ft^2 to a newly seeded 10,000 ft^2 lawn?

This fertilizer is 48% P_2O_5 by weight. The problem specifies a rate of 2 lb P_2O_5, not P. If it asked for P, this would be a different problem. In this case, the problem is worked in exactly the same way as the N problems.

Step 1

First, determine how much 0-48-0 would be needed to obtain 2 lb P_2O_5. The material is 48% P_2O_5. As with N, setting up the problem correctly and thinking of the right wording to solve the problem will be key to working it. Set the problem up as follows: 48% of something (x) will provide the 2 lb of P_2O_5.

$$(x)(0.48) = 2 \text{ lb } P_2O_5$$

Some amount (x) of 0-48-0 will provide 2 lb P_2O_5. That amount of 0-48-0 can be determined as follows:

$$x = \frac{2}{0.48}$$

$$x = 4.2 \text{ lb } 0\text{-}48\text{-}0$$

The application 4.2 lb of 0-48-0 fertilizer to 1,000 ft^2 will provide 2 lb P_2O_5.

Step 2

Next, determine how much 0-48-0 will be needed to treat the 10,000 ft^2 seedbed.

The problem is arranged as follows:

$$\frac{4.2 \text{ lb}}{1,000 \text{ ft}^2} = \frac{x \text{ lb}}{10,000 \text{ ft}^2}$$

Say the following words: If 4.2 lb 0-48-0 are to be applied to 1,000 ft^2, how much will be needed to treat 10,000 ft^2? Then, cross-multiply and divide to solve the problem.

$$(x)(1{,}000) = (4.2)(10{,}000)$$

$$x = \frac{42{,}000}{1{,}000}$$

$$x = 42 \text{ lb of 0-48-0 will be needed to treat the seedbed}$$

SOLUTION

Forty-two pounds of 0-48-0 will provide P_2O_5 at the rate of 2 lb/1,000 ft^2 to a lawn area measuring 10,000 ft^2.

The previous example involved the use of a single nutrient analysis fertilizer (0-48-0). When a mixed nutrient analysis fertilizer is used, more than one element is being applied but the rate recommendation may still relate to only one element. It is useful to be able to determine how much of those other elements are applied simultaneously and examples of this type of problem follow.

> **Fertilizer Fact**
> A *Single Nutrient Fertilizer Analysis* refers to a fertilizer product containing only one of the three major elements present.
> A *Mixed Nutrient Fertilizer Analysis* refers to a fertilizer product containing more than one of the three major elements present.
> A *Complete Fertilizer Analysis* refers to a *mixed nutrient* fertilizer product that contains all three of the major elements (N-P-K).
> An *Incomplete Fertilizer Analysis* refers to a *mixed nutrient* fertilizer product that is missing one or more of the three major elements.

EXAMPLE 5-8

If a 20-15-20 fertilizer product is applied at the rate 1 lb of N/1,000 ft^2, how many pounds of P_2O_5 are being applied at the same time?

Step 1

Determine how many pounds of 20-15-20 fertilizer product are required to deliver 1 lb of N by using a proportion.

$$\frac{20 \text{ lb N}}{100 \text{ lb 20-15-20}} = \frac{1 \text{ lb N}}{x \text{ lb 20-15-20}}$$

Cross-multiply.

$$20x = 100$$

Isolate and solve for x.

$$\frac{20x}{20} = \frac{100}{20}$$

$$x = \frac{100}{20}$$

$$x = 5 \text{ lb 20-15-20}$$

Five pounds of 20-15-20 will deliver 1 lb of N.

Step 2

Determine how many pounds of P_2O_5 are found in five pounds of 20-15-20 by setting up a proportion.

$$\frac{15 \text{ lb } P_2O_5}{100 \text{ lb } 20\text{-}15\text{-}20} = \frac{x \text{ lb } P_2O_5}{5 \text{ lb } 20\text{-}15\text{-}20}$$

Cross-multiply.

$$100x = 15 \times 5$$

Isolate and solve for x.

$$\frac{100x}{100} = \frac{75}{100}$$

$$x = \frac{75}{100}$$

$$x = 0.75 \text{ lb } P_2O_5$$

SOLUTION

When 20-15-20 is applied at the rate of 1 lb N/1,000 ft^2, a total of 0.75 lb of P_2O_5/1,000 ft^2 also is applied.

Practice Problem Set 5-5   Determining How Much Fertilizer to Apply When Rate Is Expressed as the Amount of the Oxide Form (P_2O_5 or K_2O) of a Nutritional Element per Unit Area

1. How much 5-10-5 fertilizer is needed to apply 2 lb P_2O_5/1,000 ft^2 to a 3,000 ft^2 perennial flower bed?
2. How many pounds N/1,000 ft^2 is applied with the application in question number one?
3. How much 0-0-60 fertilizer is needed to apply 2.0 lb K_2O/1,000 ft^2 to a 10,000 ft^2 lawn?

Calculations Specific to Liquid Fertilizer Products

Liquid fertilizers are commonly used in professional lawn care and are becoming more widely used in other areas of the horticultural industry. Fertilizers are also available in the liquid form; home owners can find this in garden store outlets.

The methodology used in solving liquid fertilizer problems is similar to those used in solving problems for granular fertilizers. The difference is that liquid fertilizers are purchased and applied in volume measurements, such as ounces and gallons. This leads to differences in the thought process involved in working the problems.

The analysis that appears on the label of a liquid has the same meaning as it does on the label of a granular fertilizer. It provides the percentage by weight of N-P_2O_5-K_2O. Some confusion may arise, however, because this is the percentage by weight of a liquid material. The label of most liquid fertilizers will provide the analysis but will also provide information on the number of pounds of nitrogen and other elements per gallon.

If the number of pounds of nitrogen is not listed on the label, it can be determined by weighing a gallon of the fertilizer and multiplying that weight by the percent of nitrogen. To determine the weight of the liquid, weigh a gallon container of the product and subtract the weight of the empty container.

See the following example, if a gallon of 15-2-4 weighs 9.6 lb:

$$(9.6)(0.15) = 1.44 \text{ lb N/gal}$$

Again, the number of pounds of nitrogen per gallon, as well as the pounds per gallon of the other nutrients, will generally be provided on the label or will be available in the written information provided with the product.

EXAMPLE 5-9

A lawn care company is planning to use an 18-2-3 liquid fertilizer to apply 1 lb N/1,000 ft^2 to their customers' lawns. The tank on the truck contains 500 gal when fully filled. The spray system is calibrated to apply a total volume (fertilizer plus water) of 2 gal/1,000 ft^2. The label states that the fertilizer contains 1.8 lb N/gal.

1. How much 18-2-3 needs to be applied to 1,000 ft^2 to achieve a rate of 1 lb N/1,000 ft^2?
2. How many gallons of liquid fertilizer need to be added to the 500 gal tank?
3. How many ft^2 of lawn can be treated with the 500 gal tank?

First, determine how much 18-2-3 is needed for every 1,000 ft^2 of lawn. Rule number one for solving liquid problems is to *forget about the analysis*. If the analysis is used to solve the problem in the same way it is used in dry fertilizer problems, it will lead to considerable errors. The key piece of information in working this type of problem is that the fertilizer contains 1.8 lb N/gal.

Step 1

Begin with a very logical question. If 1 gal contains 1.8 lb N, how many gallons (x) are needed to apply 1 lb N? Arrange the problem as a proportion and solve it as follows:

$$\frac{1 \text{ gal}}{1.8 \text{ lb N}} = \frac{x \text{ gal}}{1 \text{ lb N}}$$

$$1.8x = (1)(1)$$

$$x = \frac{1}{1.8}$$

$$x = 0.56 \text{ gal}$$

SOLUTION FOR #1

Every 1,000 ft^2 of area should receive 0.56 gal of 18-2-3 fertilizer to achieve a rate of 1 lb N/1,000 ft^2.

Helpful Hint: To avoid the major errors that can sometimes occur when numbers are placed in a calculator, it may be helpful to do a mental estimate of how much fertilizer is needed before any calculation is performed. In the problem above, it would be useful to ask the following question: If a gallon of this product is applied to 1,000 ft^2, will that be too much or not enough? Remember that the product contains 1.8 lb N/gal. Therefore, if an entire gallon were applied per 1,000 ft^2, 1.8 lb N would be applied. The desired rate is 1 lb N/1,000 ft^2 and 1.8 lb N is too much. A one-half gallon would provide 0.9 lb N and would not be quite enough. Therefore, the answer will be a little more than one-half gallon. If the calculation is performed and the answer is not slightly more than one-half gallon, a mistake has been made. In this case, it is apparent that 0.56 is a reasonable answer.

Step 2

Next comes a very practical problem. How is the 500 gal of liquid to be mixed?

It is known that the spray wand is calibrated to deliver 2 gal of total solution/1,000 ft^2. Of that 2 gal, 0.56 gal must be 18-3-4 liquid fertilizer. Therefore, if there is 0.56 gal of fertilizer in every 2 gal of solution, how much fertilizer (x) should there be in 500 gal of solution? Review the following:

$$\frac{0.56 \text{ gal}}{2 \text{ gal}} = \frac{x \text{ gal}}{500 \text{ gal}}$$

$$2x = (0.56)(500)$$

$$x = \frac{280}{2}$$

$$x = 140 \text{ gal of 18-2-3}$$

SOLUTION FOR #2

For every 500 gal tank of spray that is mixed, 140 gal of 18-2-3 should be placed in the tank. Place approximately 200 gal of water in the tank, start the circulating pump, put 140 gal of liquid fertilizer in the tank, and finally bring it to the 500 gal mark with additional water.

Every 2 gal of this solution that is applied to 1,000 ft^2 will contain 0.56 gal of liquid fertilizer, which contains 1 lb N.

Step 3

How many 1,000 ft^2 will a 500 gal tank of fertilizer solution treat?

Every 1,000 ft^2 will receive 2 gal of solution, and there are 500 gal of solution. Therefore, if 1,000 ft^2 will be treated with 2 gal of solution, how many 1,000 ft^2 can be treated with 500 gal of solution?

$$\frac{1,000 \text{ ft}^2}{2 \text{ gal}} = \frac{x \text{ ft}^2}{500 \text{ gal}}$$

$$2x = (1,000)(500)$$

$$x = \frac{500{,}000}{2}$$

$$x = 250{,}000 \text{ ft}^2$$

SOLUTION FOR #3

A total of 250,000 ft^2 can be treated at 1 lb N/1,000 ft^2 with a 500 gal tank.

Practice Problem Set 5-6 — Calculations Involving Liquid Fertilizer

1. How many gallons of a 12-0-4 liquid fertilizer containing 1.2 lb N/gal and 0.33 lb K/gal will be needed to apply 0.50 lb N/1,000 ft^2 to a 60,000 ft^2 sports field?
2. How much potassium would be applied per 1,000 ft^2 using the product and the product rate in question number one?
3. A hand-operated sprayer that holds 3 gal of spray is to be used to apply a liquid fertilizer to a flower bed. The fertilizer solution is an 8-1-3 that contains 0.74 lb N/gallon. The fertilizer is to be applied at a rate of 0.50 lb N/1,000 ft^2 with a total spray volume of 1 gal of water plus fertilizer. How much fertilizer should be placed in the sprayer to mix 3 gal of solution?

Calculations Based on Rate Expressed in Metric Units

Those who use the metric system for fertilizer calculations will find it to be easier than the U.S. Customary system because everything is based on a factor of 10. The thought process involved in working the metric problems will be the same as discussed earlier. A particularly useful set of conversions for fertilizer calculations are as follows:

$$1 \text{ lb}/1{,}000 \text{ ft}^2 \cong 0.5 \text{ kg}/100 \text{ m}^2 = 5 \text{ g}/\text{m}^2$$

These conversions are not exact, but they are close enough for practical situations. They allow for quick mental conversions between the metric and English systems.

Other conversions that will commonly be used for both dry and liquid fertilizer problems are found in Table 5-1.

TABLE 5-1 • METRIC CONVERSION FOR USE IN FERTILIZER CALCULATIONS

1 pound (lb)	=	454 g		
1 square meter (m^2)	=	10.76 ft^2		
1 hectare (ha)	=	10,000 m^2	=	2.47 ac
1 kilogram (kg)	=	1,000 g	=	2.2 lb
1 lb/acres (ac)	=	1.12 kg/ha		
1 kg/ha	=	0.89 lb/ac		
1 liter (L)	=	1,000 ml	=	0.264 gal
1 gallon (gal)	=	3785 ml	=	3.785 L

EXAMPLE 5-10

A Kentucky bluegrass school yard is to be treated with 5 g N/m². The area is 10,000 m². How much 20-2-10 fertilizer in kilograms will be needed to make the application?

An analysis of 20-2-10 has the same meaning as it does in the U.S. Customary system. This fertilizer is 20% by weight N, 2% by weight P_2O_5, and 10% by weight K_2O. To apply 5 g N/m² with this fertilizer would require the following:

$$(x)(0.20) = 5$$

$$x = \frac{5}{0.20}$$

$$x = 25 \text{ g of 20-2-10 fertilizer}$$

If 25 g of 20-2-10 fertilizer are applied to each m², a rate of 5 g N/m² (1 lb N/1,000 ft²) will be achieved.

If 25 g of 20-2-10 fertilizer are to be applied to each m², how much will be needed to treat 10,000 m²?

$$\frac{25 \text{ g}}{1 \text{ m}^2} = \frac{x \text{ g}}{10,000 \text{ m}^2}$$

$$x = (25)(10,000)$$

$$x = \frac{250,000}{1}$$

$$x = 250,000 \text{ g or 250 kg of 20-2-10 fertilizer}$$

A total of 250 kg of 20-2-10 fertilizer needs to be applied to the school yard.

EXAMPLE 5-11

How many kilograms of 10-20-10 fertilizer will be needed to apply 1 kg P_2O_5/100 m² to a 1 ha seed bed?

First, determine how many kilograms of 10-20-10 fertilizer must be applied to provide a rate of 1 kg P_2O_5/100 m². The fertilizer contains 20% P_2O_5 by weight. Therefore, 0.20 of some amount of 10-20-10 fertilizer is needed to provide 1 kg P_2O_5. This is written mathematically as follows:

$$(x)(0.20) = 1 \text{ kg } P_2O_5$$

$$x = \frac{1}{0.20}$$

$$x = 5 \text{ kg 10-20-10/100 m}^2$$

An application of 5 kg of 10-20-10 on 100 m² will provide 1 kg P_2O_5/100 m².
Next, determine how much 10-20-10 is needed for a hectare.

There are 10,000 m² in a hectare. If 5 kg of 10-20-10 fertilizer are to be applied to 100 m², how many kilograms are needed to treat 10,000 m²?

$$\frac{5\,kg}{100\,m^2} = \frac{x\,kg}{10{,}000\,m^2}$$

Cross-multiply and divide.

$$(100)(x) = (5)(10{,}000)$$

$$100x = 50{,}000$$

$$x = \frac{50{,}000}{100}$$

$$x = 500\,kg\ of\ 10\text{-}20\text{-}10/ha$$

To treat the 1 ha seed bed with 1 kg P_2O_5/100 m² requires 500 kg of 10-20-10.

EXAMPLE 5-12

An 18-0-3 liquid fertilizer solution contains 215 g N/liter (L). How many liters of the fertilizer would be needed to apply 5 g N/m² to 3,500 m² soccer field?

If 5 g of N are to be applied per m², how many grams will be needed for 3,500 m²?

$$\frac{5\,g}{1\,m^2} = \frac{x\,g}{3{,}500\,m^2}$$

$$x = (5\,g/m^2)(3{,}500\,m^2)$$

$$x = 17{,}500\,g\ of\ N\ are\ needed.$$

If 1 L contains 215 g of N, how many liters will be needed to apply 17,500 g of N?

$$\frac{1\,L}{215\,g} = \frac{x}{17{,}500\,g}$$

$$(x)(215) = (17{,}500)(1)$$

$$x = \frac{17{,}500}{215}$$

$$x = 81.4\ liters\ of\ fertilizer$$

If 81.4 L of this liquid fertilizer are applied to 3,500 m², a rate of 5 g N/m² will be applied.

Fertilizer Economics: Comparing Fertilizer Products Based on Cost per Pound of Nitrogen

◈ Practice Problem Set 5-7   ◈ Calculation Based on Rate Expressed in Metric Units

1. How many kilograms of a 15-3-10 fertilizer would be needed to apply 5 g N/m^2 to an 800 m^2 lawn?
2. How many liters of a 14-2-3 liquid fertilizer with 168 g N/L would be needed to apply 0.50 kg N/100 m^2 to a 930 m^2 lawn?

Fertilizer Economics: Comparing Fertilizer Products Based on Cost per Pound of Nitrogen

When making economic decisions on which fertilizer to buy, an important criterion is the cost-per-unit weight of the desired element, not on the cost-per-unit weight of the package.

Nitrogen will usually be the element on which the most money is spent. There are different sources of nitrogen and this will affect cost. Where the nitrogen source is the same, however, the evaluation should be made on the cost per pound of nitrogen. Start the process by placing a $ sign over lb N to prevent confusion, as follows:

$$\frac{\$}{\text{lb N}}$$

EXAMPLE 5-13

The following fertilizers are available from a local distributor, and the nitrogen sources are the same. Which is the best buy?

Fertilizer A	Fertilizer B	Fertilizer C
8-2-6	22-3-6	11-2-3 liquid with 1.1 lb N/gal
40 lb bag	50 lb bag	5 gal can
$8.60/bag	$12.40/bag	$27.60/5 gal can

Begin the process by calculating how much nitrogen is being purchased in the following situations:

1. Fertilizer A contains 8% N and comes in a 40 lb bag.

$$(40)(0.08) = 3.2 \text{ lb N/bag}$$

2. Fertilizer B contains 22% N and comes in a 50 lb bag.

$$(50)(0.22) = 11 \text{ lb N/bag}$$

3. Fertilizer C contains 1.1 lb N/gal and comes in a 5 gal can.

$$(5)(1.1) = 5.5 \text{ lb N/can}$$

Remember that the number of interest is the cost per pound of nitrogen. The next step is to put the cost/container over the amount of nitrogen in the container.

4. Fertilizer A contains 3.2 lb N and costs $8.60.

$$x = \frac{\$8.60}{3.2 \text{ lb N}}$$

$$x = \$2.69/\text{lb N}$$

5. Fertilizer B contains 11 lb N and costs $12.40.

$$x = \frac{\$12.40}{11 \text{ lb N}}$$

$$x = \$1.13/\text{lb N}$$

6. Fertilizer C contains 5.5 lb N and costs $27.60.

$$x = \frac{\$27.60}{5.5 \text{ lb N}}$$

$$x = \$5.02/\text{lb N}$$

While Fertilizer A appears to be the least expensive material, it is actually more than twice the cost per pound of nitrogen of Fertilizer B. Fertilizer C is by far the most expensive on a cost-per-pound basis. This is often the case with liquids because of the shipping costs of the water used to make the solution.

Practice Problem 5-8   Fertilizer Economics: Comparing Fertilizer Products Based on Cost per Pound of Nitrogen

What is the cost per pound of nitrogen for a 50-lb bag of a 12-5-9 fertilizer that sells for $12.00/bag?

Chapter 6: Pesticide and Plant Growth Regulator Calculations

Introduction

The mathematical calculations associated with pesticide and Plant Growth Regulator (PGR) applications are among the most challenging for the green industry professional. Unfortunately, the impact of a minor misstep in a calculation can lead to disastrous results. Plants, people, and the environment all can be harmed as a result of a pesticide or PGR application made after a mathematical mistake. Knowledge of pesticide and PGR products and product formulations, understanding how to read and interpret a pesticide or PGR label, and mastering the mathematical skills to calculate the correct amount of product for an application are essential skills for the green industry professional. This chapter will focus on the skills necessary to read a pesticide or PGR label, interpret a rate of application, and calculate the proper amount of product for a given application.

Pesticide and Plant Growth Regulator Terminology

Working with pesticide and PGR calculations requires an understanding of terminology used in the crop protection industry. Terms that are most important in understanding how to read a label and perform pesticide and PGR calculations are included in this chapter.

Active Ingredient

Active ingredient or a.i. is the portion of a pesticide or PGR product that acts on the plant or plant pest and provides the desired outcome. The active ingredient is identified on the product label first by the common scientific name and second, by the full scientific name.

Percent Active Ingredient

Percent active ingredient or % a.i. is described in terms of percent by weight of active ingredient found in a product. If the percentage of active ingredient and percentage of other ingredients are summed they equal 100%. The percent by weight of active ingredient in a product is found adjacent to the common scientific and the scientific names that identify the active ingredient on a product label.

Formulation

Pesticide and PGR product formulations are numerous and diverse. Formulations are broken out into two broad categories: dry and liquid. Within each of the broad categories are a number of formulations (Table 6-1). Why does the industry have such a broad range of formulations? Pesticide and PGR formulations represent the most innovative segment of the industry. Improvements in pesticide and PGR formulations offer many advantages such as: increased safety in handling of the product, increased efficacy of the product, decreased incidence of phytotoxicity (non-target plant damage), decreased persistence in the environment, and overall reduced environmental impact.

It is important to note that some dry products are intended to be applied in the dry form. These would include the dusts (D), the granules (G), and the pellets (P or PS). All of the other dry products listed are either dissolved or suspended in water and are sprayed or drenched onto the targeted crop or pest. Most of the liquid products are either diluted by or suspended in water and are sprayed or drenched onto the targeted crop or pest.

TABLE 6-1 • PESTICIDE FORMULATIONS

Dry Pesticide Formulations	Liquid Pesticide Formulation
Dusts (D)	Aqueous Flowables (AF)
Dry Flowables (DF)	Aqueous Suspensions (AS)
Dry Solubles (DS)	Cation Liquids (CL)
Granules (G)	Emulsifiable Concentrates (EC or E)
Pellets (P or PS)	Emulsifiable Solutions (ES)
Water Dispersible Granules (WDG, WG)	Emulsions in Water (EW)
Water Soluble Bags (WSB)	Flowables (F or FL)
Water Soluble Granules (WSG)	Liquids (L)
Water Dispersible Powders (WS)	Micro Emulsions (ME)
Wettable Powders (WP or W)	Slurries (SL)
Soluble Powders (SP or S)	Solutions (S)
Soluble Granules (SG)	Suspo-emulsions (SE)
	Suspension Concentrates (SC)
	Oil Soluble Liquids (OL)
	Ultra-Low Volume Concentrates (ULV)

Rate

Rate or application rate of pesticide and PGR products is expressed in two primary formats. Rate is commonly expressed on a product label either in terms of the amount of product to be applied or the amount of active ingredient to be applied. The amount of product to be applied is described in units of weight for dry products and in units of volume for liquid products.

Rate Expression for Dry Products

If the product is dry and is intended to be applied in the dry or original state, rate is expressed as amount of product by weight per unit area or amount of active ingredient by weight per unit area.

$$\frac{\text{Amount of Product by Weight}}{\text{Unit Area}} \quad \text{or} \quad \frac{\text{Amount of Active Ingredient by Weight}}{\text{Unit Area}}$$

If the product is dry and is intended to be applied dissolved or suspended in water then applied as a spray or drench, rate is expressed as the amount of product by weight per volume of spray or drench solution. The label also will indicate the volume of spray or drench solution to be applied per unit area. Ultimately, the intention is to apply the correct amount of product or active ingredient by weight per unit area via the spray or drench.

$$\frac{\text{Amount of Product by Weight}}{\text{Unit Area}} \quad \text{or} \quad \frac{\text{Amount of Active Ingredient by Weight}}{\text{Unit Area}}$$

Rate Expression for Liquid Products

If the product is liquid, it is diluted or suspended in water and is applied as a spray or a drench. Rate is expressed as the amount of product by volume per volume of spray or drench solution. The label also will indicate the volume of spray or drench solution to be applied per unit area. Ultimately, the intention is to apply the correct amount of product by volume per unit area or the correct amount of active ingredient by weight per unit area.

$$\frac{\text{Amount of Product by Volume}}{\text{Unit Area}} \quad \text{or} \quad \frac{\text{Amount of Active Ingredient by Weight}}{\text{Unit Area}}$$

How to Read a Pesticide or PGR Label

Pesticide and PGR labels are multipage documents that detail the contents of the package and all of the appropriate information required for safe and effective use of the product. For the purposes of comparing dry and liquid formulations and for performing calculations involving these products, the focus will be on the front of the package or the first page of the pesticide or PGR label.

130 Part Two — Chapter 6 Pesticide and Plant Growth Regulator Calculations

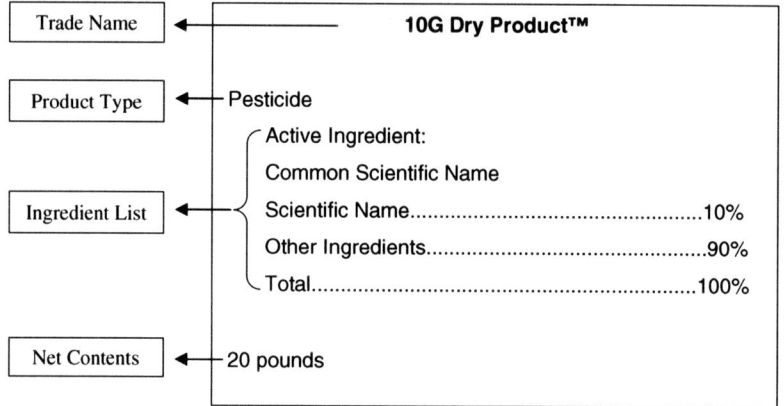

Figure 6-1

Illustration of a dry formulation product label

Dry Formulations

A dry formulation product label is illustrated in Figure 6-1. At the top of the label is the product trade name. Included in the name is the abbreviation for the formulation and a number. The number 10 in the trade name indicates that the product contains 10% by weight active ingredient. The letter, a G for granule in this example, is the abbreviation for the formulation. See Table 6-1 for additional dry formulations. The product type is listed next. Sometimes a brief description is included here as well. The ingredient list includes the active ingredient identified by common scientific and scientific names. The percent active ingredient by weight is indicated on the active ingredient line. When summed, the active and inactive or other ingredients will total 100%. The weight of net contents is listed last. The instructions for use, including recommended rates of application, are found on subsequent pages.

Liquid Formulations

A liquid formulation product label is illustrated in Figure 6-2. At the top of the label is the product trade name. Included in the name is the abbreviation for the formulation and a number. The number two in the trade name indicates that the product contains 2 pounds of active ingredient per gallon of product. The letter, an L for liquid, in this example, is the abbreviation for the formulation. See Table 6-1 for additional liquid formulations. On the product label, the product type is listed next. Sometimes a brief description is included here as well. The ingredient list includes the active ingredient identified by common scientific and scientific names. The percent active ingredient by weight is indicated on the active ingredient line. *Notice that this number (22%) is not useful in liquid formulation calculations and does not agree with the 2 L identifier or the statement regarding pounds of active ingredient per gallon of product.* When summed, the active and inactive or other ingredients will total 100%. Located below the ingredient list is a statement that tells the amount of active ingredient by weight per unit volume found in the product. The weight of net contents is listed last. The instructions for use, including recommended rates of application, are found on subsequent pages.

Pesticide and PGR Calculations **131**

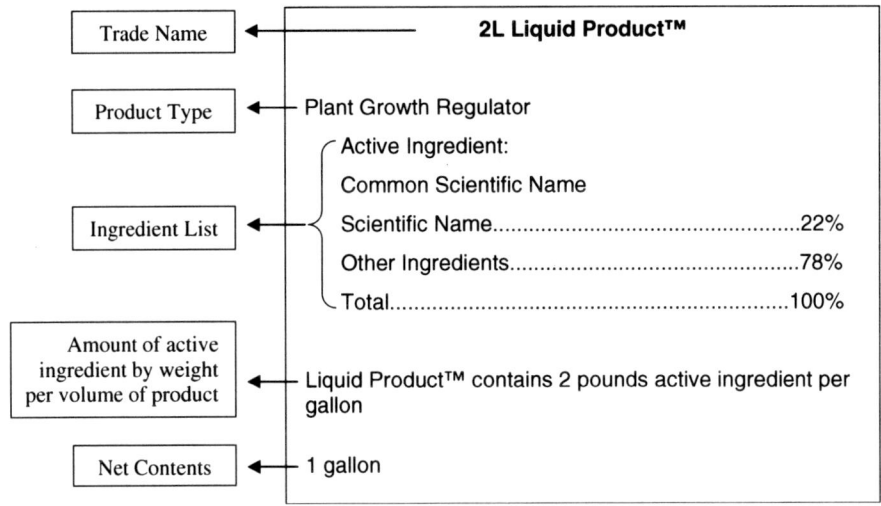

Figure 6-2
Illustration of a liquid formulation product label

❧ Helpful Hint ❧

For *dry formulations*, the product identifier 5G means:

$$5\% \text{ active ingredient by weight or } \frac{5 \text{lb a.i.}}{100 \text{ lb of product}}$$

For *liquid formulations*, the product identifier 5L means:

$$\frac{5 \text{lb a.i.}}{\text{gal of product}}$$

Pesticide and PGR Calculations

Dry Formulations Applied Dry

Rate Expressed as the Amount of Product per Unit Area

The example below shows how to calculate the amount of product needed for a typical application of a dry product when the rate is expressed in pounds of product per unit area.

In addition, this example demonstrates the proper way to convert the rate of application to a rate described in pounds of product per acre and in pounds of active ingredient per acre. Why make this transformation? Most states require that pesticide applicators keep records regarding their applications with the records showing pounds of active ingredient.

Finally, the amount of active ingredient applied for this specific application will be calculated for the purposes of pesticide-use reporting.

EXAMPLE 6-1

A 10G herbicide is to be applied at a rate of 2 lb of product/1,000 ft² to 12,000 ft² of lawns. How much of the herbicide is needed?

DETERMINE THE AMOUNT OF PRODUCT NEEDED FOR THIS APPLICATION

Step 1

Determine how many pounds of product are required to treat 12,000 ft² at the rate of 2 lb of product/1,000 ft² by setting up a proportion.

$$\frac{2 \text{ lb product}}{1,000 \text{ ft}^2} = \frac{x \text{ lb product}}{12,000 \text{ ft}^2}$$

Cross-multiply.

$$1,000x = 2 \times 12,000$$

Isolate and solve for x.

$$\frac{1,000x}{1,000} = \frac{24,000}{1,000}$$

$$x = \frac{24,000}{1,000}$$

$$x = 24 \text{ lb product}$$

SOLUTION

A total of 24 lb of this 10G herbicide product is required for this application.

CONVERT THE APPLICATION RATE TO POUNDS PER ACRE

Step 1

Convert the recommended rate from pounds of product per 1,000 ft² to pounds of product per acre by setting up a proportion. Note that one acre is equivalent to 43,560 ft².

$$\frac{2 \text{ lb product}}{1,000 \text{ ft}^2} = \frac{x \text{ lb product}}{43,560 \text{ ft}^2}$$

Cross-multiply.

$$1,000x = 2 \times 43,560$$

Isolate and solve for x.

$$\frac{1,000x}{1,000} = \frac{87,120}{1,000}$$

$$x = \frac{87,120}{1,000}$$

$$x = 87.12 \text{ lb product}$$

SOLUTION
The converted rate is 87.12 lb of product per ac.

CONVERT THE APPLICATION RATE TO POUNDS OF ACTIVE INGREDIENT PER ACRE

Step 1
Convert the recommended rate from pounds of product per acre to pounds of active ingredient per acre. Remember that the product is a 10G, and for a dry product, that means the product is 10% active ingredient by weight. A proportion or the percent expressed as a decimal number can be used for this calculation.

As a proportion:
$$\frac{10 \text{ lb a.i.}}{100 \text{ lb product}} = \frac{x \text{ lb a.i.}}{87.12 \text{ lb product}}$$

Cross-multiply.
$$100x = 10 \times 87.12$$

Isolate and solve for x.
$$\frac{100x}{100} = \frac{871.2}{100}$$
$$x = \frac{871.2}{100}$$
$$x = 8.7 \text{ lb a.i.}$$

Using percent active ingredient expressed as a decimal number, active ingredient is calculated as a percent of the weight of the product or:
$$0.10 \times 87.12 = 8.7 \text{ lb a.i.}$$

SOLUTION
The converted rate is 8.7 lb a.i./ac.

HOW MUCH ACTIVE INGREDIENT WAS APPLIED IN THIS APPLICATION?
This application used 24 lb of a 10G product. Calculate the equivalent number of pounds of active ingredient.

Step 1
This is a 10G product, so it is 10% by weight active ingredient. Using the percent active ingredient written as a decimal, calculate the amount of active ingredient used in this application.
$$0.10 \times 24 \text{ lb} = 2.4 \text{ lb a.i.}$$

134 Part Two ⌒ Chapter 6 Pesticide and Plant Growth Regulator Calculations

SOLUTION

This application used 2.4 lb of active ingredient.

Dry Formulations Applied Dry

Rate Expressed as the Amount of Active Ingredient per Unit Area

The following example shows how to calculate the amount of product needed for a typical application of a dry product that is applied dry when the rate is expressed in pounds of active ingredient per unit area. The first step is to calculate the pounds of active ingredient needed for this application, and the second step is to convert the pounds of active ingredient to pounds of product. It is useful to calculate the pounds of active ingredient for the application first because knowing how many pounds of active ingredient applied is useful for pesticide-use reporting.

EXAMPLE 6-2

A 5G pre-emergence herbicide is to be applied at 1.5 lb a.i./ac. to 10,000 ft^2 of perennial flower beds. How much of the 5G herbicide will be needed?

Step 1

Using a proportion, calculate the amount of active ingredient needed for 10,000 ft^2 when the product is to be applied at the rate of 1.5 lb a.i./ac. Note that one acre is equivalent to 43,560 ft^2.

$$\frac{1.5 \text{ lb a.i.}}{43,560 \text{ ft}^2} = \frac{x \text{ lb a.i.}}{10,000 \text{ ft}^2}$$

Cross-multiply.

$$43,560x = 1.5 \times 10,000$$

Isolate and solve for x.

$$\frac{43,560x}{43,560} = \frac{15,000}{43,560}$$

$$x = \frac{15,000}{43,560}$$

$$x = 0.344 \text{ lb a.i.}$$

A total of 0.344 pounds of active ingredient will be used in this application. This amount is useful in pesticide-use reporting.

Step 2

Using a proportion, calculate the equivalent pounds of product for 0.344 pounds of active ingredient for a 5G formulation. The product is 5% active ingredient.

$$\frac{5 \text{ lb a.i.}}{100 \text{ lb product}} = \frac{0.344 \text{ lb a.i.}}{x \text{ lb of product}}$$

Cross-multiply.
$$5x = 100 \times 0.344$$

Isolate and solve for x.
$$\frac{5x}{5} = \frac{34.4}{5}$$
$$x = \frac{34.4}{5}$$
$$x = 6.88 \text{ lb product}$$

SOLUTION

A total of 6.9 lb of 5G product is required for this application to 10,000 ft^2 of perennial beds.

Dry Formulations Applied as a Spray

Rate Expressed as the Amount of Product per Unit Area

The example below shows how to calculate the amount of product needed for a typical application of a dry product that is applied as a spray when the rate is expressed in pounds of product per unit area. The first step is to calculate the pounds of product needed for this application. The second step is to convert the pounds of product used in this application to pounds of active ingredient. It is useful to calculate the pounds of active ingredient for the application first because knowing how many pounds of active ingredient applied is useful for pesticide-use reporting. Just for fun, we have converted pounds of active ingredient for the application to pounds of active ingredient per acre. This is a standardized way of reporting that is useful for comparisons.

Since this product is applied as a spray, the product manufacturer recommends a specific spray volume described in gallons per unit area. This example also shows how to calculate the appropriate spray volume for the application. In other words, this example shows how to calculate the total amount of water to put in the spray tank to which the dry product is added.

EXAMPLE 6-3

A 25WP plant growth regulator is to be used at the rate of 0.50 oz of product/1,000 ft^2 to control clippings on a commercial property measuring 20 acres. The label indicates that the spray volume should be 1 gal/1,000 ft^2 of treated area. How many ounces of product are needed for this application? How many ounces of active ingredient are found in the applied amount? How many gallons of spray solution are needed for this application? Convert the application rate to ounces of active ingredient per acre.

HOW MANY OUNCES OF PRODUCT ARE NEEDED FOR THIS APPLICATION?

Step 1

Using a proportion, convert 20 acres to the equivalent square feet. Note that one acre is equivalent to 43,560 ft².

$$\frac{1 \text{ acre}}{43,560 \text{ ft}^2} = \frac{20 \text{ acres}}{x \text{ ft}^2}$$

Cross-multiply.

$$x = 43,560 \times 20$$

Solve for x.

$$x = 871,200 \text{ ft}^2$$

Step 2

Using a proportion, determine the amount of product needed for an application to 20 acres when the application rate is 0.50 oz of product/1,000 ft².

$$\frac{0.50 \text{ oz product}}{1,000 \text{ ft}^2} = \frac{x \text{ oz product}}{871,200 \text{ ft}^2}$$

Cross-multiply.

$$1,000x = 0.50 \times 871,200$$

Isolate and solve for x.

$$\frac{1,000x}{1,000} = \frac{435,600}{1,000}$$

$$x = \frac{435,600}{1,000}$$

$$x = 435.6 \text{ oz of product}$$

SOLUTION

A total of 435.6 ounces of product are needed for this application.

HOW MANY OUNCES OF ACTIVE INGREDIENT ARE FOUND IN THE APPLIED AMOUNT?

Step 1

Convert ounces of a 25WP product to ounces of active ingredient by using the percent active ingredient as a decimal number.

$$0.25 \times 435.6 \text{ ounces} = 108.9 \text{ ounces of active ingredient}$$

SOLUTION

A total of 108.9 ounces of active ingredient will be applied to this 20 acre commercial property.

Pesticide and PGR Calculations **137**

HOW MANY GALLONS OF SPRAY SOLUTION ARE NEEDED FOR THIS APPLICATION?

Step 1
Using a proportion, calculate the number of gallons of spray volume needed for this 20-acre application when the rate is 1 gal of spray volume/1,000 ft². Remember that 20 acres is equivalent to 871,200 ft².

$$\frac{1 \text{ gal}}{1,000 \text{ ft}^2} = \frac{x \text{ gal}}{871,200 \text{ ft}^2}$$

Cross-multiply.
$$1,000x = 871,200$$

Isolate and solve for x.
$$\frac{1,000x}{1,000} = \frac{871,200}{1,000}$$
$$x = \frac{871,200}{1,000}$$
$$x = 871.2 \text{ gal}$$

Solution
A total of 871.2 gallons of spray solution are needed for this application.

CONVERT THE APPLICATION RATE TO OUNCES OF ACTIVE INGREDIENT PER ACRE

Step 1
Using a proportion, convert the application rate of 0.50 oz of product/1,000 feet to a rate of ounces of product per acre. Note that an acre is equivalent to 43,560 ft².

$$\frac{0.50 \text{ oz product}}{1,000 \text{ ft}^2} = \frac{x \text{ oz product}}{43,560 \text{ ft}^2}$$

Cross-multiply.
$$1,000x = 0.50 \times 43,560$$

Isolate and solve for x.
$$\frac{1,000x}{1,000} = \frac{21,780}{1,000}$$
$$x = \frac{21,780}{1,000}$$
$$x = 21.78 \text{ oz of product}$$

The rate converts to 21.78 ounces of product per acre.

Step 2

Using the percent active ingredient in the 25 WP product as a decimal number, convert 21.78 oz of product/ac to oz a.i./ac.

$$0.25 \times 21.78 \text{ oz of product} = 5.5 \text{ oz a.i.}$$

SOLUTION

The rate converts to 5.5 oz a.i./ac.

Rate Expressed as the Amount of Active Ingredient per Unit Area

The example below shows how to calculate the amount of product needed for a typical application of a dry product that is applied as a spray when the rate is expressed in pounds of active ingredient per unit area. The first step is to calculate the pounds of active ingredient needed for this application, and the second step is to convert the pounds of active ingredient to pounds of product. It is useful to calculate the pounds of active ingredient for the application first because knowing how many pounds of active ingredient applied is useful for pesticide-use reporting.

Since this product is applied as a spray, the product manufacturer recommends a specific spray volume described in gallons per unit area. This example also shows how to calculate the appropriate spray volume for the application. In other words, this example shows how to calculate the total amount of water to put in the spray tank to which the dry product is added.

EXAMPLE 6-4

The label of a 60 WP insecticide recommends a rate of 2 lb a.i./ac. The product needs to be mixed with water and applied at the rate of 2 gal/1,000 ft^2. How many gallons of water are needed for this application? How much of this insecticide is needed to treat a 35,000 ft^2 sports field?

HOW MANY GALLONS OF WATER ARE NEEDED FOR THIS APPLICATION?

Step 1

Using a proportion, determine how many gallons of water are needed for 35,000 ft^2 of sports turf when used at the rate of 2 gal/1,000 ft^2.

$$\frac{2 \text{ gal}}{1,000 \text{ ft}^2} = \frac{x \text{ gal}}{35,000 \text{ ft}^2}$$

Cross-multiply.

$$1,000x = 2 \times 35,000$$

Isolate and solve for x.

$$\frac{1,000x}{1,000} = \frac{70,000}{1,000}$$

$$x = \frac{70,000}{1,000}$$

$$x = 70 \text{ gal}$$

SOLUTION

A total of 70 gallons of spray solution are needed for this application.

HOW MUCH PESTICIDE IS NEEDED FOR THIS APPLICATION?

Step 1—Calculate Pounds of Active Ingredient

Using a proportion, determine the pounds of active ingredient needed for 35,000 ft^2 of turf when applied at the rate of 2 lb a.i./ac. Note that one acre is equivalent to 43,560 ft^2.

$$\frac{2 \text{ lb a.i.}}{43,560 \text{ ft}^2} = \frac{x \text{ lb a.i.}}{35,000 \text{ ft}^2}$$

Cross-multiply.

$$43,560x = 2 \times 35,000$$

Isolate and solve for x.

$$\frac{43,560x}{43,560} = \frac{70,000}{43,560}$$

$$x = \frac{70,000}{43,560}$$

$$x = 1.6 \text{ lb a.i.}$$

A total of 1.6 pounds of active ingredient are needed for 35,000 ft^2 when the product is applied at the rate of 2 lb a.i./ac. This number is useful for pesticide-use reporting.

Step 2—Calculate Pounds of Product

Using a proportion, determine how many pounds of this 60 WP product yields 1.6 lb of active ingredient.

$$\frac{60 \text{ lb a.i.}}{100 \text{ lb product}} = \frac{1.6 \text{ lb a.i.}}{x \text{ lb product}}$$

Cross-multiply.

$$60x = 100 \times 1.6$$

Isolate and solve for x.

$$\frac{60x}{60} = \frac{160}{60}$$

$$x = \frac{160}{60}$$

$$x = 2.7 \text{ lb of product}$$

SOLUTION

A total of 2.7 pounds of a 60 WP product are needed for an application to 35,000 ft^2 of turf when applied at the rate of 2 lb a.i./ac.

Practice Problem Set 6-1 — Calculations Involving Dry Pesticide Formulations

1. A 2G insecticide is to be applied at a rate of 1 lb a.i./ac to a 30,000 ft^2 lawn. How much of the insecticide will be needed for this application?
2. A 50DF fungicide is to be applied at a rate of 0.25 lb a.i./ac to 5 acres of sports fields. How much of this fungicide will be needed for the application?
3. A 50WDG insecticide is to be applied at the rate of 0.25 oz of product/1,000 ft^2 for insect control on 20,000 ft^2 planting of ornamental shrubs. How much product is needed to treat the 20,000 ft^2 planting?
4. In question number three, how much active ingredient was applied per acre?
5. In question number three, how much total active ingredient was applied to 20,000 ft^2?

Liquid Formulations Applied as a Spray

Rate Expressed as the Amount of Product per Unit Area

The example below shows how to calculate the amount of product needed for a typical application of a liquid product that is applied as a spray when the rate is expressed in volume of product per unit area. The first step is to calculate the volume of product needed for this application. The second step is to convert the volume of product used in this application to pounds of active ingredient.

It is not a straightforward task to convert the amount of product used (on a volume basis) to the amount of active ingredient expressed in units of weight. The percent active ingredient listed on the label is expressed as a percent by weight, and since the product is a liquid, this number is not useful without some mathematical conversions that employ the product specific gravity (density). The pesticide or PGR label can reveal the relationship of pounds of active ingredient per volume of product. One way to find this information is to look at the trade name for an indication of the amount of active ingredient found in a formulation; for example, a formulation of a 4FL indicates that each gallon of product contains 4 pounds of active ingredient. If the product trade name does not include any numbers, the amount of active ingredient by weight per volume of product often can be found in small print on the label.

Pesticide and PGR Calculations **141**

It is useful to calculate the pounds of active ingredient for the application because knowing how many pounds of active ingredient applied is useful for pesticide-use reporting. We have converted pounds of active ingredient for the application to pounds of active ingredient per acre. This is a standardized way of reporting that is useful for comparisons.

Since this product is applied as a spray, the product manufacturer recommends a specific spray volume described in gallons per unit area. Since this skill has been demonstrated in the previous two examples, it has been omitted here.

> **◄ Helpful Hint ►**
>
> Do not confuse *ounces* (ounces by weight) and *fluid ounces* (ounces by volume) or significant errors will occur.
>
> **16** *ounces* **are equivalent to one pound**
>
> and
>
> **128** *fluid ounces* **are equivalent to one gallon**

EXAMPLE 6-5

A 1ME fungicide is to be applied at the rate of 44 fl oz of product/ac to a sports complex that measures 100,000 ft². How much product is needed for this application? How many pounds of active ingredient were applied during this application? Convert the application rate of 44 fl oz of product/ac to pounds of active ingredient per acre.

HOW MUCH PRODUCT IS NEEDED FOR THIS APPLICATION?

The fact that this is a 1ME product does not enter into this first calculation. The label simply states that 44 fluid ounces of product are needed per acre.

Step 1

Using a proportion, determine how many fluid ounces of product are needed for this 100,000 ft² application when the product is applied at the rate of 44 fl oz/ac. Note that an acre is equivalent to 43,560 ft².

$$\frac{44 \text{ fl oz}}{43,560 \text{ ft}^2} = \frac{x \text{ fl oz}}{100,000 \text{ ft}^2}$$

Cross-multiply.

$$43,560x = 44 \times 100,000$$

Isolate and solve for x.

$$\frac{43,560x}{43,560} = \frac{4,400,000}{43,560}$$

$$x = \frac{4{,}400{,}000}{43{,}560}$$

$$x = 101 \text{ fl oz}$$

SOLUTION

A total of 101 fluid ounces of product are needed for this application.

HOW MANY POUNDS OF ACTIVE INGREDIENT WERE APPLIED DURING THIS APPLICATION?

This product is a 1ME product. For a liquid product, this means that there is one pound of active ingredient per gallon of product.

Step 1

Using a proportion, determine how many pounds of active ingredient are found in 101 fluid ounces of this 1ME liquid product. Note that one gallon is equivalent to 128 fluid ounces.

$$\frac{1 \text{ lb a.i.}}{128 \text{ fl oz of product}} = \frac{x \text{ lb a.i.}}{101 \text{ fl oz of product}}$$

Cross-multiply.

$$128x = 101$$

Isolate and solve for x.

$$\frac{128x}{128} = \frac{101}{128}$$

$$x = \frac{101}{128}$$

$$x = 0.79 \text{ lb a.i.}$$

SOLUTION

A total of 0.79 pounds of active ingredient were applied during this application.

CONVERT THE APPLICATION RATE OF 44 FLUID OUNCES OF PRODUCT PER ACRE TO POUNDS OF ACTIVE INGREDIENT PER ACRE.

Step 1

Using a proportion, determine how many pounds of active ingredient are equivalent to 44 fl oz of a 1ME product. Remember that 1ME means that the product contains 1 lb a.i./gal and that one gallon is equivalent to 128 fluid ounces.

$$\frac{x \text{ lb a.i.}}{44 \text{ fl oz of product}} = \frac{1 \text{ lb a.i.}}{128 \text{ fl oz of product}}$$

Cross-multiply.

$$128x = 44$$

Isolate and solve for x.

$$\frac{128x}{128} = \frac{44}{128}$$

$$x = \frac{44}{128}$$

$$x = 0.34 \text{ lb a.i.}$$

SOLUTION

The application rate of 44 fluid ounces per acre is equivalent to 0.34 pounds active ingredient per acre.

Rate Expressed as the Amount of Active Ingredient per Unit Area

The example below shows how to calculate the amount of product needed for a typical application of a liquid product that is applied as a spray when the rate is expressed in pounds of active ingredient per unit area. The first step is to calculate the amount of active ingredient needed for the application. The second step is to convert the amount of active ingredient to the volume of liquid product that can deliver the desired amount of active ingredient.

Since this product is applied as a spray, the product manufacturer recommends a specific spray volume described in gallons per unit area. Since this skill has been demonstrated in previous examples, it has been omitted here.

EXAMPLE 6-6

A 4L fungicide is to be applied at the rate of 1.8 lb a.i./ac. to 30,000 ft^2 of golf course greens. How much material, in fluid ounces, will be needed to treat the golf course greens?

Step 1

Using a proportion, determine how many pounds of active ingredient, applied at the rate of 1.8 lb a.i./ac., are needed to treat the 30,000 ft^2 of golf greens. Note that one acre is equivalent to 43,560 ft^2.

$$\frac{1.8 \text{ lb a.i.}}{43,560 \text{ ft}^2} = \frac{x \text{ lb a.i.}}{30,000 \text{ ft}^2}$$

Cross-multiply.

$$43,560x = 1.8 \times 30,000$$

Isolate and solve for x.

$$\frac{43,560x}{43,560} = \frac{54,000}{43,560}$$

$$x = \frac{54{,}000}{43{,}560}$$

$$x = 1.24 \text{ lb a.i.}$$

A total of 1.24 pounds of active ingredient are needed to treat the golf greens.

Step 2

The designation of 4L indicates that this is a liquid formulation that contains 4 pounds of active ingredient per gallon of product.

Using a proportion, determine the number of gallons of 4L product needed for this application when 1.24 pounds of active ingredient are needed.

$$\frac{4 \text{ lb a.i.}}{1 \text{ gal of product}} = \frac{1.24 \text{ lb a.i.}}{x \text{ gal of product}}$$

Cross-multiply.

$$4x = 1.24$$

Isolate and solve for x.

$$\frac{4x}{4} = \frac{1.24}{4}$$

$$x = \frac{1.24}{4}$$

$$x = 0.31 \text{ gal of product}$$

Convert gallons of product to fluid ounces of product to allow for easy measuring by setting up a proportion. Note that 128 fluid ounces are equivalent to a gallon.

$$\frac{x \text{ fl oz}}{0.31 \text{ gal}} = \frac{128 \text{ fl oz}}{1 \text{ gal}}$$

Cross-multiply and solve for x.

$$x = 0.31 \times 128$$

$$x = 39.7 \text{ fl oz}$$

SOLUTION

This fungicide treatment requires 39.7 fluid ounces of product.

Practice Problem Set 6-2 Calculations Involving Liquid Pesticide Formulations

1. A 2EC herbicide is to be applied at 1.2 lb a.i./ac. How many gallons will be needed to treat 6 acres of turf?

2. If 1.6F insecticide is to be applied at 3.8 fl oz of product/ac., how much is needed to treat 20,000 ft^2?
3. In question number two, how much active ingredient in pounds per acre is being applied?

Making Economic Decisions

The same active ingredient often is available in several different formulations. The decision of which product to buy may depend on the equipment available for application and on the skill and knowledge of the applicator. When the flexibility exists to apply the material in any of the available formulations, the least expensive material may be the right choice. The least expensive material per container may not be the best buy. The decision should always be based on the cost-per-unit weight of active ingredient, not on the cost per package.

Other factors that should be considered when selecting a product are: the application interval or duration of control, the acceptability of the reentry interval (REI), and the advantages of enhanced formulations that may include increased safety in handling of the product, increased efficacy of the product, decreased incidence of phytotoxicity, decreased persistence in the environment, and decreased overall environmental impact.

Although economic decisions can be complex, comparison based on cost-per-unit weight of active ingredient is an appropriate place to begin.

EXAMPLE 6-7

The same active ingredient can be purchased as 1G or as 10G. The carrier for the two materials is the same and both are to be applied at 1.5 lb a.i./ac. If a 50 lb bag of 1G sells for $45.00, and a 50 lb bag of 10G sells for $105.00, which is the best buy?

Remember that the goal is to calculate the cost per pound of active ingredient.

$$\frac{\$}{\text{lb a.i.}}$$

1G: (50 lb)(0.01) = 0.50 lb a.i.

10G: (50 lb)(0.1) = 5 lb a.i.

Then, calculate the cost per pound of active ingredient.

1G: $\dfrac{\$45}{0.50 \text{ lb a.i.}}$ = $90 per lb of a.i.

10G: $\dfrac{\$105}{5 \text{ lb a.i.}}$ = $21 per lb of a.i.

The 1G appears to be the cheapest material on a cost-per-bag basis but is much more expensive on a cost-per-active ingredient basis.

EXAMPLE 6-8

The same insecticide is available as a 0.5G, a 75WP, and a 2F. A 30 lb bag of a 0.5G sells for $73, the 75WP sells for $125 for a 6.4 oz bag, and 1 gal of the 2F sells for $730/gal. Which formulation costs the least per pound of active ingredient?

Begin by determining how many pounds of active ingredient are in each container:

0.5G

There is 0.5% active ingredient by weight, and the bag weighs 30 lb.

$$(30)(0.005) = 0.15 \text{ lb a.i./bag}$$

75WP

The material is 75% a.i. by weight, and the package weighs 6.4 oz.

$$(6.4)(0.75) = 4.8 \text{ oz a.i./bag}$$

There are 16 oz of dry material in 1 lb.

$$\frac{4.8 \text{ oz}}{16 \text{ oz/lb}} = 0.3 \text{ lb a.i./bag}$$

2F

The 2F by definition contains 2 lb a.i/gal. The question is, what is the cost per pound of active ingredient?

$$\frac{\$}{\text{lb a.i.}}$$

In the case of the 0.5G, $73 buys 0.15 lb a.i.

$$\frac{\$73}{0.15 \text{ lb a.i.}} = \$487/\text{lb a.i.}$$

The 75WP costs $125 per bag and each bag contaains 0.3 lb a.i.

$$\frac{\$125}{0.3 \text{ lb a.i.}} = \$417/\text{lb a.i.}$$

There are 2 lb a.i. in a gal of the 2F, and it sells for $730.

$$\frac{\$730}{2 \text{ lb a.i.}} = \$365/\text{lb a.i.}$$

The 2F costs the least per pound of active ingredient.

Practice Problem Set 6-3 — Making Economic Decisions

1. A fungicide is available in a 6AS for $75/gal, an 82.5WDG for $50/5 lb bag, and a 5G for $36/25 lb bag. What is the cost per pound of active ingredient for each material?
2. A 50DF fungicide sells for $75/10 lb bag. The same active ingredient in a 2G sells for $40/50 lb bag. Which formulation is the most expensive?

Dry and Liquid Pesticide Calculations in Metric Units

As was the case for fertilizer calculations, pesticide calculations are performed in the same way in the metric system as they are in the U.S. Customary system. Again, careful attention must be paid to the definitions and formulations.

Dry materials will generally be labeled with the percentage by weight of a.i. A 50WDG would be a wettable, dispersible granule with 50% by weight active ingredient. Liquids can be confusing. The label of pesticides marketed in countries that use the metric system will list the grams of active ingredient per liter (g a.i./L). The label may also list a number and formulation. An example would be an emulsifiable concentrate with 250 g a.i./L listed as a 250E. The way in which the formulation is listed on the label is not consistent, and the number of grams of active ingredient per liter should always be confirmed and used in the calculations.

The following conversion factors will be very useful when comparing application rates between the two systems:

$$1 \text{ lb}/1{,}000 \text{ ft}^2 \cong 0.5 \text{ kg}/100 \text{ m}^2 = 5 \text{ g/m}^2$$

$$1 \text{ lb/ac} = 1.12 \text{ kg/ha}$$

$$1 \text{ kg/ha} = 0.89 \text{ lb/ac}$$

Metric Calculations for Dry Formulations

EXAMPLE 6-9

A 50WP fungicide is to be applied at 2.8 kg a.i./ha to 8,500 m² of golf greens. How many kilograms of material need to be applied to the greens?

The 50WP fungicide is by definition 50% a.i. by weight. To apply 2.8 kg a.i./ha would require the following:

$$(x)(.50) = 2.8 \text{ kg a.i.}$$

$$x = \frac{2.8}{0.50}$$

$$x = 5.6 \text{ kg of 50 WP fungicide}$$

If 5.6 kg of this fungicide is needed per hectare (10,000 m^2), how many grams are needed for 8,500 m^2 of greens?

$$\frac{5.6\,\text{kg}}{10,000\,\text{m}^2} = \frac{x\,\text{kg}}{8,500\,\text{m}^2}$$

$$10,000x = (5.6)(8,500)$$

$$x = \frac{47,600}{10,000}$$

$$x = 4.76\,\text{kg fungicide}$$

A total of 4.76 kg (4,760 g) of 50WP insecticide is needed for 8,500 m^2 of greens.

Practice Problem 6-4 Metric Calculations for Dry Formulations

A 0.5G insecticide is to be applied to 0.22 kg a.i./ha for ornamental beds. How many grams are needed to treat a 200 m^2 bed?

Metric Calculations for Liquid Formulations

EXAMPLE 6-10

A 300F insecticide contains 300 g a.i./L. The product is to be applied at 0.56 kg a.i./ha. How many milliliters of product are needed to treat 15,000 m^2?

The product contains 300 g a.i./1,000 ml and is to be applied at 0.56 kg (560 g) a.i./ha (10,000 m^2). If 1,000 ml contains 300 g, how many milliliters are needed to apply 560 g a.i.?

$$\frac{1,000\,\text{ml}}{300\,\text{g}} = \frac{x\,\text{ml}}{560\,\text{g}}$$

$$300x = (1,000)(560)$$

$$x = \frac{560,000}{300}$$

$$x = 1,867\,\text{ml}$$

An application of 1,867 ml of the insecticide applied to 1 ha is equal to a rate of 560 g or 0.56 kg/ha.

If 1,867 ml are needed for 10,000 m^2, how much is needed to treat 15,000 m^2?

$$\frac{1,867\,\text{ml}}{10,000\,\text{m}^2} = \frac{x\,\text{ml}}{15,000\,\text{m}^2}$$

$$10,000x = (1,867)(15,000)$$

$$x = \frac{28{,}005{,}000}{10{,}000}$$

$$x = 2{,}801 \text{ ml}/15{,}000 \text{ m}^2$$

An application of 2,801 ml (2.8 L) to 15,000 m² would provide 0.56 kg a.i./ha of this insecticide.

✐ Practice Problem Set 6-5 ✐ *Metric Calculations for Liquid Formulations*

1. A 250EC fungicide contains 250 g a.i./L. How many liters are needed to treat 20 ha at a rate of 0.125 kg a.i./ha?
2. A 480E herbicide that contains 480 g a.i./L is to be applied at 0.6 ml/m². At this rate of application, how many kilograms of active ingredient per hectare are being applied?

Chapter 7: Calibration of Application Equipment

Introduction

The accuracy of calculations that determine the rate of materials such as fertilizer, pesticides, and seeds are very important, but the best calculations are of little value if the equipment used to apply the material is not properly calibrated.

Equipment calibration requires knowledge of the calculations involved in determining the rates of pesticides and fertilizers, and those unfamiliar with those procedures should review earlier sections of this book before proceeding with this chapter. This chapter discusses the methods of calibrating spreaders for dry materials and sprayers for liquids.

Spreader Calibration

Spreaders are used for the application of dry fertilizers, pesticides, and seeds. Spreaders vary in size from handheld models to large units that must be attached to tractors and other equipment. They can be drop spreaders that release material straight down from the bottom of a hopper (Figure 7-1), or they can be broadcast spreaders that release a fan of material from spinning blades (Figure 7-2).

The principle involved in calibrating all spreaders is the same. The process includes a trial and error approach in which the setting of the spreader is adjusted until the proper rate of material is consistently released.

Drop Spreaders

Drop spreaders are used for applying a variety of material in the landscape. They are an effective means of applying fertilizer to lawns and other landscape areas. Many pesticides are formulated as granular materials and can be applied with drop spreaders. Drop spreaders are particularly useful when very accurate applications are required. Seed of grasses and flowers can also be spread with drop spreaders. They are the preferred method of applying very light seed that can be blown by wind. Drop spreaders release the seed from the bottom of the hopper directly to the soil surface.

Figure 7-1

Drop spreaders release material from the bottom of the hopper on the soil surface

Figure 7-2

Broadcast spreaders release a fan of material from a spinning disk

Spreader Calibration 153

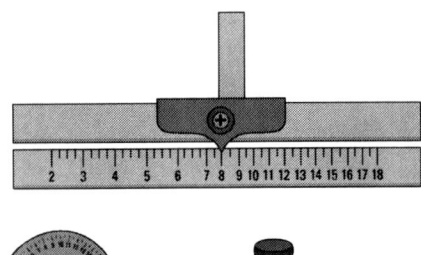

Figure 7-3
Spreaders have varying mechanisms for adjusting the release rate of material

There are many makes and models of drop spreaders, but their means of operation is generally quite similar. They have a lever that opens and closes the spreader, which allows material to be released from the base of the hopper. Remember that once the spreader is open, material will continue to flow whether the spreader is moving or still. It is best to begin moving the spreader and to open it when the desired operating speed has been reached. It should then be turned off before the spreader is stopped. It takes practice, and test runs should be made before the process of calibration begins.

The amount of material released from the hopper can be adjusted. The mechanism for adjusting flow varies among models, but the principle is the same. These set mechanisms will have either letters or numbers (Figure 7-3). The higher the number or letter on the set mechanism, the more material is released. To reduce flow, set the mechanism to a lower setting. The objective of calibration is to set the flow-control mechanism to release the right amount of material per unit area. When the proper setting has been achieved, write down the setting for future use. However, it is important to recheck the calibration each time the spreader is used with new material.

The first step in calibration is the calculation of the amount of material that should be released on a given area when the unit is properly calibrated. Once that amount of material is determined, the spreader is adjusted to the right setting. This takes some guess work initially, and it may involve arbitrarily selecting a setting to begin. A trial and error approach is used to adjust the setting.

Finally, a test area is established for the calibration process. For instance, the area of a 25 ft strip will be the length times the width. In this case, 25 ft is multiplied by the width of the spreader in ft.

The actual process of determining how much material is being released from the spreader can be done in a variety of ways. If a large enough scale is available, the spreader is filled with product and weighed. The spreader is then operated over the test area and weighed again to determine how much material was released. An alternate process is to operate the spreader over a known area on a clean shop floor. Then, sweep up the material after spreading and weigh it. A more accurate method is to place a plastic sheet over the test area and collect the material. Catch trays also can be made that collect the material from the base of the hopper as the spreader is operated (Figure 7-4).

Figure 7-4
Catch tray for catching material released by a drop spreader

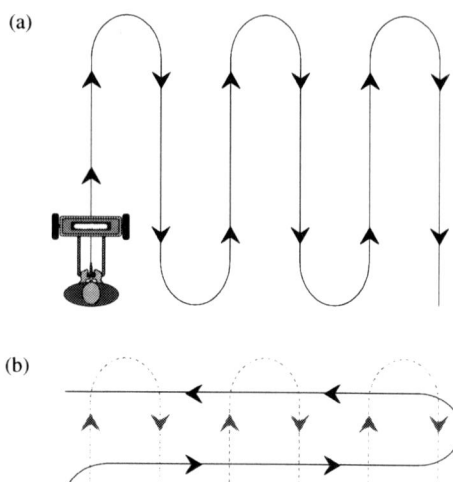

Figure 7-5

Material can be applied in one direction (a) at a full rate, or in two directions at right angles to one another (b) at one-half rate in each direction

Whatever the method used, the concept is the same. The spreader is operated at the desired speed over the test area and the amount of material released is determined. The spreader setting is then adjusted until the desired rate is achieved.

Always double check the calibration when the applications begin. For instance, if it has been calculated that 10 lb of product is to be applied to every 1,000 ft^2 of area, approximately one 50 lb bag will be needed for a 5,000 ft^2 area. If this is not the case, recheck the calibration.

A skilled applicator is able to calibrate a drop spreader and make the application with a single application over the area (Figure 7-5a). This is difficult to do and may lead to skips and misses on the area. The best way to avoid this is to calibrate the spreader to apply one-half of the desired rate of material in a single direction and then to apply the remainder of the material at a right angle to the first (Figure 7-5b). This ensures a more uniform application, although it will take more time to make the application. This is particularly important when applying seed to ensure a uniform stand of plants. Be sure to define exactly how the application will be made before the process of calibration begins.

EXAMPLE 7-1

A 12-3-8 fertilizer is to be applied at a rate of 1 lb N/1,000 ft^2 to a lawn. A drop spreader with a 36 in. hopper width is to be calibrated to make the application. The spreader will be calibrated to apply the 1 lb N in a single pass. The lawn measures 8,000 ft^2.

Spreader Calibration

The spreader is to be calibrated by operating it over a test strip on a concrete floor. The material will be collected with a broom and dust pan and weighed after each pass. An accurate scale will be needed that can hold approximately 0.1 oz of material. Begin with a couple of test runs to establish a reasonable walking speed. Remember that a uniform speed will need to be maintained uniformly throughout the application process. It is important to set a pace that can be sustained. Any change in speed during the application process will affect the application rate.

Measure a 25 ft strip on the concrete floor and mark the ends with a piece of tape and calculate the area of the test strip. Since the spreader is 36 in. wide, convert the measurement to feet.

$$\frac{36 \text{ in.}}{12 \text{ in./ft}} = 3 \text{ ft}$$

The area of the test strip is as follows:

```
┌─────────────────────────────────────┐
│ 3 ft            75 ft²              │
└─────────────────────────────────────┘
                  25 ft
```

$$(3 \text{ ft}) \times (25 \text{ ft}) = 75 \text{ ft}^2$$

Each time the spreader is operated over the test strip, fertilizer will be released on an area of 75 ft².

Next, determine how much fertilizer will be released on the 75 ft² test strip when the spreader is properly set. To do this, first determine how much fertilizer needs to be applied to 1,000 ft² to achieve an application rate of 1 lb N.

$$(x)(0.12) = 1 \text{ lb N}$$

$$x = \frac{1}{0.12}$$

$$x = 8.33 \text{ lb of 12-3-8 fertilizer}$$

If 8.33 lb of 12-3-8 fertilizer are applied per 1,000 ft², the desired rate of 1 lb N/1,000 ft² will have been achieved.

Set up a proportion to determine how much fertilizer would be released on a 75 ft² test strip. If 8.33 lb of 12-3-8 fertilizer are to be applied to 1,000 ft² to achieve a rate of 1 lb N/1,000 ft², how much should be released by the spreader on a 75 ft² test area when it is properly set?

$$\frac{8.33 \text{ lb}}{1,000 \text{ ft}^2} = \frac{x \text{ lb}}{75 \text{ ft}^2}$$

$$(x)(1,000) = (8.33)(75)$$

$$(x)(1,000) = 624.75$$

$$x = \frac{624.75}{1,000}$$

$$x = 0.625 \text{ lb}$$

There are 16 oz in 1 lb and the amount of fertilizer in ounces is as follows:

$$(16 \text{ oz/lb})(0.625 \text{ lb}) = 10 \text{ oz of 12-3-10 fertilizer}$$

When the spreader is properly set, 10 oz of fertilizer will be collected from the 75 ft² test area.

The next part of the process will involve some guess work. Pick a setting on the spreader and operate it over the test strip. If more than 10 oz is released, lower the setting, and if less than 10 oz is released, increase it. This process will be repeated until the spreader consistently releases 10 oz of fertilizer on the test strip.

Notice that small changes in the setting will result in large changes in the release of product. For instance, if a setting of 6 releases 5 oz of product and 10 oz is desired, adjust the setting to 7 and try again. If the setting is doubled to 12, the result will be a very large release of product that will be several times the desired amount.

A final check of the setting will come during treatment of the 8,000 ft² lawn. If 8.33 lb of product are needed to treat 1,000 ft², 8,000 ft² will require:

$$(8.33)(8) = 66.6 \text{ lb fertilizer}$$

If the amount of actual material applied to the lawn during the application is not very close to 67 lb, the calibration needs to be adjusted before the next application to the lawn.

Practice Problem 7-1   Calibration of a Drop Spreader

A 42 in.-wide drop spreader is to be calibrated to apply 1 lb N/1,000 ft² using an 8-4-4 fertilizer. A 20 ft test strip will be used for the calibration. Place a plastic sheet over the test strip, operate the spreader over the plastic sheet, and collect the material by lifting the sheet. The spreader will be calibrated to apply 0.50 lb N/1,000 ft² and two passes at right angles to one another will be used to make the application (Figure 7-5b). How many ounces of 8-4-4 fertilizer will be collected from the test strip when the spreader is properly calibrated at 0.50 lb N/1,000 ft²?

EXAMPLE 7-2

A 1G insecticide is to be applied at a rate of 3.5 lb a.i./ac with a 54 in.-wide drop spreader. A scale large enough to weigh the spreader filled with insecticide is available, and the amount of release from the spreader will be determined by weighing the spreader before and after operating it over a 25 ft test strip. The treatment will be made with a single pass, and the spreader will be calibrated to release the full rate in a single pass over the area. How many ounces of the 1G insecticide will be released over the 25 ft strip when the spreader is properly calibrated?

The material comes in 50 lb bags. When the spreader is properly calibrated, how many square feet will a bag cover?

The first step will be to determine how much product must be applied per unit area, in this case one acre. Once that is known, it can be determined how much should be released on the test strip when the spreader is properly calibrated.

This is done as follows:

$$(x)(0.01) = 3.5 \text{ lb a.i./ac}$$

$$x = \frac{3.5}{0.01}$$

$$x = 350 \text{ lb insecticide/ac}$$

An application of 350 lb of the insecticide to 1 ac is required to apply 3.5 lb a.i.

The test strip is 25 ft and the width of the spreader is 54 in. (4.5 ft). The area of the test strip is as follows:

$$(25)(4.5) = 112.5 \text{ ft}^2$$

How much of the insecticide should be applied to a 112.5 ft² test area? An acre is 43,560 ft². If 350 lb are needed to cover 43,560 ft², how much would be needed to treat 112.5 ft²?

$$\frac{350 \text{ lb}}{43,560 \text{ ft}^2} = \frac{x \text{ lb}}{112.5 \text{ ft}^2}$$

$$(x)(43,560) = (350)(112.5)$$

$$(x)(43,560) = 39,375$$

$$x = \frac{39,375}{43,560}$$

$$x = 0.9 \text{ lb of insecticide}$$

There are 16 oz/lb, therefore:

$$(16)(0.9) = 14.4 \text{ oz}$$

There will be 14.4 oz of the insecticide applied to the test area when the spreader is properly calibrated.

How much area will a 50 lb bag cover when the spreader is properly calibrated?

If 350 lb of insecticide are to be applied to an acre (43,560 ft²), how many square feet will 50 lb cover?

$$\frac{43,560 \text{ ft}}{350 \text{ lb}} = \frac{x \text{ ft}}{50 \text{ lb}}$$

$$(350)(x) = (50)(43,560)$$

$$(350)(x) = 2,178,000$$

$$x = \frac{2,178,000}{350}$$

$$x = 6,223 \text{ ft}^2$$

A 50 lb bag will cover 6,223 ft² of area. This can be used to check the calibration as applications are made. If the first bag does not cover this much area (approximately), the calibration will need to be checked before continuing.

Practice Problem 7-2   Drop-Spreader Calibration Using a Granular Herbicide

A 2G herbicide is to be applied to 2.0 lb a.i./ac using a 42 in.-wide spreader. A 30 ft test strip will be used and the material will be collected in a catch tray attached to the base of the spreader. The spreader will be calibrated at 1 lb a.i./ac, and two passes will be used to make the 2 lb treatment.

How many ounces of material will be released on the test strip when the spreader is properly calibrated?

How many square feet will a 40 lb bag of the material cover when the spreader is properly calibrated?

Broadcast Spreaders

Broadcast spreaders disperse material released from the hopper with a spinning disk that broadcasts granular materials in a semicircular pattern ahead of the machine (Figure 7-6). These spreaders are also known as rotary spreaders because of the rotary action of the spinning disk. The broadcast spreader has an advantage over drop spreaders in that they can cover large areas much more quickly. They are generally a little less accurate, although a skilled applicator can become quite accurate if proper methods of application are used.

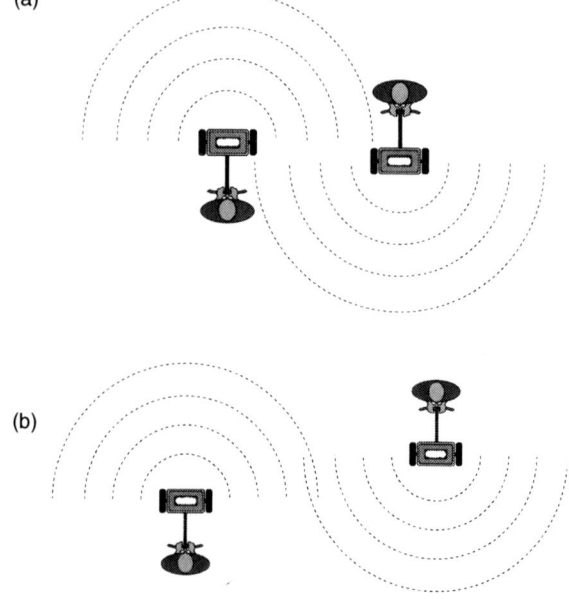

Figure 7-6
Broadcast spreaders can be operated wheel track to wheel track (a) or can be operated to the edge of the fan of material or (b)

The method of calibration for broadcast spreaders is basically the same as for drop spreaders. The release rate of material on a test area will be measured and the setting will be adjusted until the proper calibration is achieved. There are some important differences, however. The broadcast spreader throws the granular materials much further; it is not practical to try to calibrate them on a concrete floor. A large plastic sheet can be used, but again, the materials are thrown so far that collecting all of the released materials can be difficult. A known weight of granular material can be placed in the hopper and can be emptied and reweighed after spreading the material on the test strip, but this is slow. The quickest way to calibrate broadcast spreaders is to have a scale large enough to weigh the spreader and material before and after the test run.

There are also catch trays made specifically for the calibration of broadcast spreaders (Figure 7-7). Either a large scale or a catch tray is recommended for those who find it necessary to calibrate several different products each season.

Another difference from drop spreaders is that the width of the fan of material spread by the spinning disk will vary from material to material. The heavier the granule, the farther it will be thrown and the wider the effective spread width will be for that product. The effective spread width is defined as the width of the fan of material where most of the product is being released. A few particles will bounce beyond the fan of material, but these should be disregarded in the measuring process. It is important to measure the effective spread width for the material being applied before calculations are made. The effective spread width for various materials can vary by several feet. The area of the test strip will be the effective spread width times the length of the test strip. For instance a 12 ft effective spread width applied over a 40 ft test strip will have the following area:

Figure 7-7
Specialized catch tray to collect material from a broadcast spreader

$$(12\,\text{ft})(40\,\text{ft}) = 480\,\text{ft}^2$$

There are several ways to apply fertilizer with a broadcast spreader, and you will need to make a decision about which technique to use before calibration begins. The spread pattern can be from wheel track to wheel track (Figure 7-6a), or it can be to the edge of the fan (Figure 7-6b). In addition to that, the material can be applied in a single pass or at half rates applied at right angles to one another (Figure 7-5).

If the application is to be made to the edge of the fan of material and in a single direction, the full rate of material should be used for the calibration. For instance, if 10 lb of product is needed per 1,000 ft^2, the spreader should be calibrated to release the full 10 lb of material. If the spread pattern is from wheel track to wheel track, the spreader will need to be calibrated at one-half of the desired rate because of the overlap. For instance, if 10 lb of product are desired per 1,000 ft^2, the spreader should be calibrated at 5 lb/1,000 ft^2. If the spreader is to be operated wheel track to wheel track (Figure 7-6a) and the double application technique is to be used (7-5b), the spreader should be calibrated at one-quarter of the rate, or 2.5 lb/1,000 ft^2. This would provide 5 lb with the overlap on the first pass and an additional 5 lb/1,000 ft^2 on the second application at a right angle to the first. It is very important to establish the method of application before beginning the calibration process.

Remember that walking speed (or driving speed in the case of a spreader towed by a machine) also has an impact of application rate. Speed of operation must be established first and maintained during the application process.

EXAMPLE 7-3

A broadcast spreader is to be calibrated to apply 1 lb N/1,000 ft². The spreader will be operated so that the edge of the application fan overlaps on each pass (Figure 7-6b), and the application will be made in two passes at right angles to one another (Figure 7-5b). The fertilizer is an 18-3-9. The spreader has an effective spread width of 10 ft. A 40 ft test strip is to be used. How much fertilizer will be released from the spreader on the 40 ft test strip when it is properly calibrated?

How many square feet will a 50 lb bag of the 18-3-9 fertilizer cover?

The first step is to determine the area of the test strip. The effective spread width is determined to be 10 ft and the length of the strip is 40 ft. The area on which fertilizer will be released is as follows:

$$(10\,\text{ft})(40\,\text{ft}) = 400\,\text{ft}^2$$

The spreader will be operated so that the fans of material just overlap and the application will be made in two directions with 50% of the material released on each pass; therefore, the spreader must be calibrated at one-half the desired application rate, or 0.50 lb N/1,000 ft². The first pass will release 0.50 lb N/1,000 ft². The second application applied at the same rate at a right angle to the first will result in a total application of 1 lb N/1,000 ft².

The next part of the process is to determine how much 18-3-9 fertilizer will be needed to apply 0.50 lb N/1,000 ft².

$$(x)(0.18) = 0.50\,\text{lb N}$$

$$x = \frac{0.50}{0.18}$$

$$x = 2.8\,\text{lb of 18-3-9}$$

An application of 2.8 lb of 18-3-9 applied to 1,000 ft² will provide 0.50 lb N.

Now determine how much 18-3-9 fertilizer will be released on the 400 ft² test strip when the spreader is properly calibrated.

If 2.8 lb are to be applied to 1,000 ft², how much will be applied to 400 ft²?

$$\frac{2.8\,\text{lbs}}{1000\,\text{ft}^2} = \frac{x\,\text{lb}}{400\,\text{ft}^2}$$

$$(1{,}000)(x) = (2.8)(400)$$

$$(1{,}000)(x) = 1{,}120$$

$$x = \frac{1{,}120}{1{,}000}$$

$$x = 1.12\,\text{lb}$$

There are 16 oz/lb, therefore:

$$(16\,oz/lb)(1.12\,lb) = 17.9\,oz$$

A total of 17.9 oz of fertilizer is released on the test strip when the spreader is properly calibrated. For practical purposes, this can be rounded to 18 oz.

A comfortable walking speed needs to be established. Remember that this walking speed will need to be maintained for an extended period of time. If the speed starts fast and slows later in the application, the rate will change. The person who actually makes the application should establish the walking speed.

For this calibration, a floor scale is available that is large enough to weigh the spreader and the fertilizer together. Fill the spreader to approximately the three-quarter level and weight it. Pick a setting and operate it over the test strip. The loss of weight is the amount of fertilizer released on the 40 ft test strip. If it is more than 18 oz, lower the setting. If it is less than 18 oz, raise the setting. Continue this process until a consistent 18 oz of product is released over the test strip.

The spreader is now set to apply 0.50 lb N/1,000 ft^2 per pass. For two passes over the area at right angles to one another, 1 lb N/1,000 ft^2 should be applied uniformly over the area.

Finally, determine how much area a 50 lb bag will cover. This will provide a quick check to see if the right amount of material is being applied.

A bag of 18-3-9 contains the following:

$$(50)(0.18) = 9\,lb\ of\ N$$

If the application rate is 1 lb N/1,000 ft^2, 9 pounds will cover 9,000 ft^2. If 50 lb of fertilizer is not treating approximately 9,000 ft^2, the calibration needs to be rechecked before continuing.

Practice Problem 7-3 Calibration of a Broadcast Spreader 1

A broadcast spreader that is designed to overlap from wheel track to wheel track is to be used to apply a 1 lb N/1,000 ft^2 using a 20-2-8 fertilizer. The application will be made in two passes at a right angle to one another. A test strip of 50 ft is to be used, and the effective spread width has been determined to be 12 ft. How much product should be released on the test strip when the spreader is properly calibrated?
How many square feet will a 40 lb bag cover?

EXAMPLE 7-4

A 5G herbicide is to be applied to a 38,000 ft^2 sports field at a rate of 2 lb a.i./ac with a broadcast spreader. The spreader is designed to overlap slightly at the edge of the fan of material (Figure 7-6b). All of the material will be applied in a single direction over the field. A 50 ft test strip will be used for the calibration. The effective spread width with this material is 10 ft.

How much material will be needed to treat the entire field?

The first step is to determine the amount of product that will be needed per acre. The material is a 5G and the application rate is 2 lb a.i./ac.

$$(x)(0.05) = 2 \text{ lb a.i.}$$

$$x = \frac{2}{0.05}$$

$$x = 40 \text{ lb of product}$$

The application of 40 lb of product on 1 ac (43,560 ft^2) will provide the recommended rate of 2 lb a.i. of the herbicide.

The test strip is 50 ft long and 10 ft wide for a total of 500 ft^2 when multiplied.

If 40 lb of product is to be applied to 43,560 ft^2, how much would be applied to a 500 ft^2 test strip?

$$\frac{40 \text{ lb}}{43,560 \text{ ft}^2} = \frac{x \text{ lb}}{500 \text{ ft}^2}$$

$$(43,560)(x) = (40)(500)$$

$$x = \frac{20,000}{43,560}$$

$$x = 0.459 \text{ lb}$$

A total of 0.459 lb, or (16 oz/lb)(0.459 lb) equals 7.4 oz of material, should be applied on the test strip when the spreader is properly calibrated.

Using trial and error, adjust the setting until approximately 7.4 oz of product is released on the test strip.

Now determine how much material will be needed on the 38,000 ft^2 field.

$$\frac{40 \text{ lb}}{43,560 \text{ ft}^2} = \frac{x \text{ lb}}{38,000 \text{ ft}^2}$$

$$(43,560)(x) = (40)(38,000)$$

$$(43,560)(x) = (1,520,000)$$

$$x = \frac{1,520,000}{43,560}$$

$$x = 34.9 \text{ lb}$$

A total of 34.9 lb will be needed to treat the 38,000 ft^2 field.

Practice Problem 7-4   Calibration of a Broadcast Spreader 2

A soccer complex that includes 180,000 ft^2 of turf is to be treated with a 2G insecticide at 1.6 lb a.i./ac. A broadcast spreader is to be calibrated to apply the material with overlap at the edge of the fan of material (Figure 7-6b). One-half of the material will be applied in each of two applications at right angles to one another (Figure 7-5b). A 100 ft test strip will be used. The effective spread width is 12 ft. How much material will be released on the test strip when the spreader is properly set?

If the material comes in 50 lb bags, how many bags will be needed for the application?

Sprayer Calibration

The calibration of sprayers takes a clear understanding of how the equipment works. Begin by reading the owner's manual and familiarize yourself with operation of the equipment. There are a number of variables that will affect the application rate. These include the type of nozzle on the boom, the speed of operation of the unit, the number of revolutions per minute (RPM) at which the engine is operated, and the pressure generated by the pump. Once these variables have been determined, the method of calibration is relatively simple. The first step is to determine the amount of time required for the unit to cover an acre, followed by the determination of how much liquid is released in that amount of time.

Nozzle wear can also affect the rate of application of liquids. Nozzles often wear unevenly, which will result in an inaccurate application of material. Check the flow rate of each nozzle first. If there are significant variations in flow rate among nozzles, they should be replaced with new nozzles.

Modern sprayers will come with tables that show application rates given well-defined operating conditions. There are also advanced sprayers that are controlled by computers. These sprayers can even adjust release rates as the speed changes. All sprayers, including the most advanced sprayers with computer controls, should have their calibration checked (as described in the following section) on a regular basis to be sure that proper rates are being applied.

EXAMPLE 7-5

A standard boom sprayer mounted on a maintenance cart (Figure 7-8) has a 12 ft boom with 10 nozzles. Determine the number of gallons per acre applied by the unit.

First, as mentioned, read the instruction manual. Determine the proper number of RPMs of operation and the proper pump pressure for application. Remember that these settings should be maintained as much as possible during the calibration and application process.

Check the flow rate from each nozzle to determine how uniform the release rate is from each nozzle. If they vary 10% or more from one another, replace them with new nozzles before calibration.

Next, determine the amount of time required to cover one acre with this unit when it is operated at the chosen speed of operation. Do not trust the speedometer. Set up a test strip and measure the number of seconds required to travel that distance. Use a stop watch and run several tests until a uniform speed has been established.

Figure 7-8
Boom sprayer mounted on a maintenance cart

In this case, a 100 ft test strip will be used and the required time to travel 100 ft is 15 seconds (sec).

The width of the boom is 12 ft, and it is now determined that 15 sec are required to treat an area that measures 12 ft by 100 ft, or 1,200 ft².

100 ft

12 ft | 1200 ft²

Once it has been determined that this unit covers 1,200 ft² in 15 sec, the amount of time required for the unit to treat 1 ac (43,560 ft²) can easily be calculated. If it takes 15 sec to cover 1,200 ft², how long does it take to cover 43,560 ft²?

$$\frac{15 \text{ sec}}{1,200 \text{ ft}^2} = \frac{x \text{ sec}}{43,560 \text{ ft}^2}$$

$$1,200x = (15)(43,560)$$

$$1200x = (653,400)$$

$$x = \frac{653,400}{1,200}$$

$$x = 545 \text{ sec}$$

It takes 545 sec for this sprayer to cover 1 ac.

By simply determining how much liquid is released from the 12 ft boom in 545 sec the application rate per acre can be determined. This is done by establishing the number of RPMs and the pump pressure that will be used for the application. The cart will remain parked during this part of the calibration. Then, measure the flow rate from the nozzles during a preset time of operation. This time of operation can be any amount of time, but it will be easier if the amount of time required to cover the test strip is used. In this case, that was 15 sec.

The spray can be collected from the nozzles in a number of ways. The easiest is to use graduated containers that show the amount of liquid in the container following collection from the nozzle (Figure 7-9).

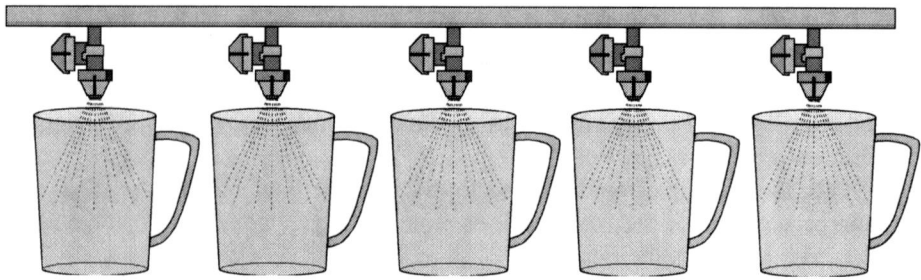

Figure 7-9

Liquid is collected from each nozzle

In this example, 14.1 oz of spray were collected from each of the 10 nozzles for a total release of 141 oz [(14.1)(10) = 141 oz] from the boom in 15 sec. If 141 oz of spray are released in 15 sec from the nozzles on the boom, how much would be released in the time it takes the sprayer to cover 1 ac (43,560 ft²), which in this case is 545 sec?

$$\frac{141 \text{ oz}}{15 \text{ sec}} = \frac{x \text{ oz}}{545 \text{ sec}}$$

$$15x = (141)(545)$$

$$15x = 76,845$$

$$x = \frac{76,845}{15}$$

$$x = 5,123 \text{ oz}$$

There are 128 oz/gal, therefore:

$$\frac{5,123 \text{ oz}}{128 \text{ oz/gal}} = 40 \text{ gal}$$

A total of approximately 40 gal of liquid/ac would be released.

Practice Problem 7-5   Calibration of a Boom Sprayer 1

A 15 ft-wide boom sprayer with 12 nozzles was timed to cover a 150 ft test strip in 21 sec. If each nozzle releases 27.5 oz in 21 sec, what is the application rate of this spreader in gallons per acre?

EXAMPLE 7-6

A 17 ft-wide boom sprayer with 11 nozzles is operated at 6 miles per hour (mph). In a flow rate test, each nozzle released an average of 68 oz/min. What is the application rate in gallons per acre for this sprayer?

When an accurate measurement of the speed in mph is known, the problem can be worked as follows: In 60 min, or one hour, this sprayer will travel 6 miles. If one mile, or 5,280 ft, is multiplied by 6 miles, the total is 31,680 ft [(6)(5,280) = 31,680]. The sprayer is 17 ft wide; therefore, in 60 min of operation, this sprayer will cover a total of 538,560 ft².

```
              31,680 ft
      ┌─────────────────────────┐
17 ft │                         │
      └─────────────────────────┘
       (17 ft)(31,680 ft) = 538,560 ft²
```

If 60 min are required to cover 538,560 ft², how many minutes are required to treat 1 ac (43,560 ft²)?

$$\frac{60 \text{ mins}}{538,560 \text{ ft}^2} = \frac{x \text{ min}}{43,560 \text{ ft}^2}$$

$$538,560 x = (60)(43,560)$$

166 Part Two ⬩ Chapter 7 Calibration of Application Equipment

$$538,560\,x = (2,613,600)$$
$$x = \frac{2,613,600}{538,560}$$
$$x = 4.85\ \text{min}$$

There are 17 nozzles and each nozzle releases 68 oz/min. The boom will release a total of 1,156 oz/min [(68)(17) = 1,156 oz/min]. If 1,156 oz are released in 1 min, how many ounces are released in the amount of time to cover 1 ac, which is 4.85 min?

$$\frac{1,156\ \text{oz}}{1\ \text{min}} = \frac{x\ \text{oz}}{4.85\ \text{min}}$$
$$x = (1,156)(4.85)$$
$$x = 5,607\ \text{oz}$$

There are 128 oz/gal, therefore:

$$\frac{5,607}{128} = 43.8\ \text{gal}$$

The application rate of this sprayer is 43.8 gal/acre, or approximately 1 gal/1,000 ft².

⬩ Practice Problem 7-6 ⬩ Calibration of a Boom Sprayer 2

An 18 ft-wide boom sprayer with 12 nozzles is to be operated at 7 mph. Each nozzle releases an average of 82 oz in 60 sec. What is the application rate in gallons per acre for this sprayer?

EXAMPLE 7-7

A boom sprayer is calibrated to release 50 gal/ac. It has a 200 gal tank. A sports field is to be treated with an 18-2-3 liquid fertilizer that contains 1.8 lb N/gal at 0.50 lb N/1,000 ft² with a 60 WDG herbicide at 1.5 lb a.i./ac. How much fertilizer in gallons and how much herbicide in pounds would need to be added to a 200 gal tank? How many acres will the 200 gal tank cover?

First, determine how much area a 200 gal tank will cover. If the sprayer is calibrated so that 1 ac (43,560 ft²) is to receive 50 gal, how many acres will a 200 gal tank cover?

$$\frac{43,560\ \text{ft}^2}{50\ \text{gal}} = \frac{x\ \text{ft}^2}{200\ \text{gal}}$$
$$50x = (43,560)(200)$$
$$50x = 8,712,000$$

$$x = \frac{8{,}712{,}000}{50}$$

$$x = 174{,}240 \text{ ft}^2$$

$$\frac{174{,}240 \text{ ft}^2}{43{,}560 \text{ ft}^2/\text{ac}} = 4 \text{ ac}$$

A tank will cover 174,240 ft², or 4 ac.

In the next step, determine how much liquid fertilizer needs to be placed in the tank. The 18-2-3 fertilizer contains 1.8 lb N/gal; how many gallons should be applied per 1,000 ft² to achieve a rate of 0.50 lb N?

$$\frac{1 \text{ gal}}{1.8 \text{ lb N}} = \frac{x \text{ gal}}{0.50 \text{ lb N}}$$

$$1.8x = (1)(0.50)$$

$$x = \frac{0.50}{1.8}$$

$$x = 0.28 \text{ gal}$$

If 0.28 gal are to be applied per 1,000 ft², how much will be needed to treat 174,240 ft²?

$$\frac{0.28 \text{ gal}}{1{,}000 \text{ ft}^2} = \frac{x \text{ gal}}{174{,}240 \text{ ft}^2}$$

$$1{,}000\,x = (0.28)(174{,}240)$$

$$1{,}000\,x = (48{,}787)$$

$$x = \frac{48{,}787}{1{,}000}$$

$$x = 48.8 \text{ gal}$$

As the tank is filled to 200 gal, 48.8 gal of 18-2-3 fertilizer needs to be added to the tank.

Next, determine how much 60 WDG herbicide needs to be added to the tank. The herbicide is 60% a.i.; therefore, 0.60 times some amount of herbicide (x) will provide 1.5 lb a.i.

$$(x)(0.60) = 1.5 \text{ lb a.i.}$$

$$x = \frac{1.5}{0.60}$$

$$x = 2.5 \text{ lb herbicide}$$

Each acre treated needs to receive 2.5 lb of product. If 1 ac requires 2.5 lb product, how much is needed to treat 4 ac—the number of acres that a full tank will treat?

$$(4 \text{ ac})(2.5 \text{ lb/ac}) = 10 \text{ lb product}$$

As the tank is filled, 48.8 gal of 18-2-3 fertilizer and 10 lb herbicide need to be added to the tank. Be sure to start the pump for proper mixing.

EXAMPLE 7-8

A back-pack sprayer with a 3 gal tank is calibrated to apply 1 gal of liquid/1,000 ft^2. A 4EC herbicide is to be applied at 1.8 lb a.i./ac with the sprayer to a 6,000 ft^2 lawn. How many ounces of product need to be placed in the tank when it is filled to 3 gal? The product contains 4 lb a.i./gal. How much is needed per 1,000 ft^2?

$$\frac{1 \text{ gal}}{4 \text{ lb a.i.}} = \frac{x \text{ gal}}{1.8 \text{ lb a.i.}}$$

$$4x = (1)(1.8)$$

$$x = \frac{1.8}{4}$$

$$x = 0.45 \text{ gal/ac}$$

There are 128 oz in 1 gal and 0.45 gal is as follows:

$$(128)(0.45) = 57.6 \text{ oz}$$

If 57.6 oz is needed for 1 ac (43,560 ft^2), how much is needed for 1,000 ft^2?

$$\frac{57.6 \text{ oz}}{43,560 \text{ ft}^2} = \frac{x \text{ oz}}{1,000 \text{ ft}^2}$$

$$43,560x = (57.6)(1,000)$$

$$43,560x = (57,600)$$

$$x = \frac{57,600}{43,560}$$

$$x = 1.3 \text{ oz}/1,000 \text{ ft}^2$$

The sprayer will treat 3,000 ft^2; therefore, 3.9 oz of product needs to be added to each tank [(1.3)(3) = 3.9 oz]. The lawn is 6,000 ft^2 and two tanks will need to be mixed.

Practice Problem 7-7   *Calibration of a Small Lawn Sprayer*

A small lawn and garden sprayer that is designed to be pulled behind a lawn tractor has a 5 gal tank and two nozzles on a rear boom that release spray over an area 5 ft wide. When operated over a 50 ft test strip, it was found to release 64 oz of spray material from the tank. How much spray material would be applied per 1,000 ft^2 using this sprayer, and how many square feet will a full sprayer cover?

If this sprayer were to be used to apply a 4EC herbicide to a lawn at 1.5 lb a.i./ac, how many ounces of the herbicide should be placed in the 5 gal tank as it is filled to capacity?

Chapter 8: Mathematical Applications for the Turfgrass Industry

Introduction

The turfgrass industry is a diverse industry. It includes home lawns, commercial properties, parks, general turfgrass areas, and sod production. This chapter covers common calculations used in the turfgrass industry that are not covered in the other chapters.

The three main categories of calculations covered in this chapter are establishment calculations, topdressing calculations and water-use calculations. Establishment calculations are core to all turfgrass uses, while topdress and water calculations may target sports fields and sod producers respectively.

Establishment Calculations

There are several ways to establish turfgrass and each has some unique calculations. Both seed and vegetative establishments are included.

Seeding Calculations

The calculation of seeding rate begins with the information on the seed label (Figure 8-1). Understanding what goes into the seed mixture will help in selecting the right seed mixture. Seed containers are required by law to list on a label an analysis containing important information about the seed. Key pieces of information contained on the seed label are the amount of pure seed and the percent of germination. These values are used to calculate pure live seed.

> **◄ Definition ►** *Pure Live Seed (PLS)*
>
> **Pure Live Seed** is defined as the pure seed in the lot that can be expected to germinate under laboratory test conditions. It is calculated by multiplying the percentage of pure seed listed on the seed label by the percentage of germination, as follows:
>
> (% Pure Seed)(% Germination) = Pure Live Seed

Figure 8-1

Turfgrass seed label

```
Seed Mixture Analysis
Fine Textured Grasses            Pure Seed      Germination %
Midnight II Kentucky bluegrass      62%            96.25%
Blacksburg Kentucky bluegrass       36%            98.50%
Other Ingredients
Weed Seed                           0.1%
Other Crop Seed                     0.0%
Inert Matter                        1.9%
No Noxious Weeds
        Seed Lot# 66666
        Tested 11/2007
        100 lb Net Weight

                                             RST Seed Company
                                              Somewhere, USA
```

A seed lot with 90% pure seed and 90% germination would have a PLS of:

$$(0.90)(0.90) = 0.81 \text{ PLS}$$

PLS can also be expressed as a percentage, where 0.81 PLS is the same as saying the seed lot contains 81% PLS.

Once the PLS is determined, the pounds of pure live seed (lb PLS) can be determined in a seed lot. The seed lot above has a PLS of 0.81. If the lot weighs 100 lb, the weight of PLS would be calculated by multiplying the weight of the seed by the calculated PLS.

$$(100 \text{ lb seed})(0.81 \text{ PLS}) = 81 \text{ lb of PLS}$$

The 100 lb seed lot contains 81 pounds of pure live seed that should germinate under good growing conditions.

EXAMPLE 8-1

A landscaper has a 50 lb bag of seed that contains 3 cultivars of turfgrass (Figure 8-2). The landscaper wants to determine the following:

1. Total pounds of pure live seed for each cultivar.
2. Total pounds of pure live seed in a 50 lb bag of seed

Step 1

Determine the PLS for each individual seed cultivar. This is done separately for each grass cultivar because the percent germination differs for all three.

Establishment Calculations **171**

```
Seed Mixture Analysis
Fine Textured Grasses           Pure Seed      Germination %
Citation III perennial ryegrass    32%           92.75%
Seville II perennial ryegrass      32%           91.50%
Blacksburg Kentucky bluegrass      34%           98.50%
Other Ingredients
Weed Seed                          0.1%
Other Crop Seed                    0.0%
Inert Matter                       1.9%
```

Figure 8-2
Turfgrass seed label

The PLS for Citation III perennial ryegrass is the following:

$$(0.32)(0.9275) = 0.2968 \text{ PLS}$$

The PLS for Seville II perennial ryegrass is the following:

$$(0.32)(0.915) = 0.2928 \text{ PLS}$$

The PLS for Blacksburg Kentucky bluegrass is the following:

$$(0.34)(0.985) = 0.3349 \text{ PLS}$$

Step 2

Calculate the total PLS for all 3 seed cultivars combined. This is calculated by adding the individual PLS values together.

$$0.2968 + 0.2928 + 0.3349 = 0.9245 \text{ PLS}$$

or

92.45% of the seed in the seed lot is pure live seed.

Step 3

Determine the pounds of PLS for each individual seed cultivar.
The pounds of PLS for Citation III perennial ryegrass are calculated as follows:

$$(0.2968)(50) = 14.84 \text{ lb PLS}$$

The pounds of PLS for Seville II perennial ryegrass are calculated as follows:

$$(0.2928)(50) = 14.64 \text{ lb PLS}$$

The pounds of PLS for Blacksburg Kentucky bluegrass are calculated as follows:

$$(0.3349)(50) = 16.75 \text{ lb PLS}$$

Step 4

The total pounds of PLS in the 50 lb seed bag are calculated in one of two ways.

Option 1

The individual cultivar pounds of PLS can be added together to calculate total pounds of PLS in the 50 lb bag of seed.

$$14.84 + 14.64 + 16.75 = 46.23 \text{ lb PLS in a 50 lb bag of seed}$$

Option 2

The PLS of the total seed lot can be multiplied by the weight of the seed lot.

$$(0.9245 \text{ PLS})(50 \text{ lb seed}) = 46.23 \text{ lb PLS in a 50 lb bag of seed}$$

SOLUTION

There are 46.23 lb of PLS in the 50 lb seed lot.

Knowing the pounds of PLS in a seed lot is used to determine how much actual seed should be sown during establishment. Actual seeding rates should be altered to account for the amount of PLS in a seed lot. Table 8-1 provides recommended seeding rates for four turfgrass species. A turfgrass manager uses both PLS and a recommended seeding rate to determine the actual seeding rate.

EXAMPLE 8-2

The landscaper is to apply the seed described in Example 8-1 to 20,000 ft^2 at a rate of 3.5 lb of PLS/1,000 ft^2. How much seed needs to be purchased if the seed lot has a PLS of 0.9245 and there are 46.23 lb PLS in a 50 lb bag?

Step 1

The first step is to determine how much PLS will be needed for the area. This is done by asking the following question: If 3.5 lb of PLS is needed for 1,000 ft^2, then how many pounds of PLS will be needed for 20,000 ft^2?

$$\frac{3.5 \text{ lb PLS}}{1,000 \text{ ft}^2} = \frac{x \text{ lb PLS}}{20,000 \text{ ft}^2}$$

$$x \text{ lb PLS} = \frac{(3.5 \text{ lb PLS})(20,000 \text{ ft}^2)}{1,000 \text{ ft}^2}$$

$$x \text{ lb PLS} = \frac{70,000}{1,000}$$

$$x = 70 \text{ lb PLS}$$

Step 2

The next step is to determine how much seed needs to be purchased to apply 70 lb of PLS to the 20,000 ft^2 area. This is done by asking the following question: If the seed lot contains 0.9245 PLS, then how many pounds of this seed are needed to deliver 70 lb of PLS? A PLS of 0.9245 is the same as saying 92.45 lb out of 100 lb is pure live seed.

Establishment Calculations

TABLE 8-1 •

Turfgrass Species	Seeding Rate lb/1,000 ft²
Kentucky bluegrass (*Poa pratensis*)	1.0–1.5
Perennial ryegrass (*Lolium perenne*)	7.0–9.0
Fine Leaf fescues	3.5–4.5
Chewing fescue (*Festuca rubra* ssp *falax*)	
Hard fescue (Festuca brevipila)	
Creeping red fescue (*Festuca rubra* ssp *rubra*)	
Tall fescue (*Festuca arundinacea*)	7.0–9.0

From: Christians, N. E. (2007). *Fundamentals of Turfgrass Management*, 3rd edition, page 81.

Therefore:

$$\frac{100 \text{ lb seed}}{92.45 \text{ lb PLS}} = \frac{x \text{ lb seed}}{70 \text{ lb PLS}}$$

Cross multiply and divide to achieve the answer.

$$(92.45 \text{ lb PLS})(x \text{ lb seed}) = (70 \text{ lb PLS})(100 \text{ lb seed})$$

$$x \text{ lb seed} = \frac{(70 \text{ lb PLS})(100 \text{ lb seed})}{92.45 \text{ lb PLS}}$$

$$x \text{ lb seed} = \frac{7{,}000}{92.45}$$

$$x = 75.7 \text{ lb of seed}$$

SOLUTION

A total of 75.7 lb of seed is required to deliver 70 lb of PLS to 20,000 ft².

When shopping for seed, the quality can differ greatly. Therefore, it is important to compare seed lot based on seed contamination and PLS. Contaminated seed should be avoided. Select from seed lots that offer the turfgrass species and cultivars that are desired. Percent PLS can be used to compare the cost of seed lots that are of comparable quality.

EXAMPLE 8-3

A 100 lb lot of creeping red fescue contains 95% pure seed and lists a germination of 89%. The price of the lot is $190. The cost per pound of seed without adjusting for PLS is $1.90. What is the cost per pound of PLS?

Step 1

The first step is to calculate the percent of PLS by multiplying the percent of pure seed by the percent of germination, as follows:

$$(0.95)(0.89) = 0.8455 \text{ PLS}$$

Step 2

Calculate the pounds of PLS by multiplying the percent of PLS by the total pounds of seed.

$$(0.8455 \text{ PLS})(100 \text{ lb}) = 84.55 \text{ lb PLS}$$

Step 3

To determine the cost of seed per pound of PLS, divide the cost (dollars) by the pounds of PLS.

$$\$/\text{lb PLS} = \frac{\text{Total cost of bag}}{\text{Total lb PLS in bag}}$$

In this problem, there are 84.55 lb PLS that sell for $190. The cost per pound of PLS is then determined by dividing $190 by 84.55 lb PLS as follows:

$$\$2.25/\text{lb PLS} = \frac{\$190}{84.55 \text{ lb PLS}}$$

SOLUTION

On the surface without calculating PLS, the cost of the seed is $1.90/lb. After calculating PLS, the real cost of the seed is $2.25/lb PLS.

Vegetative Turfgrass Establishment Calculations

Turfgrass can be established vegetatively with sod, plugs, and stolons. The calculations used for vegetative establishment are based on determining how much product (sod, plugs, or stolons) you should purchase to establish a turfgrass area like a lawn or sports field. In order to know how much product to purchase, one must first determine the area of the site to be established with turfgrass. Area calculations are found in Chapters 3 and 4.

Sod

Sod provides an instant lawn by covering the entire area to be established. Sod can be purchased as either a slab or a roll. A slab is a rectangular piece of sod that can easily be handled by hand. Sod rolls are separated into both small and big rolls. Small rolls are light enough to be handled by hand, while big rolls can easily weigh 750 lb or more, thus requiring special equipment to move it. The dimensions of the sod can vary by sod producers. Slab sizes vary in size from widths of 16–36 in. and lengths of 21–36 in. The sizes of a small roll vary from widths of 16–24 in. and lengths of 48–72 in. Big rolls can vary from 42 in. by 7 ft up to 4 ft by 100 ft. Purchase sod in sizes that are appropriate for the installation.

Depending upon the sod grower, sod is sold by the square foot or square yard. Prices will vary depending upon the quantity purchased. Single rolls will normally demand a premium price, while discounts are typically given to large volume purchases.

Establishment Calculations **175**

EXAMPLE 8-4

An area is to be established using sod. If the sod rolls measure 24 in. by 60 in., assuming a 5% waste due to trimming, how much needs to be purchased to sod an area that measures 75 ft by 175 ft? The cost of the sod has been priced at $0.25/ft^2.

Step 1
Determine the area of the site to be established plus 5% waste. The area in this example is a rectangle. The formula for the area of a rectangle is length × width. Therefore:

$$\text{Area} = 175 \text{ ft} \times 75 \text{ ft}$$
$$\text{Area} = 13{,}125 \text{ ft}^2$$

Five percent waste is calculated by multiplying the total area by 0.05.

$$5\% \text{ waste} = 13{,}125 \text{ ft} \times 0.05$$
$$5\% \text{ waste} = 656.25 \text{ ft}^2$$

The total square footage to order is then determined by adding the value for the area and the value for 5% waste together.

$$\text{Total sod square footage} = 13{,}125 \text{ ft}^2 + 656.25 \text{ ft}^2$$
$$\text{Total sod square footage} = 13{,}781 \text{ ft}^2$$

Step 2
Determine the amount of sod to purchase by calculating the area of each sod piece. Sod rolls measure 24 in. by 60 in. This is the same as saying they measure 2 ft by 5 ft, since 12 in. = 1 ft. The area of 1 sod roll is calculated by multiplying length by width.

$$\text{Area of one sod roll} = (5 \text{ ft})(2 \text{ ft})$$
$$\text{Area of one sod roll} = 10 \text{ ft}^2$$

Step 3
Divide the total sod square footage by the area of each sod piece to determine the number of rolls needed to sod the area.

$$\text{Number of sod rolls} = \frac{13{,}781 \text{ ft}^2}{10 \text{ ft}^2/\text{roll}}$$
$$\text{Number of sod rolls} = 1{,}378 \text{ rolls}$$

Step 4
Determine the cost of the sod. Set up a proportion as follows:

$$\frac{\$}{1 \text{ ft}^2} = \frac{\$x}{\text{total ft}^2}$$

In this problem, the turfgrass manager is purchasing 1,378 sod rolls that are 10 ft² each.

$$(1{,}378 \text{ rolls})(10 \text{ ft}^2/\text{roll}) = 13{,}780 \text{ ft of sod to purchase}$$

The cost of the sod is $0.25/ft². Returning to the proportion above:

$$\frac{\$0.25}{1 \text{ ft}^2} = \frac{\$x}{13{,}780 \text{ ft}^2}$$

$$x = (\$0.25)(13{,}780 \text{ ft}^2)$$

$$x = \$3{,}445.00$$

SOLUTION

The turfgrass manager will purchase 1,378 sod rolls at a cost of $3,445.

Plugs

Plugs rely on the capability of a grass species to spread into an area for total coverage. Plugs are cut from sod and typically measure 2 or 4 in.². Small quantities can be purchased in trays, while larger projects utilize special machines that cut sod into plugs and plant the plugs. The math involved is similar to that used in calculating sod needs.

EXAMPLE 8-5

Using the area in Example 8-4 (13,125 ft²), determine how many plugs are needed to establish a zoysia grass lawn using 2 in. square plugs that are planted on 8 in. centers (Figure 8-3).

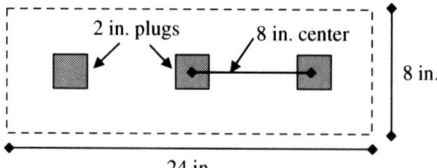

Figure 8-3
Spacing of Zoysia grass plugs

Step 1

Determine the number of plugs per square foot. A plug that is planted on 8 in. centers means that each plug occupies an area 8 in. by 8 in., or 64 in.².

$$\frac{1 \text{ plug}}{64 \text{ in.}^2} = \frac{x \text{ plug}}{144 \text{ in.}^2}$$

$$64x = 144$$

$$x \text{ plug} = \frac{144 \text{ in.}^2}{64 \text{ in.}^2}$$

$$x = 2.25 \text{ plugs/ft}^2$$

Step 2

If 2.25 plugs are planted per 1 ft², how many plugs are needed to plant 13,125 ft²?

$$\frac{2.25 \text{ plugs}}{1 \text{ ft}^2} = \frac{x \text{ plugs}}{13,125 \text{ ft}^2}$$

$$x = (2.25 \text{ plugs})(13,125 \text{ ft}^2)$$

$$x = 29,532 \text{ plugs}$$

SOLUTION

A total of 29,532 plugs are needed to establish a 13,125 ft² turfgrass area.

Stolons

The last method of establishing a turfgrass area is through the use of individual stolons. Stolons are aboveground stems that have nodes that contain reproductive buds. These buds develop new roots and shoots, thereby creating new plants. Stolons can be planted by either sprigging or by stolonizing. Sprigging is the placement of stolons in narrow furrows that are spaced six to nine inches apart, whereas stolonizing is the process of spreading the stolons on the soil surface. Stolonizing requires up to 10 bushels of stolons/1,000 ft², while sprigging may take as little as 2 bushels of stolons/1,000 ft². The mathematics involved are the same with both methods.

EXAMPLE 8-6

Using the area in Example 8-4 (13,125 ft²), determine how many bushels of stolons are needed to establish a Bermuda grass lawn using 10 bushels of stolons/1,000 ft².

Step 1

Set up the proportion, as follows:

$$\frac{10 \text{ bushels}}{1,000 \text{ ft}^2} = \frac{x \text{ bushels}}{13,125 \text{ ft}^2}$$

Step 2

Isolate and solve for x.

$$1,000 \, x = (10)(13,125)$$

$$x \text{ bushels} = \frac{(10 \text{ bushel})(13,125 \text{ ft}^2)}{1,000 \text{ ft}^2}$$

$$x = 132 \text{ bushels}$$

SOLUTION

A total of 132 bushels of Bermuda grass stolons are needed to establish 13,125 ft² of lawn.

Practice Problem Set 8-1   Establishment

1. Determine the pounds of PLS for the following seed lots:
   a. 1 ton of tall fescue seed that contains 91% pure seed and 86% germination rate.
   b. 100 lb of Kentucky bluegrass that contains 95% pure seed and 96% germination rate.
2. A turf site that is 1.5 ac needs to be overseeded with perennial ryegrass. How much of the following seed lots would need to be applied to deliver 1.5 lb of PLS/1,000 ft²?
   a. Lot A: 98% pure seed and 92% germination
   b. Lot B: 89% pure seed and 92% germination
   c. Lot C: 85% pure seed and 90% germination
3. What is the cost per pound of PLS for the following seed lots?
   a. Lot A: 98% pure seed and 96% germination. Cost $120 for 50 lb
   b. Lot B: 90% pure seed and 90% germination. Cost $110 for 50 lb
   c. Lot C: 84% pure seed and 88% germination. Cost $105 for 50 lb
4. A lawn that measures 6,400 ft² is to be sodded with sod slabs that measure 18 in. by 24 in.
   a. What is the square footage of each piece of sod?
   b. Assuming a 5% waste, how many sod pieces need to be purchased?
   c. If the sod cost $1.90/yd², what is the total cost of the sod?
5. A lawn that measures 5,000 ft² is to be established.
   a. How many 4 in. plugs would need to be purchased if planting on 10 inch centers?
   b. How many bushels of stolons would need to be purchased if sprigging at a rate of 3 bushels/1,000 ft²?

Soil Modification and Topdressing Calculations

Soil modification is the process of adding amendments that improve the root zone of a turfgrass area. While this is more common on athletic fields, the practice can be used on any turfgrass area. Topdressing is the practice of spreading a thin layer of soil, sand, compost or other material on a turfgrass area. It is a practice that can be used on athletic fields and other turf area to control thatch, level out uneven area, fill in aeration holes and assist in overseeding. Soil modification differs from topdressing because any additional soil amendments are incorporated into the root zone, while topdress spreads the amendments on top of the turf surface.

The mathematical calculations used for topdress and soil modification calculations relate to the volume of the soil amendments. While many suppliers of topdressing and soil amendments sell their product by the ton, it is important to make your decision based on volume, not weight. Therefore, it is best to purchase topdress or soil amendments by the cubic yard or to convert weight to volume measurements.

Soil Modification

The process of modifying soils will differ between turfgrass sites. Athletic fields can be modified by removing existing soil and replacing it with a new media over a compacted subgrade. In this case, the soil should be modified off-site prior to establishment. Home lawns may be modified by incorporating topsoil and/or compost to existing soil. This is commonly done by tilling the soil amendment into the soil profile. The mathematical calculations for these are quite similar.

EXAMPLE 8-7

The sod of a football field has been stripped off and the existing soil is removed to an 18 in. depth. A new root-zone mix that contains 85% sand, 5% soil, and 10% organic matter will be used as a replacement. The root-zone mix will be prepared off-site adjacent to the football field. How much sand, organic matter, and soil are needed to produce an 85-5-10 root-zone mix for the football field?

Step 1

Determine the surface area that needs to be modified. The area of a football field has a total surface area of 57,600 ft^2.

Step 2

First convert 18 inches into feet. If there are 12 in. in 1 ft, then there are x feet in 18 in.

$$\frac{1 \text{ ft}}{12 \text{ in.}} = \frac{x \text{ ft}}{18 \text{ in.}}$$

$$x \text{ ft} = \frac{18 \text{ in.}}{12 \text{ in./ft}}$$

$$x = 1.5 \text{ ft}$$

Therefore, there are 1.5 ft in 18 in.

Step 3

Determine the volume of root-zone mix that is needed for the football field. The area is 57,600 ft^2 and the soil depth is 1.5 ft. To determine the volume of the root-zone mix needed for the football field, multiply the area (ft^2) by the depth (ft) to yield cubic feet (ft^3).

$$(57,600 \text{ ft}^2)(1.5 \text{ ft}) = 86,400 \text{ ft}^3$$

A total of 86,400 ft^3 of root-zone mix is needed for the football field.

Step 4

Determine the volume of sand, soil, and organic matter that is needed to make the root-zone mix. The total root-zone volume has been determined to be 86,400 ft^3, and the mix is 85% sand, 5% soil, and 10% organic matter. To calculate the volume of each root-zone-mix component, multiply the total root-zone volume by the percent of the component. This is done as follows:

Volume of sand in cubic feet:

$$(86,400 \text{ ft}^3)(0.85) = 73,440 \text{ ft}^3$$

Volume of soil in cubic feet:

$$(86,400 \text{ ft}^3)(0.05) = 4,320 \text{ ft}^3$$

Volume of organic matter in cubic feet:

$$(86,400 \text{ ft}^3)(0.10) = 8,640 \text{ ft}^3$$

Step 5

Since bulk soil, sand, and organic matter is sold and delivered by cubic yard (yd^3), it is important to convert cubic feet to cubic yard. There are 27 ft^3 in 1 yd^3; therefore, it is necessary to divide the total volume of each root-zone-mix component by 27 ft^3/yd^3 to derive the total cubic yard of each.

Volume of sand in cubic yards:

$$\frac{1 \text{ yd}^3}{27 \text{ ft}^3} = \frac{x \text{ yd}^3}{73,440 \text{ ft}^3}$$

$$x \text{ yd}^3 = \frac{73,440 \text{ ft}^3 \text{ of sand}}{27 \text{ ft}^3/\text{yd}^3}$$

$$x = 2,720 \text{ yd}^3 \text{ of sand}$$

Volume of soil in cubic yards:

$$\frac{1 \text{ yd}^3}{27 \text{ ft}^3} = \frac{x \text{ yd}^3}{4,320 \text{ ft}^3}$$

$$x \text{ yd}^3 = \frac{4,320 \text{ ft}^3 \text{ of sand}}{27 \text{ ft}^3/\text{yd}^3}$$

$$x = 160 \text{ yd}^3 \text{ of soil}$$

Volume of organic matter in cubic yards:

$$\frac{1 \text{ yd}^3}{27 \text{ ft}^3} = \frac{x \text{ yd}^3}{8,640 \text{ ft}^3}$$

$$x \text{ yd}^3 = \frac{8{,}640 \text{ ft}^3 \text{ of sand}}{27 \text{ ft}^3/\text{yd}^3}$$

$$x = 320 \text{ yd}^3 \text{ of organic matter}$$

SOLUTION

A total of 2,270 yd³ of sand, 160 yd³ of soil, and 320 yd³ of sand need to be used to produce 85-5-10 root-zone mix for the football field.

When houses and building complexes are built, it is common for the topsoil to be stripped off as part of the construction process. It is important to place topsoil back over the subgrade prior to establishing the turfgrass.

Practice Problem 8-2   Soil Modification

A 10,000 ft² lawn is to be established. The topsoil was stripped off prior to construction, and it needs to be replaced. It has been determined that 6 in. of amended topsoil will be placed on the lawn. How much amended topsoil needs to be ordered?

Topdressing

Topdressing calculations are very similar to those described for soil modification. The main difference is the depth and how it is applied. Topdressing is the process of adding amendments on top of the turfgrass. The depth of the topdressing is significantly less than that described in soil modification; instead of 6 to 18 inches, topdressing is usually less than one-half inch in depth.

EXAMPLE 8-8

How much topdressing material would need to be purchased to topdress an infield of a baseball field to a depth of $\frac{1}{4}$ in.? Bases in this field are spaced at 90 ft as shown in Figure 8-4.

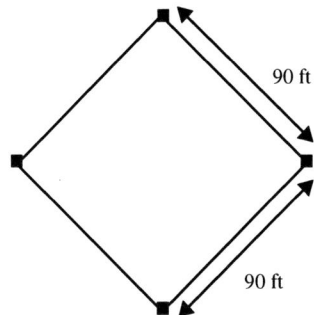

Figure 8-4

Baseball diamond

Step 1

Determine the area to be topdressed. The bases are spaced 90 ft apart, and each base path is set at a 90° angle; therefore, the area can be calculated using the formula for the area of a square. The formula for a square is: Area = s², where s = side.

$$\text{Area} = s^2$$
$$\text{Area} = 90^2$$
$$\text{Area} = 8,100 \text{ ft}^2$$

Step 2

To determine volume, convert the topdress depth from inches to feet.
 The topdress depth is $\frac{1}{4}$ in., or 0.25 in.

$$\tfrac{1}{4} = 0.25 \text{ in.}$$

To convert 0.25 in. into ft, divide by 12 because there are 12 in. in 1 ft.

$$\frac{1 \text{ ft}}{12 \text{ in.}} = \frac{x \text{ ft}}{0.25 \text{ in.}}$$

$$x \text{ ft} = \frac{0.25 \text{ in.}}{12 \text{ in./ft}}$$

$$x = 0.02083 \text{ ft}$$

Step 3

Volume is determined by multiplying the depth (in feet) by the square footage of the infield.

$$\text{Volume} = (8,100 \text{ ft}^2)(0.02083 \text{ ft})$$
$$\text{Volume} = 168.7 \text{ ft}^3$$

Step 4

Topdress will be purchased in cubic yards; therefore, convert cubic feet to cubic yards. Set up the proportion by stating the following: If 1 yd³ = 27 ft³, then (*x*) yd³ = 168.7 ft³.

$$\frac{1 \text{ yd}^3}{27 \text{ ft}^3} = \frac{x \text{ yd}^3}{168.7 \text{ ft}^3}$$

$$x \text{ yd}^3 = \frac{(168 \text{ ft}^3)(1 \text{ yd}^3)}{27 \text{ ft}^3}$$

$$x = 6.25 \text{ yd}^3$$

SOLUTION

It will take 6.25 yd³ to topdress the infield to a $\frac{1}{4}$ in. depth.

Soil Modification and Topdressing Calculations **183**

Another common topdressing calculation question relates to storage. Often, topdress material is stored in bins or piles without any idea of the volume on hand in storage. If the topdress material is to be used, it becomes necessary to calculate the volume of the stored topdress material to determine how much area can be topdressed.

EXAMPLE 8-9

How many square feet can be topdressed to a $\frac{3}{8}$ in. depth with a quantity of topdress material that is stored in a 25 ft by 18 ft rectangular bin that measures 6 ft high? To solve this problem, first determine the best method for determining volume. This is a rectangular-shaped stockpile of topdress material, thus, the calculation of a cube should suffice to determine the volume of the bin of topdress material.

Step 1
Determine the surface area of the storage bin.

$$(25 \text{ ft})(18 \text{ ft}) = 450 \text{ ft}^2$$

Step 2
Determine the volume of the bin of topdress material in cubic feet. The surface area of the storage bin is 450 ft² and the height of the topdress material is 6 ft. To calculate volume, multiply the base area by the height.

$$x \text{ ft}^2 \text{ of topdress material} = (450 \text{ ft}^2)(6 \text{ ft})$$
$$x = 2{,}700 \text{ ft}^3 \text{ of topdress material}$$

Step 3
Convert 2,700 ft³ to cubic yards. Set up the proportion as shown in Example 8-8.

$$\frac{1 \text{ yd}^3}{27 \text{ ft}^3} = \frac{x \text{ yd}^3}{2{,}700 \text{ ft}^3}$$
$$x \text{ yd}^3 = \frac{2{,}700 \text{ ft}^3}{27 \text{ ft}^3/\text{yd}^3}$$
$$x = 100 \text{ yd}^3$$

Step 4
Determine how many cubic yards are needed to topdress 1 ft² to a $\frac{3}{8}$ in. depth. The topdress depth is $\frac{3}{8}$ in., or 0.375 in.

$$\tfrac{3}{8} = 0.375 \text{ in.}$$

To convert 0.375 in. into ft, divide by 12 because there are 12 in. in 1 ft.

$$\frac{1 \text{ ft}}{12 \text{ in.}} = \frac{x \text{ ft}}{0.375 \text{ in.}}$$

$$x \text{ ft} = \frac{0.375 \text{ in.}}{12 \text{ in./ft}}$$

$$x = 0.03125 \text{ ft}$$

Volume for 1 ft² is determined by multiplying the area (1 ft²) by the height (0.03125 ft).

$$x = (0.03125 \text{ ft})(1 \text{ ft}^2)$$

$$x = 0.03125 \text{ ft}^2$$

Convert cubic feet to cubic yards.

$$\frac{1 \text{ yd}^3}{27 \text{ ft}^3} = \frac{x \text{ yd}^3}{0.03125 \text{ ft}^3}$$

$$x \text{ yd}^3 = \frac{0.03125 \text{ ft}^3/\text{ft}^2}{27 \text{ ft}^3/\text{yd}^3}$$

$$x = 0.001157 \text{ yd}$$

Step 5

Determine how much surface area the pile of topdress material will cover. Set up the proportion by asking the following: If 0.001157 yd³ is needed to cover 1 ft², then how many x ft² will be covered by 100 yd³?

$$\frac{1 \text{ ft}^2}{0.001157 \text{ yd}^3} = \frac{x \text{ ft}^2}{100 \text{ yd}^3}$$

$$x \text{ ft}^2 = \frac{100 \text{ yd}^3}{0.001157 \text{ yd}^3/\text{ft}^2}$$

$$x = 86,430 \text{ ft}^2$$

SOLUTION

A storage bin that measures 25 ft by 18 ft by 6 ft contains 100 yd³ of topdress material that will cover 86,430.4 ft² when applied at a $\frac{3}{8}$ in. depth.

Practice Problem Set 8-3 Topdressing

1. A landscape company is going to topdress a 6,000 ft² seedbed with compost amended soil to a depth of $\frac{1}{2}$ in. How many cubic yards of compost amended soil needs to be ordered?

2. A sports turf supply company has a quantity of topdress media that is stored in a cone-shaped pile. The pile has a base diameter of 30 ft and a height of 18 ft.
   a. How many cubic yards are in the pile of topdress (see Chapter 3 for the formula for determining the volume of a cone).
   b. How many square feet will the pile cover if the topdress is applied to a $\frac{5}{16}$ in. depth?

Water Use Calculations

Water is a very important resource that needs to be protected and used wisely. It is also an important component to maintain high-quality turfgrass areas. This section will cover calculations involving irrigation of turf that are the responsibility of the turfgrass manager. These problems include total water use, the cost of irrigation water, and the capacity of storage ponds.

Measuring Water Use

> **Definition** *Acre-foot (ac-ft)*
> One *acre-foot* is the amount of water that is needed to cover one acre (43,560 ft^2) to the depth of one foot. The conversion factor used to determine the number of gallons in an acre-foot is as follows:
> $$1 \text{ ac-ft} = 325{,}851.4 \text{ gal}$$

Irrigation water is measured in terms of acre-feet (ac-ft). This is a very useful conversion factor to use when calculating the capacity of an irrigation pond.

Irrigation applied to turfgrass is typically measured in terms of inches rather than feet. Therefore, it is important to calculate how much water is in one acre-inch (ac-in.).

> **Definition** *Acre-inch (ac-in.)*
> One *acre-inch* is the gallon of water that is needed to cover one acre (43,560 ft^2) one inch deep. It is calculated by dividing 1 ac-ft (325,851.4 gal) by 12.
> $$\frac{325{,}851.4 \text{ gal}}{12 \text{ in.}} = \frac{x \text{ gal}}{1 \text{ in.}}$$
> $$x \text{ gal} = \frac{325{,}851.4 \text{ gal}}{12 \text{ in.}}$$
> $$1 \text{ ac-in.} = 27{,}154.3 \text{ gal}$$

Total Water Use

Understanding how much water is used with one irrigation event is the first step for the turfgrass manager.

EXAMPLE 8-10

If 5 ac are to be irrigated with 1 in. of water, how many gallons would be applied to the area?

Step1

There are 5 ac, each receiving 1 ac-in. There are 27,154.3 gallons of water in an acre-inch.

$$x \text{ gal} = (27,154.3 \text{ gal/ac-in.})(5 \text{ ac})$$
$$x = 135,771.5 \text{ gal}$$

SOLUTION

A total of 135,771.5 gallons of water would be needed to apply 1 in. of water to 5 ac.

Many turfgrass areas are reported in square feet, rather than acres. It is important to convert them to acres prior to completing the calculation.

EXAMPLE 8-11

A football field is to be irrigated with 0.75 in. of water. How many gallons would be applied to the area, if the area of a football field is 57,600 ft².

Step 1

Convert 57,600 ft² to acres. Begin by setting up the proportion.

$$\frac{1 \text{ ac}}{43,560 \text{ ft}^2} = \frac{x \text{ ac}}{57,600 \text{ ft}^2}$$
$$x \text{ ac} = \frac{57,600 \text{ ft}^2}{43,560 \text{ ft}^2}$$
$$x = 1.32 \text{ ac}$$

Step 2

Determine how much water is in 0.75 ac-in.

$$\frac{27,154.3 \text{ gal}}{1 \text{ ac-in.}} = \frac{x \text{ gal}}{0.75 \text{ ac-in.}}$$
$$x \text{ gal} = (27,154.3 \text{ gal})(0.75 \text{ ac-in})$$
$$x = 20,365.7 \text{ gal in } 0.75 \text{ ac-in.}$$

Step 3
There are 1.32 acres, each receiving 0.75 ac-in. There are 20,365.7 gallons of water in 0.75 ac-in.

$$\frac{20{,}365.7 \text{ gal}}{1 \text{ ac}} = \frac{x \text{ gal}}{1.32 \text{ ac}}$$

$$x \text{ gal} = (20{,}365.7 \text{ gal})(1.32 \text{ ac})$$

$$x = 26{,}882.7 \text{ gal}$$

SOLUTION

A total of 26,882.7 gal of water would be needed to apply 0.75 in. of water to a football field.

Another helpful calculation is to determine the total water use for the year. To do these types of calculations, it is important to know the typical water use rates for a particular locale. Good estimates of water use rates usually can be obtained from the local supplier of irrigation equipment.

EXAMPLE 8-12

A property manager is considering the addition of an irrigation system for a 10-ac property. The manager wants the following information:

1. How much water would have to be supplied to the property in a peak irrigation week in midsummer. For this problem, it will be assumed that a total of 1.4 in. of irrigation, including adjustments for evaporation, will be needed in a peak irrigation week.
2. An estimate of the total water cost for the season. For this problem, assume that 15 in. of irrigation water will be required.

PART A: TOTAL WATER USE FOR A PEAK IRRIGATION WEEK

There are 10 ac of irrigated turfgrass that are to receive 1.4 in. of water during a peak week of irrigation. It already has been calculated that 1 in. of water on 1 ac will contain 27,154.3 gal.

Step 1
Calculate total water use for 1 acre.

$$\frac{27{,}154.3 \text{ gal}}{1 \text{ in. H}_2\text{O}} = \frac{x \text{ gal.}}{1.4 \text{ in H}_2\text{O}}$$

$$(27{,}154.3 \text{ gal})(1.4 \text{ in. H}_2\text{O}) = (x \text{ gal})(1 \text{ in. H}_2\text{O})$$

$$x \text{ gal} = (27{,}154.3 \text{ gal})(1.4 \text{ in. H}_2\text{O})$$

$$x = 38{,}016 \text{ gal to apply } 1.4 \text{ in. of water to 1 ac}$$

Step 2

Next, determine how much water would need to be applied to 10 ac.

$$x \text{ gal} = (38{,}016 \text{ gal/ac})(10 \text{ acres})$$
$$x = 380{,}160 \text{ gal}$$

SOLUTION

A total of 380,160 gal of water would need to be applied to irrigate 10 ac with 1.4 in. of water.

PART B: COST OF WATER FOR ONE SEASON

To determine the cost of water, first calculate the total water use for the season. It has already been calculated that 380,160 gal of water will be needed to apply 1.4 in. of water to 10 ac. How much water will be needed to apply 15 in.?

Step 1

Calculate the total water use. If 380,160 gal deliver 1.4 in., then x gal will deliver 15 in. of irrigation.

$$\frac{380{,}160 \text{ gal}}{1.4 \text{ in.}} = \frac{x \text{ gal}}{15 \text{ in.}}$$

$$(380{,}160 \text{ gal})(15 \text{ in.}) = (x \text{ gal})(1.4 \text{ in.})$$

$$x \text{ gal} = \frac{(5{,}702{,}400 \text{ gal})}{1.4 \text{ in.}}$$

$$x = 4{,}073{,}143 \text{ gal}$$

A total of 4,073,143 gal is needed to irrigate 10 ac for 1 season.

Step 2

The cost of water is based on a charge of $0.02/ft^3. Since water cost is based on cubic feet (ft^3) instead of gallons, it is necessary to convert gallons into cubic feet. There are 7.480519 gallons in 1 ft^3. The question is, how many cubic feet are in 4,073,143 gal?

$$\frac{1 \text{ ft}^3}{7.480519 \text{ gal}} = \frac{x \text{ ft}^3}{4{,}073{,}143 \text{ gal}}$$

$$x \text{ ft}^3 = \frac{4{,}073{,}143 \text{ ft}^2/\text{gal}}{7.480519 \text{ gal}}$$

$$x = 544{,}500 \text{ ft}^3$$

A total of 544,500 ft^3 of water is the projected volume needed.

Step 3

Calculate the total cost of water, where the base cost is $0.02/ft^3.

$$x \text{ \$} = (544{,}500 \text{ ft}^3)(\$0.02/\text{ft}^3)$$

$$x = \$10{,}890.00$$

SOLUTION

The projected cost to irrigate the property for one season is $10,890.00.

Capacity of Storage Ponds

It is common to irrigate turfgrass areas by pulling the water out of storage or holding ponds. Knowing the capacity of a storage pond allows turfgrass managers to make informed decisions.

EXAMPLE 8-13

Calculate the amount of water in a storage lake that has a surface area of 500,000 ft^2 and an average depth of 18 ft. How many gallons of water are in the lake?

Step 1

Determine the surface area of the lake in acres. If 1 ac equals 43,560 ft^2, how many acres are in 500,000 ft^2?

$$\frac{1 \text{ ac}}{43,560 \text{ ft}^2} = \frac{x \text{ ac}}{500,000 \text{ ft}^2}$$

$$x \text{ ac} = \frac{500,000 \text{ ft}^2}{43,560 \text{ ft}^2}$$

$$x = 11.5 \text{ ac of surface area}$$

Step 2

Determine the amount of water in the lake as acre-feet. The average depth of the lake has been determined to be 18 ft and the surface area is 11.5 ac. Therefore:

$$x \text{ ac-ft} = (11.5 \text{ ac})(18 \text{ ft})$$

$$x = 207 \text{ ac-ft}$$

There are 207 ac-ft of water in the lake.

Step 3

Convert acre-feet to gallons of water. There are 325,851.4 gal in 1 ac-ft, therefore:

$$\frac{325,851.4 \text{ gal.}}{1 \text{ ac-ft}} = \frac{x \text{ gal}}{207 \text{ ac-ft}}$$

$$x \text{ gal} = (325,851.4 \text{ gal/ac-ft})(207 \text{ ac-ft})$$

$$x = 67,451,239 \text{ gal}$$

SOLUTION

There are 67,451,239 gallons of water in the storage lake.

Once the capacity of a storage lake is determined, the irrigation potential can be calculated.

EXAMPLE 8-14

There are 200 acres of turf on a sod farm that are to be irrigated from the storage lake in Example 8-12. How many inches of water could be applied to the 200 acres without any recharge? Evaporation loss will not be considered in this calculation.

Step 1

Calculate the gallons of water needed to irrigate 200 ac with 1 in. of water. There are 27,154.3 gal in 1 ac-in. This is done by multiplying gallons per acre-inch by total acres.

$$x \text{ gal} = (27,154.3 \text{ gal/ac-in.})(200 \text{ ac})$$

$$x = 5,430,860 \text{ gal}$$

A total of 5,430,860 gal are needed to apply 1 in. of water to 200 ac of sod.

Step 2

Calculate the total inches of water that can potentially be applied to 200 ac of sod. There are 67,451,239 gal in the lake and irrigating the sod with 1 in. of water delivers 5,430,860 gallons. Therefore:

$$x \text{ in.} = \frac{67,451,239 \text{ gal.}}{5,430,860 \text{ gal.}}$$

$$x = 12.4 \text{ in. of water}$$

SOLUTION

The lake is capable of supplying 12.4 in. of water to the 200 ac without any recharge.

Practice Problem Set 8-4 — Water Use Calculations

1. A total of 125 ac of sod are to be irrigated with 2.5 in. of water over the next 2 weeks. How much water will be needed to make this application?

2. The department of natural resources is monitoring the amount of irrigation water used by a sports complex. The complex irrigates 16 ac with 1.3 in. of water/week. Total irrigation for the year is 15.6 in. of water. How many cubic feet of water does the sports complex use for the year?

3. An office complex draws its irrigation water from a small retention pond. The surface area of the pond is 95,832 ft^2 and the depth is 10 ft. There are 5.5 ac of turf that needs to be irrigated. They want to determine how many inches of water can be applied to irrigate 5.5 ac during one season.

Chapter 9

Mathematical Applications for the Landscape Industry

Introduction

This chapter covers common calculations used in the landscape installation and maintenance industry. The three main categories of calculations include: determining slopes as part of site grading, landscape installation costs including material and labor costs and production rates, and landscape maintenance costs including labor costs and production rates. Many of the calculations will build on concepts covered earlier in this text.

Site Grading

Almost all landscape installation projects require some amount of site grading. Grading requires the excavation and moving of existing soil or the addition of new soil to change the contours of a site. Since calculations for large-scale grading plans must be done by a licensed professional (civil engineer or landscape architect), only smaller residential landscape examples will be discussed in this chapter.

Calculating Slope

Slopes are used in landscapes to create interest, add spatial definition, and ensure a landscape site will drain properly. The term slope is often used to describe the measurement of the steepness, incline, or gradient of a straight line. A higher slope value indicates a steeper incline. In landscape construction, slope is commonly expressed in two ways: as a ratio or a percent.

The slope is defined as the ratio of the "rise" divided by the "run" between two points on a line. In other words, the ratio of the altitude change to the horizontal distance between any two points on the line. For example, a ratio of 1:20 indicates a 1 ft change vertically for every 20 ft of horizontal length. This slope can also be expressed as a percent and is 5% (Figure 9-1).

It is essential to use the same units for the vertical and horizontal distances. For example, do not calculate a slope using a 10 in. rise and a 30 ft run, unless you have converted them both to either feet or inches.

192 Part Two · Chapter 9 Mathematical Applications for the Landscape Industry

Percent slope

$$= \frac{1}{20} = 0.05$$

$$= (0.05)(100) = 5\%$$

Figure 9-1
A 1:20 ratio or a 5% slope

Rise = 1 ft

Run = 20 ft

EXAMPLE 9-1

Calculate the ratio and percent slope based on the following site information: The dimensions of the Smith property are shown in Figure 9-2.

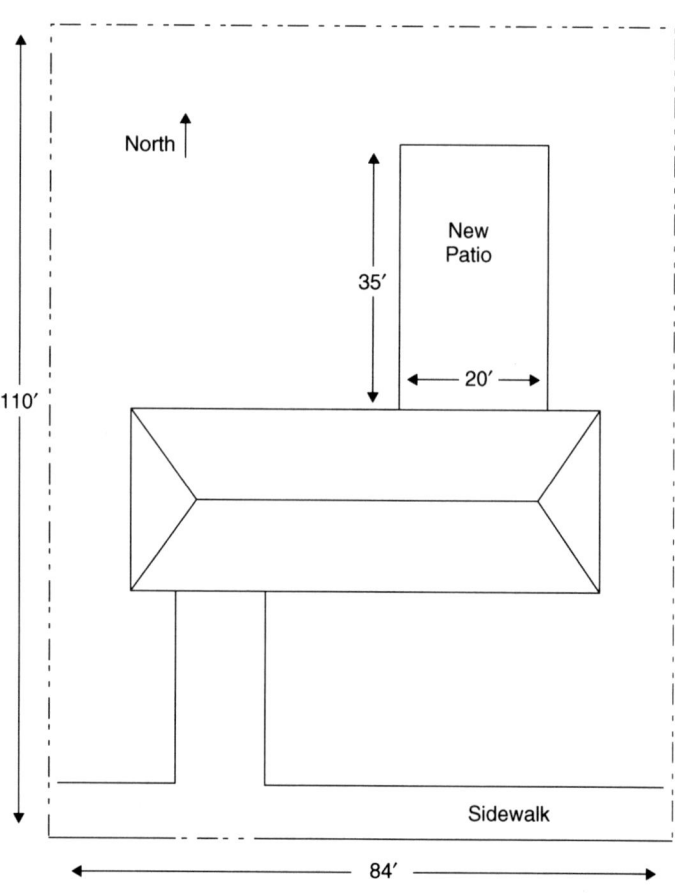

Figure 9-2
The Smith property

From the north property line to the sidewalk along the south property line, the lot rises 8 ft in elevation. What is the slope from north to south?

SOLUTION

$$\text{Ratio} = \text{rise:run}$$

$$8\,\text{ft}:110\,\text{ft}$$

$$\text{Percent slope} = \frac{\text{rise}}{\text{run}}$$

$$\frac{8\,\text{ft}}{110\,\text{ft}} = 0.0727$$

$$(0.0727)(100) = 7.27\%\ \text{slope}$$

From the west boundary of the property to the east boundary there is a drop of 2 ft in elevation. What is the slope from west to east?

SOLUTION

$$\text{Ratio} = \text{rise:run}$$

$$2\,\text{ft}:84\,\text{ft}$$

$$\text{Percent slope} = \frac{\text{rise}}{\text{run}}$$

$$\frac{2\,\text{ft}}{84\,\text{ft}} = 0.0238$$

$$(0.0238)(100) = 2.38\%\ \text{slope}$$

Many common landscape features have recommended slopes (Table 9-1). These slope guidelines ensure proper site drainage and good vehicular and pedestrian circulation around the site.

TABLE 9-1 • SLOPE GUIDELINES

Area	Maximum	Desirable	Minimum
Street	17%	1–8%	0.5%
Parking	6%	1–4%	0.5%
Walks			
Bldg. Approach	4%	2–3%	1%
Major Walk	5%	2–4%	0.5%
Ramp	8.33%	6–8%	0.5%
Paved Pedestrian Areas (seating and plazas)	2%	1.5%	1%
Lawn (mowed)	25%	4–10%	2%
Unmowed bank	50%	33%	—
Swales	5–6%	1–4%	0.5%

Adapted from Ingels (2004) and Rubenstein (1996) and VanDerZanden and Rodie (2008).

Practice Problem Set 9-1

The Smiths plan to install a new paver patio adjacent to the north side of their house. The patio dimensions are 20 ft E-W and 35 ft N-S (see Figure 9-2). They have calculated a drop of 8.4 in. over 35 ft from the patio edge adjacent to the house to the north edge of the patio. Answer the following:

1. What percent of slope are they using on this patio?
2. Is it within the slope guidelines for a seating area?

Estimating Landscape Installation Costs

The estimating technique used for landscape installation projects is often based on a company's business model. Since this and numerous other factors impact landscape estimating, this chapter only will cover the basic calculations associated with a landscape estimate. This discussion will not detail the different approaches to calculate things such as equipment costs, overhead, and profit. References detailing estimating methods, including systems used to account for equipment costs, overhead, and profit are included at the end of this chapter.

Proficiency in landscape estimating is critical to having a successful landscape installation business. Costs associated with a typical landscape installation project include: materials, labor, and equipment. Occasionally the cost of using a subcontractor also needs to be included. In addition to these costs, overhead (direct and indirect) and profit need to be included in estimates so the landscape company is a viable business. The key to creating a thorough landscape installation estimate is to take one step at a time and to think through the process carefully.

A backyard renovation of the Peters' residence (Figure 9-3) will be used throughout Example 9-2, which is a landscape installation estimate. The estimate includes installing new plantings (trees, shrubs, perennials, and groundcovers), as well as a new concrete paver patio. Area measurements have been calculated for the project and include the following:

Total Planting Bed Area	214 ft^2
(Groundcover Area 46 ft^2)	
Total Lawn Area	180 ft^2
Total Paver Patio Area	121 ft^2
Total Area	515 ft^2

Materials Costs

> **Definition** — *Materials Costs*
> *Materials costs* are the actual costs of the materials used to create the landscape.

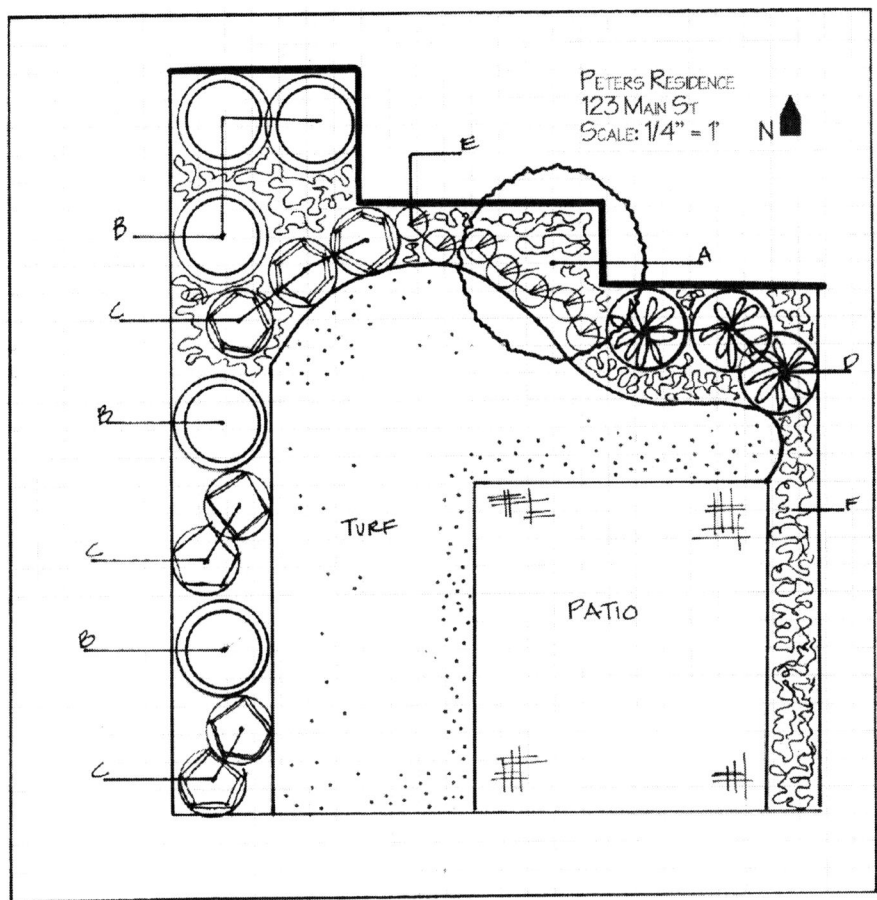

Figure 9-3
The Peters' residence

This cost includes unit items like plants, pavers, or natural stone and bulk items such as topsoil, mulch, and crushed limestone for a patio base. Depending on the material, you can either take a direct count (e.g., count each shrub drawn on a landscape plan) or calculate the amount based on area and volume (e.g., the area of planting beds to be covered with three inches of mulch).

Materials Costs Calculation

Using the new landscape plan for the Peters' residence, the quantity of plants, hard goods, and their associated costs can be calculated.

Quantities of Plant Materials

EXAMPLE 9-2a

The tree, shrubs, and perennials can be counted directly from the landscape plan and include the elements listed in Table 9-2:

TABLE 9-2 • PLANT MATERIALS FOR PETERS RESISDENCE

Key	Plant Name	Quantity
A	Flowering crabapple	1
B	Juniper	5
C	Dwarf spirea	7
D	Redtwig dogwood	3
E	Hosta	7
F	Groundcover	—

Often the number of groundcover plants and some annual and perennial plants need to be calculated based on the area they will cover in the design and the mature spread of each plant. The spread of a plant is defined as the area the plant will cover at maturity. Table 9-3 is an industry standard and shows how many plants are required per square foot based on different plant spacing regimes.

TABLE 9-3 • SPACING FOR GROUNDCOVERS, ANNUALS, AND PERENNIALS

Spacing between centers of plants (inches)	Plants/ft^2
4	9.00
6	4.00
8	2.30
10	1.40
12	1.00
15	0.65
18	0.45
24	0.25

Adapted from *Landscape Designer & Estimator's Guide*, 3rd edition.

The periwinkle groundcover (*Vinca minor*) in the Peters' residence design covers a total area of 46 ft^2. The planting scheme calls for the plants to be installed 18 in. on center (o.c.), meaning the center of one plant is 18 in. from the center of adjacent plants. This spacing accounts for the mature spread of each periwinkle plant.

EXAMPLE 9-2b

Calculate the number of periwinkle plants required for the design.

From Table 9-3 you can see that a spacing of 18 in. o.c. is equivalent to 0.45 plants/ft^2.

$$(\text{plants/ft}^2)(\text{total ft}^2) = \text{total periwinkle plants needed}$$

$$(0.45 \text{ plants/ft}^2)(46 \text{ ft}^2) = 20.7 \text{ or } 21 \text{ plants}$$

> **SOLUTION**
>
> Twenty-one periwinkle plants are required to cover 46 ft² when planted 18 in. o.c.

Seeding is a common way to establish a new lawn area. Seeding rates vary depending on the turf grass species. For example, Kentucky bluegrass is usually seeded at a rate of 1–1.5 lb/1,000 ft² (MSF is the abbreviation for 1,000 ft²), while perennial ryegrass is seeded at 7–9 lb/MSF. The new lawn area at the Peters' residence will be Kentucky bluegrass and seeded at a rate of 1.5 lb/MSF.

EXAMPLE 9-2c

Calculate the amount of seed needed for the lawn area.

Step 1

Convert 180 ft² to MSF

$$\frac{180 \text{ ft}^2}{1{,}000 \text{ ft}^2} = 0.180 \text{ MSF}$$

Step 2

$$(\text{rate})(\text{area}) = \text{amount needed}$$

$$(1.5 \text{ lb/MSF})(.180 \text{ MSF}) = 0.27 \text{ lb of seed}$$

or rounded up to 0.30 lb

> **SOLUTION**
>
> The new lawn area needs 0.30 lb of Kentucky bluegrass seed.

Quantities of Hard goods

Hard goods, or hardscapes, are any nonplant material in a landscape. Examples include: pavers, natural stone, retaining wall block, deck materials, mulch, gravel, and sand. The quantity of pavers, natural stone, and retaining wall block required for a project can be calculated based on the size of each unit (Example 9-2d). Bulk items (mulch, gravel, and sand) can be calculated using standard equations using area and volume measurements (Example 9-2 e–g). Cubic yard is the standard unit of measurement for many bulk items. Table 9-4 is an industry standard and shows the area (ft²) covered by one cubic yard (yd³) of bulk material at various depths.

After completing hardscape calculations, landscape estimators often increase the actual quantity ordered on certain materials to account for waste and breakage. Table 9-5 gives common industry standard overages associated with hard goods.

EXAMPLE 9-2 d

The Peters' new patio area is 11' × 11', and they selected a blended color paver mix with pavers that are 6" × 9" each. The installation pattern will be a simple pattern.

Calculate the number of pavers needed for the patio. This will require multiple steps, as follows.

TABLE 9-4 • AREA COVERED BY ONE CUBIC YARD OF BULK MATERIAL AT A SET DEPTH

Depth of Material (inches)	Area Covered (ft^2)
$\frac{1}{4}$	1296
$\frac{1}{2}$	648
1	324
2	162
3	108
4	81
6	54

TABLE 9-5 • OVERAGES ASSOCIATED WITH COMMON HARD GOODS

Hard Good Materials	Additional Overages (%)
Bulk items: soil, sand, gravel	10–15
Unit pavers and wall materials	3–5
When design calls for complex patterns	5–8
Concrete	2–3
Groundcover and sod	5

Step 1

Calculate the area covered by one paver.

$$(\text{paver length})(\text{paver width}) = \text{area of one paver}$$

$$(6 \text{ in.})(9 \text{ in.}) = 54 \text{ in.}^2/\text{paver}$$

Step 2

Calculate the number of pavers per square foot, given there are 144 in^2/ft^2.

$$\frac{144 \text{ in.}^2/\text{ft}^2}{54 \text{ in.}^2/\text{paver}} = 2.67 \text{ pavers/ft}^2$$

Step 3

Calculate the area of the patio.

$$(\text{patio length})(\text{patio width}) = \text{area of patio}$$

$$(11 \text{ ft})(11 \text{ ft}) = 121 \text{ ft}^2$$

Step 4
Calculate the number of pavers needed for the patio.

(area of patio in ft^2)(number of pavers/ft^2) = total number of pavers

(121 ft^2)(2.67 pavers/ft^2) = 323.07 or 323 pavers

Step 5
Calculate the total number of pavers needed for the patio when waste/breakage is included. The overage percentage for unit pavers ranges from 5–8% (Table 9-5). Since this is a simple paving pattern, use the lower percentage.

(number of pavers)(overage percentage) = additional pavers needed for waste/breakage

(323 pavers)(0.05) = 16.15 or 16 pavers

SOLUTION

calculated amount + overage amount = total amount

323 + 16 = 339 total pavers

339 pavers are required to install the Peters' new 121 ft^2 patio.

An alternative way to calculate the total pavers required for the installation is to convert the total patio area into square inches and divide that by the number of square inches per paver, as illustrated below.

Step 1
Calculate the total square inches of patio area given there are 144 in.^{2}/ft^2.

(patio area in ft^2)(in.^{2}/ft^2) = patio area in in.^{2}

(121 ft^2/patio)(144 in.^{2}/ft^2) = 17,424 in.^{2}/patio

Step 2
Having already determined that the area of each paver is 54 in.^{2} [(6 in.)(9 in.) = 54 in.^{2}], the next step is dividing the total patio area by the area of each paver:

$$\frac{\text{total patio area}}{\text{area of each paver}} = \text{total number of pavers}$$

$$\frac{17,424 \text{ in.}^2}{54 \text{ in.}^2/\text{paver}} = 322.07 \text{ or } 323 \text{ pavers}$$

Step 3
Overage is calculated the same as in the previous example.

(number of pavers)(overage percentage) = additional pavers needed for waste/breakage

(323 pavers)(0.05) = 16.15 or 16 pavers

SOLUTION

$$\text{calculated amount} + \text{overage amount} = \text{total amount}$$

$$323 + 16 = 339 \text{ total pavers}$$

339 pavers are required to install the Peters' new 121 ft² patio.

EXAMPLE 9-2e

Before the pavers are installed for the new patio, a 4 in. layer of crushed limestone base material must be installed and compacted. This layer is followed by a 1 in. layer of sand for the pavers to sit on.

Calculate the amount of crushed limestone using the total area of the patio (121 ft²) and information from Table 9-4. The table shows that at a depth of 4 in., 1 yd³ of a bulk material (crushed limestone in this case) covers 81 ft².

Step 1

$$\frac{\text{total area}}{\text{area covered/yd}^3} = \text{amount of bulk material needed}$$

$$\frac{121 \text{ ft}^2}{81 \text{ ft}^2/\text{yd}^3} = 1.49 \text{ yd}^3 \text{ crushed limestone}$$

Step 2

From Table 9-5 you can determine the additional overage to calculate this type of bulk item is 10–15%. Since this patio is relatively small, you can use the lower number.

$$(\text{amount of limestone})(\text{overage percent}) = \text{amount of overage}$$

$$(1.49 \text{ yd}^3)(.10) = 0.149 \text{ yd}^3$$

Step 3

$$\text{calculated amount} + \text{overage amount} = \text{total amount}$$

$$1.49 \text{ yd}^3 + 0.149 \text{ yd}^3 = 1.64 \text{ yd}^3$$

Generally, bulk items are sold in whole or half units so you will need to decide if you want to order 1.5 yd³ or round up to 2.0 yd³. Since rounding up only results in approximately 0.36 yd³ (2.0 − 1.64 = .36) of additional material, rounding up to 2.0 yd³ to account for uneven excavation areas is a good idea.

SOLUTION

The Peters' new 121 ft² patio requires 2.0 yd³ of crushed limestone.

EXAMPLE 9-2f

The sand for the patio is calculated using the same approach.

Step 1

$$\frac{\text{total area}}{\text{area covered/yd}^3} = \text{amount of bulk material needed}$$

$$\frac{121 \text{ ft}^2}{324 \text{ ft}^2/\text{yd}^3} = 0.37 \text{ yd}^3 \text{ sand}$$

Step 2

Since sand is also needed to sweep into the joints between the pavers, it is a good idea to use the higher overage percentage.

$$(\text{amount of sand})(\text{overage percent}) = \text{amount of overage}$$

$$(0.37 \text{ yd}^3)(0.15) = 0.06 \text{ yd}^3$$

Step 3

$$\text{calculated amount} + \text{overage amount} = \text{total amount}$$

$$0.37 \text{ yd}^3 + 0.06 \text{ yd}^3 = 0.43 \text{ yd}^3$$

In this case, round up to 0.50 yd^3 when ordering the sand. It always is better to have a little extra than to run short during an installation project.

SOLUTION

The Peters' new 121 ft^2 patio requires 0.50 yd^3 of sand.

EXAMPLE 9-2 g

Mulch is also a bulk item and often is purchased by the cubic yard. All of the bed area (214 ft^2) will be mulched after the plants are installed.

Calculate the amount of mulch needed to cover the bed area with a layer of mulch 3 in. deep.

Step 1

$$\frac{\text{total area}}{\text{area covered/yd}^3} = \text{amount of bulk material needed}$$

$$\frac{214 \text{ ft}^2}{108 \text{ ft}^2/\text{yd}^3} = 1.98 \text{ yd}^3 \text{ mulch}$$

SOLUTION

Round 1.98 yd^3 to 2.0 yd^3 of mulch for the Peters' project.

Practice Problem Set 9-2

The Schuhs plan to install a 10′ × 14′ paver patio using a complex paving pattern. Calculate the following:

1. The number of 4″ × 8″ pavers
2. The amount of crushed limestone required for a 4-inch layer
3. The amount of sand needed for the project

Calculating Material Costs and the Selling Prices An important part of having a successful landscape business is identifying your costs (materials and labor), overhead expenses, and profit margin. All of these are used to determine your breakeven price and selling price for goods and/or services. The breakeven price is what a product costs you. The selling price is the amount for which you will sell a particular good or service. In other words, it is what you charge the customer. The following are examples of how these are calculated:

$$\text{Breakeven price} = \text{wholesale cost} + \text{overhead markup}$$

$$\text{Selling price} = \text{wholesale cost} + \text{overhead markup} + \text{profit}$$

$$\text{Overhead markup} = \text{total annual overhead expenses} / \text{total annual direct material expenses}$$

Overhead markup and profit are usually expressed as a percentage, and the percentages for each vary depending on the company. In addition to the overhead markup, profit must be added to the wholesale cost of the good or service to determine the selling price.

EXAMPLE 9-3a

The following shows how to calculate the selling price for the Prairifire flowering crabapple tree in the design for the Peters' residence.

Prairifire Crabapple (wholesale cost; 12 in. B&B, 1 in. caliper)	$52.00
Overhead markup (60%)	
(percent overhead markup)(wholesale cost) = markup	
(0.60)($52.00) = $31.20	$31.20
Breakeven price (wholesale cost + markup)	$83.20

Profit (10%)

 (percent profit)(breakeven price) = selling price

 (0.10)($83.20) = $8.32 $8.32
                            ———

Total selling price $91.52

EXAMPLE 9-3b

The following shows how a similar calculation can be done to determine the selling price of the pavers for the Peters' new patio.

 Autumn blend cobble paver (wholesale cost/ft^2; 6 in. × 9 in.) $2.50

 Overhead markup (60%)

  (percent overhead markup)(wholesale cost) = markup

  (0.60)($2.50) = $1.50 $1.50
                            ———

 Breakeven price (wholesale cost + markup) $4.00

 Profit (10%)

  (percent profit)(breakeven price) = selling price

  (0.10)($4.00) = $0.40 $0.40
                            ———

Total selling price/ft^2 $4.40

Practice Problem 9-3

Determine the breakeven price and selling price for a 5 gal rhododendron assuming a wholesale cost of $80.00, overhead markup of 60%, and profit of 10%.

Once prices (B) for the individual components of a project are determined, they are multiplied by the quantity (A) needed for the project to determine the total materials selling price (C). Tabulating this information into a spreadsheet program such as Excel (Microsoft Products, Redmond, Wash.) makes the calculations easier and ensures accuracy. It also allows the landscape estimator to easily make changes and quickly recalculate a selling price.

EXAMPLE 9-4

Table 9-6 shows how to calculate the selling price of materials for the Peters' residence.

TABLE 9-6 • SELLING PRICE OF MATERIALS FOR PETERS' RESISDENCE

Description	Quantity or Unit (A)	Selling Price per unit (including wholesale cost, overhead, and profit) (B)	Total selling price for materials C = (A)(B)
Autumn blend paver 6 x 9 rectangle	121 ft^2	$ 4.40/ft^2	$ 532.40
Crushed limestone; 4 in. depth	2 yd^3	$19.60/yd^3	$ 39.20
Sand; 1 in. depth	0.50 yd^3	$18.60/yd^3	$ 9.30
Plastic paver edge restraint (includes 5% overage)	46 lineal feet (lf)	$ 0.75/lf	$ 34.50
Crabapple (12 in. B&B, by hand)	1	$91.52/plant	$ 91.52
Juniper (5 gal)	5	$53.00/plant	$ 265.00
Dwarf spirea (1 gal)	7	$16.00/plant	$ 112.00
Redtwig dogwood (1 gal)	3	$19.00/plant	$ 57.00
Hosta (4 in. pot)	7	$ 6.50/plant	$ 45.50
Periwinkle (4 in. pot)	21	$ 5.00/plant	$ 105.00
Bluegrass turf (1.5 lb/MSF)	0.30 lb	$ 2.75/lb	$ 0.83
Hardwood mulching; 3 in. deep	1.98 yd^3 (round up to 2.0 yd^3)	$27.00/yd^3	$ 54.00
Materials Selling Price Sub-Total			$1,346.25

Selling price/unit figures are based on industry data. Some figures were calculated earlier in this section.

Practice Problem 9-4

Determine the materials selling price for the new foundation planting at the Shoemaker residence (see Figure 9-4). The project includes: 375 ft^2 of bed area; adding a 4-in. layer of conditioned topsoil and tilling it into the bed area; 2 12-in. B&B kousa dogwood trees (labeled A); 4 3 gal-sized arrowwood viburnum shrubs (labeled B); 16 1 gal-sized dwarf barberry shrubs (labeled C); and a 350 ft^2 area covered with bugleweed groundcover spaced 10-in. on center.

Labor Costs and Selling Price

◄ Definition ► *Labor Costs*

Labor costs include wages, benefits, and labor burden (Worker's Compensation, state and federal payroll taxes, and unemployment taxes) for employees.

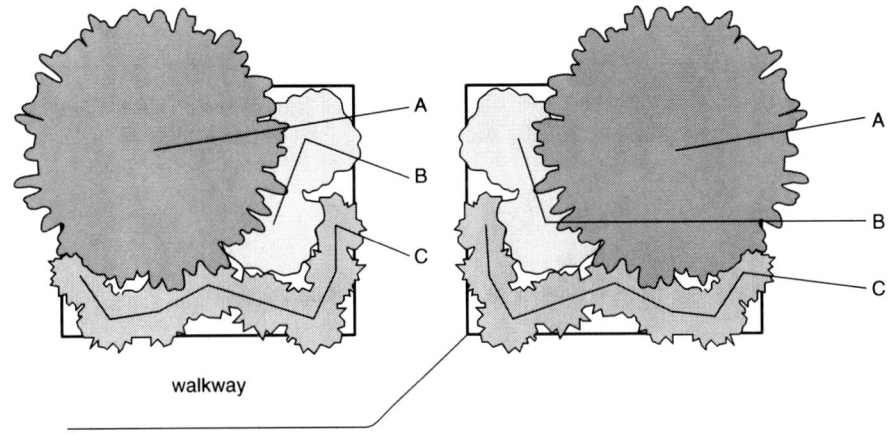

Figure 9-4
The new foundation planting for the Shoemaker's residence

Wages and benefits of the landscape crew make up the majority of labor costs for a landscaping company. Labor burden is generally 20–30% of the laborer's wage, depending on the state where he or she is employed. In addition, other labor costs such as unproductive time (e.g., lunch breaks), mandatory state and federal paid holidays, paid vacation, or sick days and, in some cases, medical insurance and pension plan contributions also need to be included. These are referred to as "labor overhead" and are often calculated as 100–125% of the hourly labor cost.

EXAMPLE 9-5

The following example shows how to calculate the total hourly labor rate for a landscape crew member:

Hourly labor rate	$8.00
Labor burden (20%)	
(percent labor burden)(hourly rate) = labor burden	
(0.20)($8.00) = $1.60	$1.60
Hourly labor cost (hourly labor rate + labor burden)	$9.60
Labor overhead (115%)	
(percent labor overhead)(hourly labor cost) = hourly labor overhead	
(1.15)($9.60) = $11.04	$11.04
Breakeven hourly labor cost (hourly labor cost + labor overhead)	$20.64
Profit (10%)	
(percent profit on labor)(breakeven hourly labor cost) = hourly labor profit	
(0.10)($20.64) = $2.06	$ 2.06
Total hourly labor rate (breakeven hourly labor cost + hourly profit)	$22.70

✏ Practice Problem 9-5 ✏

Calculate the total hourly rate for an employee who earns $10.00/hr and works for a company where labor burden is 25%, labor overhead is 120%, and profit on labor is calculated as 10%.

Production Rate

◄ **Definition** ► *Production rate*

Production rate is the time required to complete a specific task. ►

Landscape estimators use production rates found in industry-accepted reference books that contain data for every imaginable task. For instance, a production rate might specify how long it takes to plant a 4 in. potted perennial. It may also determine the average time it takes to spread bark mulch 3 in. deep over a 500 ft^2 planting bed or how long it takes to install 1,000 ft^2 of sod.

A number of resources are available that provide general production rates for the wide variety of tasks associated with landscape installation and help an estimator determine how long it will take to complete a job (see references at the end of this chapter). When the production rate for a task is multiplied by the quantity and the hourly labor rate, the estimator can determine the labor expense associated with the job (Example 9-6).

Table 9-7 provides production rates for some common plant installation tasks listed in fractions of an hour. To calculate the minutes for the task, multiply the production rate by 60 (60 min/hr). For example, calculating how many minutes it takes to plant a 3 gal rhododendron is done as follows:

(planting 3 gal container production rate)(60 min/hr) = time in minutes

$$(0.571)(60) = 34.26 \text{ min}$$

Although these figures are actual production rates from the landscape industry (reference included with the table), they are only a starting point. Variable site conditions (e.g., soil type, site accessibility) can have a significant impact on production rates.

Once the tasks to complete the job are determined, the time and associated labor price to complete each task can be calculated. In the case of the Peters' residence, the tasks for the installation include: general site grading; installing the paver patio; planting trees, shrubs, perennials, and groundcovers; seeding the turf area; mulching; and site cleanup.

Production rates (B) from Table 9-7 and the more extensive list of rates found in the references listed at the end of the chapter are multiplied by the quantity (A) needed for the project. The result is the total hours required for that portion of the project (C). It is essential that the units for the task being completed match the units of the production rate. For example, the production rate for general site grading is given in MSF (1,000 ft^2). The area to be graded at the Peters' residence (526 ft^2)

TABLE 9-7 • PRODUCTION RATES FOR COMMON LANDSCAPE INSTALLATION PROCEDURES

Description	Crew*	Quantity or Unit**	Production Rate (Quantity/ Work Hour)
Site Preparation			
Excavating topsoil, 6 in. deep, with dozer	B10B	yd^3	0.014
Spread conditioned topsoil, by hand and 6 in. deep	1 Clab	SY	0.067
Rake topsoil by hand	1 Clab	MSF	0.800
Tilling topsoil, 26 in. rototiller, by hand and 4 in. deep	1 Clab with rototiller	SY	0.008
Planting (Medium Soil)			
Container, 4 in. diameter	1 Clab	Each	0.018
Container, 1 gal	2 Clab	Each	0.271
Container, 2 gal	2 Clab	Each	0.444
Container, 3 gal	2 Clab	Each	0.571
Container, 5 gal	2 Clab	Each	0.800
B&B, 12 in. diameter, by hand	2 Clab	Each	1.231
Turf			
Bluegrass 4 lb/MSF, common push spreader	1 Clab	MSF	1.000
Shade mix 6 lb/MSF, common push spreader	1 Clab	MSF	1.000
Mulch			
Aged bark, 3 in. deep, by hand	1 Clab	SY	0.080
Stone mulch, 3 in. deep, by hand	1 Clab	SY	0.064

*Crew: 1 Clab = 1 Common Laborer; 2 Clab = 2 Common Laborer; B10B = 1 equipment operator, 0.5 laborer, 1 dozer 200 HP.

**Unit: MSF = 1,000 ft^2; SY = square yard; See Appendix for conversions.

Adapted with permission from Means Site Work and Landscape Cost Data, 2007. Copyright Reed Construction Data, Kingston, MA 781-585-7880. All rights reserved.

must be converted to MSF or 0.526 MSF. If this conversion is not made, the total hours for general site grading would be 420.8 hours instead of the actual 0.420 hours: A huge difference! In some cases, an estimator is unable to find a production rate that exactly matches the task to be completed. In these situations, he or she can select the production rate for a similar task and adjust it if necessary.

After the total hours for the particular job are calculated (C), that figure can be multiplied by the hourly labor rate (D) to calculate the total labor price (E). Again,

208 Part Two — **Chapter 9** Mathematical Applications for the Landscape Industry

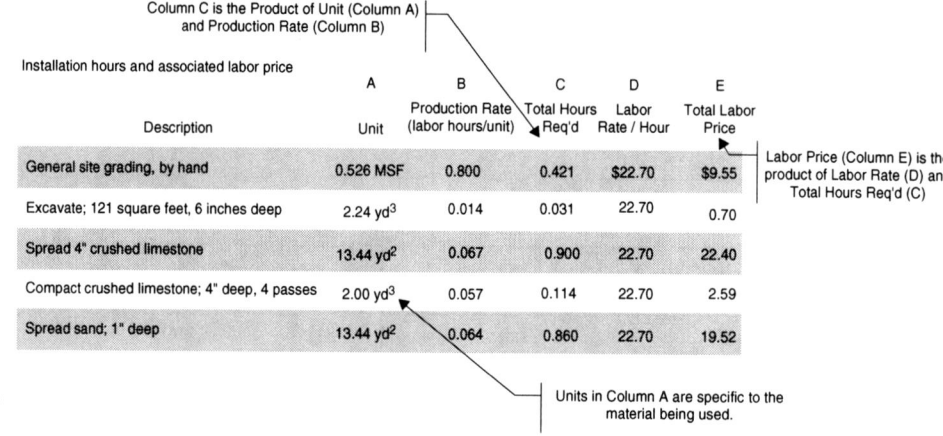

Figure 9-5

Labeled spreadsheet illustrating how the columns of information can be organized to calculate the labor price

tabulating the data into a spreadsheet makes the calculations easier and ensures accuracy provided the data is correct. Figure 9-5 is a spreadsheet showing how the columns of information can be organized to complete these calculations.

EXAMPLE 9-6

Table 9-8 shows installation hours and associated labor price for the Peters' residence.

Practice Problem 9-6

Determine the installation hours and associated labor price for the new foundation planting at the Shoemaker residence (Practice Problem 9-4 and Figure 9-4). Use an hourly labor rate of $ 29.04.

To determine the total selling price for the Peters' residence project (Example 9-7), combine the data calculated in Example 9-4 (materials prices), Example 9-5 (hourly labor price), and Example 9-6 (production rate and labor price for each component of the project). Figure 9-6 illustrates how this information can be combined in a spreadsheet.

EXAMPLE 9-7

Table 9-9 shows the total selling price for the Peters' residence.

Practice Problem 9-7

Determine the total selling price for installation of the new foundation planting at the Shoemaker residence (Practice Problem 9-4 and Figure 9-4).

Estimating Landscape Installation Costs **209**

TABLE 9-8 • INSTALLATION HOURS AND LABOR PRICE FOR PETERS' RESIDENCE

Description	Quantity or Unit (A)	Production Rate (labor hours per unit)* (B)	Total hours required for job C = (A)(B)	Total Labor rate per hour (including labor, burden, labor overhead, and profit) (D)	Total Labor Price E = (C)(D)
General site grading (by hand)	0.526 MSF	0.800	0.421	$22.70	$ 9.55
Excavate; 121 ft², 6 in. depth**	2.24 yd³	0.014	0.031	$22.70	$ 0.70
Spread 4 in. crushed limestone	13.44 SY	0.067	0.900	$22.70	$ 20.43
Compact crushed limestone; 4 in. depth, 4 passes	2.00 yd³	0.057	0.114	$22.70	$ 2.59
Spread sand; 1 in. depth	13.44 SY	0.064	0.860	$22.70	$ 19.52
Lay pavers***	121 ft²	0.060	7.26	$22.70	$164.80
Set pavers; 4 passes; (plate compactor by hand)	2.0 yd³	0.057	0.114	$22.70	$ 2.59
Patio edging; 46 lineal ft	46	0.040	1.84	$22.70	$ 41.77
Crabapple (12 in. B&B, by hand)	1	1.231	1.231	$22.70	$ 27.94
Juniper (5 gal)	5	0.800	4.000	$22.70	$ 90.80
Dwarf spirea (1 gal)	7	0.271	1.897	$22.70	$ 43.06
Redtwig dogwood (1 gal)	3	0.271	0.813	$22.70	$ 18.46
Hosta (4 in. pot)	7	0.018	0.126	$22.70	$ 2.86
Periwinkle (4 in. pot)	21	0.018	0.378	$22.70	$ 8.58
Seeding bluegrass turf (4 lb/MSF, push spreader)	0.180 MSF	1	0.180	$22.70	$ 4.09
Mulching beds (3 in. deep, by hand)	23.77 SY	0.080	1.90	$22.70	$ 43.13
Site cleanup (sweeping/raking by hand)	0.515 MSF	0.533	0.274	$22.70	$ 6.23
Time and Labor Selling Price Sub-Total			22.34 hours		$507.10

*The production rate is based on one laborer doing the work with hand tools, except as noted with multiple asterisks. This figure is in a fraction of an hour. To determine how many minutes it takes to complete a task, multiply the production rate by 60.

**Equipment operator; 0.50. laborer; 200HP dozer;

***Brick layer and brick layer's helper

Figure 9-6
Labeled spreadsheet illustrating how the columns of information can be organized to calculate the total selling price for the Peters' residence

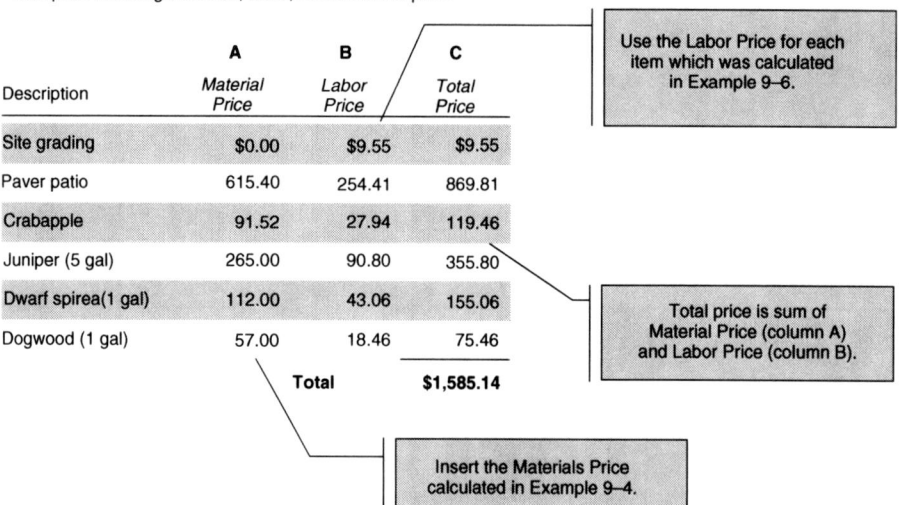

TABLE 9-9 • TOTAL SELLING PRICE FOR PETERS' RESISDENCE

Description	Materials Price (A)	Labor Price (B)	Total Price A + B = C
General site grading	0	$ 9.55	$ 9.55
Paver patio (includes all components of this part of the project)	$615.40	$252.40	$ 867.80
Crabapple (12 in. B&B; by hand)	$ 91.52	$ 27.94	$ 119.46
Juniper (5 gal)	$265.00	$ 90.80	$ 355.80
Dwarf spirea (1 gal)	$112.00	$ 43.06	$ 155.06
Redtwig dogwood (1 gal)	$ 57.00	$ 18.46	$ 75.46
Hosta (4" pot)	$ 45.50	$ 2.86	$ 48.36
Periwinkle (4" pot)	$105.00	$ 8.58	$ 113.58
Bluegrass turf (1.5#/MSF)	$ 0.83	$ 4.09	$ 4.92
Hardwood mulching; 3 in. deep	$ 54.00	$ 43.13	$ 97.13
Site cleanup	0	$ 6.36	$ 6.36
Total Price			$1,853.48

Equipment Costs

Equipment costs, either as a direct cost associated with a job or as an overhead cost, need to be accounted for in an estimate. Equipment costs are based on the type of equipment; associated costs (acquisition cost, maintenance cost, fuel cost); trade-in value; and expected lifetime hours. The *Site Work and Landscape Cost Estimating*

Data reference book includes equipment (non-hand tools) costs as part of the bare costs. Generally, the cost of hand tools (e.g., shovel, rake, wheel barrow) is $.10/hour.

Overhead and Profit

> **❦ Definition ❧** *Direct Job Overhead*
> *Direct job overhead* costs are those costs that can be directly attributed to a specific landscape project, such as a building permit fee for that site or a dumpster rental fee. ❧

These overhead costs are not actually part of the installation of the landscape, but they are still required for the project.

> **❦ Definition ❧** *General Overhead*
> *General overhead* costs are expenses associated with operating a business that cannot be directly attributed to a specific landscape project. Examples include office rental fees, electricity for the office, and business license fees. ❧

> **❦ Definition ❧** *Profit*
> *Profit* is determined when the cost of the project is subtracted from the selling price of the project. ❧

Profit is calculated as a percentage of the total costs associated with a landscape installation job. The percentage used in the profit calculation is dependent on a number of factors, including the company's business model, the local market, and competition. On average, landscape contractors realize a profit between 10–25% of all the direct and indirect costs associated with the project (Angley, et al., 2002); however, this is not a set range and is subject to much industry variability.

Accounting for overhead expenses and the desired amount of profit are an essential part of developing a landscape estimate. As mentioned at the beginning of this chapter, the scope of this book prevents a detailed discussion of these important issues. However, they must be included for a landscape company to be a viable business enterprise. In this chapter, two examples of calculating overhead and profit are included; one in the plant materials and hard goods cost section and a second in the labor cost section.

Estimating Landscape Maintenance Costs

Landscape maintenance contracts are an important source of income for many landscape companies and estimating proficiency is essential to the success of these businesses. A landscape maintenance job is estimated based on two things: the services provided and the number of times the services are provided as outlined in the contract.

Common landscape maintenance services include mowing, trimming, edging, litter patrol, fertilizing, disease, insect and weed control, pruning, and mulching. In some cases, there are also seasonal services such as spring clean up, winterization of an irrigation system, and snow removal. The frequency at which services are performed varies by the site and the client's preference.

Similarly to landscape installation, landscape maintenance services also have general production rates (see references at the end of this chapter). Table 9-10 provides production rates for some common landscape maintenance services. Keep in mind, however, that these are just a starting point. Variable site conditions can significantly impact how long it takes to complete a task.

TABLE 9-10 • LANDSCAPE MAINTENANCE PRODUCTION RATES

Description	Equipment	Quantity or unit	Production Rate* (Work Hour/Unit)
Mowing	21 in. rotary, by hand	1,000 ft^2	0.167
	25 in. rotary, riding	1,000 ft^2	0.050
Pruning	Shrub	1 plant	0.067
	Tree from ground	1 plant	0.083
Fertilizing	Rotary spreader	1,000 ft^2	0.050
	24 in. drop spreader soluble	1,000 ft^2	0.067
	by hand	18" container	0.067
Herbicide application	Rotary spreader	1,000 ft^2	0.050
	3 gal. sprayer	1,000 ft^2	0.025
Edge / trim / clean up	Power edger	100 lineal ft (clf)	0.022
Hedge trimming	5–6 in. high, with shears	100 lineal ft (clf)	0.310
Spring flower bed clean up	Prune, rake, and remove debris, by hand	1,000 ft^2	3.33
Deadheading	By hand	18 in. container	0.067
Mulching	3 in. deep	1,000 ft^2	0.500
Hard surface clean up	Sweep, by hand,	1,000 ft^2	0.500
	Blowing	100 ft^2	0.013

*The production rate is based on one laborer doing the work with hand tools, as described. This figure is in a fraction of an hour. To determine how many minutes it takes to complete a task, multiply the production rate by 60.

Sources: *Landscape Designer and Estimator's guide, 3rd edition*, (1996) National Landscape Association, Washington, DC. *Guide to Growing a Successful Landscape Maintenance Business*, (2004) Professional Landcare Network, Herndon, VA.

Estimating Landscape Maintenance Costs

To develop an estimate, the job site must be accurately measured to determine how much area is being maintained. Just as with an installation estimate, materials costs (e.g., fertilizer, herbicide, mulch) are calculated based on the quantity used, and overhead and profit are based on the company business model. Further, labor can be calculated using a similar process as outlined in Example 9-5.

Sample Landscape Maintenance Plan for the Owen Property

The Owen property (Figures 9-7 and 9-8) will be used for this landscape maintenance example (Example 9-8).

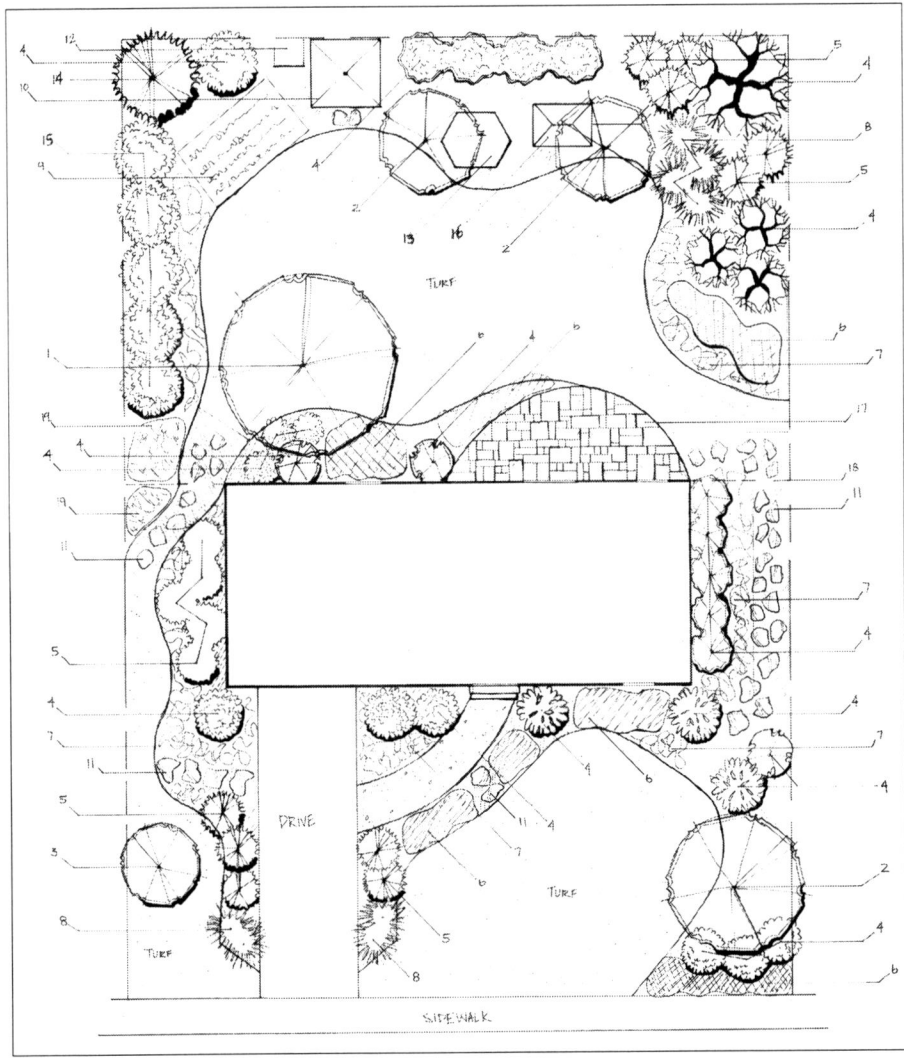

Figure 9-7

The Owen property landscape plan

Key	Description	Quantity
1	Large deciduous tree	1
2	Medium deciduous tree	3
3	Small deciduous tree	1
4	Deciduous shrub	27
5	Evergreen shrub	15
6	Herbaceous perennials	256 ft^2
7	Groundcover	660 ft^2
8	Ornamental grass	8
14	Medium evergreen tree	1
15	Trimmed deciduous hedge	36 lf
19	Evergreen perennial	54 ft^2
	Total planting bed area	4,144 ft^2
	(excluding groundcover)	
	Total turf area	3,636 ft^2

Figure 9-8

The plant key for the Owen property

The Owens requested an annual maintenance contract beginning April 1 and ending September 30 (Table 9-11).

To calculate the landscape maintenance estimate, you need to know the individual services to be provided, the frequency of each service for the annual contract, the production rate for each service, and the labor costs. Example 9-8 shows the multiple steps in calculating a landscape estimate.

The steps include multiplying the unit or quantity (A) by the associated production rate (B) to get the hours required for a single unit or quantity (C). Next, multiply the hours for the individual unit or quantity (C) by the frequency (D) that the service will be provided during the annual maintenance contract to get the total hours (E). Finally, multiply the total hours by the hourly labor rate to calculate the total labor price. Using a spreadsheet, as described earlier, for these calculations should ensure the estimating accuracy.

Estimating Landscape Maintenance Costs **215**

TABLE 9-11 • ANNUAL MAINTENANCE CONTRACT FOR OWEN PROPERTY

Maintenance Service	April	May	June	July	August	September
Turf	Fertilize/Herbicide combination 0.75 lb N/1,000 ft^2 (30-0-4)	Fertilize 0.75 lb N/1,000 ft^2 (29-3-4)			Fertilize 0.50 lb N/1,000 ft^2 (16-16-16)	Fertilize 1.0 lb N/1,000 ft^2 (22-4-14)
		Mow (weekly) ———————————————————→				
		Trim (weekly) ———————————————————→				
Trees	Fertilize (16-16-16) Corrective pruning					
Shrubs	Fertilize (16-16-16) Corrective pruning		Shear hedge			
Herbaceous perennials; ornamental grasses	Cut to the ground Fertilize (16-16-16)					
Annual flowers (6–18 in. patio containers)		Soluble fertilizer (15-30-15) (bi-weekly) —————————————→				
		Deadhead as needed (weekly) ————————————————→				
Planting beds	Apply herbicide pre-emergence Apply hardwood mulch (3 in. deep)					

EXAMPLE 9-8a

Table 9-12 shows annual hours for each landscape service performed as part of the Owen property maintenance contract.

Calculate the total labor price for the maintenance plan.

SOLUTION

(total hours)(labor rate/hour) = total labor price

(34.97 hours)($22.70/hour) = $793.82 labor price

216 Part Two • Chapter 9 Mathematical Applications for the Landscape Industry

TABLE 9-12 • ANNUAL HOURS FOR LANDSCAPE SERVICE AT OWEN PROPERTY

Maintenance Service	Unit or Quantity (A)	Production Rate* (B)	Hours / Unit C = (A)(B)	Annual Frequency** (D)	Total Annual Hours E = (C)(D)
Turf					
Mowing	3.636 MSF	0.167	0.61	23	14.03
Fertilizer/ pre-emergence herbicide combination	3.636 MSF	0.067	0.24	1	0.24
Fertilizing	3.636 MSF	0.067	0.24	3	0.72
Trim Edge	4.45 clf	0.022	0.10	23	2.30
Trees (Deciduous)					
Corrective Pruning	5	0.083	0.42	1	0.42
Shrubs (Deciduous)					
Corrective Pruning	27	0.067	1.81	1	1.81
Hedge Shearing	0.36 clf	0.31	0.11	1	0.11
Perennials					
Spring clean up; cut to ground (includes ornamental grasses)	0.256 MSF	3.33	0.85	1	0.85
Annuals in Containers					
Fertilizing: Soluble	6	0.067	0.40	10	4.00
Deadheading	6	0.067	0.40	20	8.00
Flower beds					
Fertilizing includes: trees, shrubs, groundcover, perennials	4.144 MSF	0.050	0.21	1	0.21
Herbicide: pre-emergence	4.144 MSF	0.05	0.21	1	0.21
Mulching	4.144 MSF	0.50	2.07	1	2.07
Total Hours					**34.97**

*Production Rate from Table 9-10; **Annual Frequency from Table 9-11

✎ Practice Problem 9-8 ✎

1. Based on the information in Table 9-13, determine the annual hours for each landscape service performed for the Newberry property maintenance contract.

TABLE 9-13 •

Newberry Property					
Maintenance Service	Unit or Quantity (A)	Production Rate (B)	Hours / Unit C = (A)(B)	Annual Frequency (D)	Total Annual Hours E = (C)(D)
Turf					
Mowing	1.50 MSF			20	
Fertilizing	1.50 MSF			3	
Shrubs (Deciduous)					
Corrective Pruning	15			1	
Flower beds					
Herbicide: Pre-emergent	0.50 MSF			1	
Mulching	0.50 MSF			1	
Total Hours					

2. Calculate the labor price for the maintenance contract for the Newberry property assuming a total hourly labor rate of $25.00/hour.

In addition to the labor price, the amount and price of materials must also be calculated.

Refer to Chapter 5 to review fertilizer calculations to determine the amount of lawn fertilizer needed based on the fertilizer analysis and application rate.

Example 9-8b shows how to calculate the price of the soluble fertilizer needed for the annuals in containers. Example 9-8c shows how to calculate the price of the pre-emergence herbicide used in the flower beds. And, finally, Example 9-8d shows how to calculate the price of the hardwood mulch needed for the project.

EXAMPLE 9-8b

One gallon of a dissolved soluble fertilizer (15-30-15) will be used to fertilize each of the six 18 in. containers of annual flowers every two weeks for the contract period, for a total frequency of 60 (6 pots × 10 fertilizer applications). Each application uses 1 tablespoon (Tbsp) of fertilizer per gallon of water. The product price (including overhead and profit) is $14.00 for a 5 lb box.

Calculate the total amount of fertilizer needed and the price.

Step 1

(application rate)(frequency) = total amount of fertilizer needed

(1 Tbsp/gal)(60 frequency) = 60 Tbsp

Step 2

Convert: Tablespoons to ounces; where 3 Tbsp = 1 oz

$$\frac{\text{Tbsp required}}{3 \text{ Tbsp/oz}} = \text{number of ounces}$$

$$\frac{60 \text{ Tbsp}}{3 \text{ Tbsp/ounce}} = 20 \text{ oz}$$

Step 3

Convert 20 oz to pounds because the product is sold in pounds.

$$\frac{\text{total ounces of fertilizer needed}}{16 \text{ oz/lb}} = \text{lb needed}$$

$$\frac{20 \text{ oz}}{16 \text{ oz/lb}} = 1.25 \text{ lb needed}$$

Step 4

Calculate the fertilizer price per pound.

$$\frac{\text{price of fertilizer}}{\text{size of container in pounds}} = \text{price/pound}$$

$$\frac{\$14.00}{5 \text{ lb}} = \$2.80/\text{lb}$$

Step 5

Calculate the price of the fertilizer required for the job.

(fertilizer price/pound)(number of pounds needed) = price of fertilizer needed

($2.80/lb)(1.25 lb) = $3.50

SOLUTION

The price for the soluble (15-30-15) fertilizer is $3.50

EXAMPLE 9-8c

A granular pre-emergence herbicide (Trifluralin) is applied to 4,144 ft^2 of flower bed area. The product price (including overhead and profit) is $64.25 for a 17.5 lb bag.

Step 1

Calculate how many bags of herbicide are needed. According to the product label, the 17.5 lb bag of pre-emergence herbicide covers 2,800 ft^2.

$$\frac{\text{total flower bed area}}{\text{area covered by herbicide/bag}} = \text{total bags of herbicide}$$

$$\frac{4{,}144 \text{ ft}^2}{2{,}800 \text{ ft}^2/\text{bag}} = 1.5 \text{ bags}$$

Step 2
Calculate the total amount of herbicide needed in pounds.

(number of bags)(pounds/bag) = total pounds of herbicide needed

(1.5 bags)(17.5 lb/bag) = 26.25 lb

Step 3
Calculate the price of the herbicide per pound.

$$\frac{\text{price of herbicide}}{\text{size of bag in pounds}} = \text{price/pound}$$

$$\frac{\$64.25}{17.5 \text{ lb}} = \$3.67/\text{lb}$$

Step 4
Calculate the price of the herbicide required for the job.

(herbicide price/pound)(number of pounds needed) = price of herbicide needed

($3.67/lb)(26.25 lb) = $96.34

SOLUTION

The price for the pre-emergence herbicide is $96.34.

EXAMPLE 9-8d

The flower bed needs a 3 in. application of hardwood mulch. From Table 9-4 you can find that 1 yd^3 of material that is spread 3 in. deep covers 108 ft^2. The area to be covered with mulch is 4,144 ft^2. The mulch price is $27.00/yd^3.

Step 1
Calculate how many cubic yards of mulch is needed.

$$\frac{\text{total bed area}}{\text{area covered/yd}^3} = \text{total yd}^3 \text{ needed}$$

$$\frac{4,144 \text{ ft}^2}{108 \text{ ft}^2/\text{yd}^3} = 38.4 \text{ yd}^3$$

Step 2
Calculate the price of the mulch required for the job.

(total yd^3 of mulch needed)(price/yd^3) = total price for mulch

(38.4 yd^3)($27.00/yd^3) = $1,036.80

SOLUTION

The price for the hardwood mulch is $1,036.80.

Practice Problem Set 9-9

Using Table 9-13, calculate the following:

1. Materials price for the pre-emergence herbicide
2. Materials price for the hardwood mulch

Assume the product price, including overhead and profit, is $64.25 for a 17.5 lb bag of pre-emergence herbicide (Trifluralin). Use Table 9-1 to decide how many yd^3 of mulch are needed. Assume the mulch price, including overhead and profit, is $27.00/$yd^3$.

Total materials price (C) are calculated by multiplying the total amount of material required (A) by the material price (B), which should include the wholesale cost and appropriate overhead and profit amounts. Table 9-14 show the price of materials associated with the annual maintenance contract for the Owen property.

TABLE 9-14 • PRICE OF MATERIALS FOR OWEN PROPERTY

Materials	Annual Quantity (A)	Unit Price (B)	Total Price (C) = (A)(B)
Turf			
April fertilizer/herbicide product (0.75 lb N /1,000 ft^2)(30-0-4 + pendimethalin)	9.1 lb	$ 1.00/lb	$ 9.10
May fertilizer (0.75 lb N /1,000 ft^2)(29-3-4)	9.4 lb	$ 0.95/lb	$ 8.93
June fertilizer (0.50 lb N /1,000 ft^2)(16-16-16)	11.4 lb	$ 0.88/lb	$ 10.03
September fertilizer (1.0 lb N /1,000 ft^2)(22-4-14)	16.5 lb	$ 0.88/lb	$ 14.52
Annuals			
Fertilizing: Soluble (15-30-15)	1.25 lb	$ 2.80/lb	$ 3.50
Flower beds			
Fertilizer: Includes trees, shrubs, groundcover, perennials (16-16-16)	12.9 lb	$ 0.88/lb	$ 11.35
Herbicide: Pre-emergent (Trifluralin)	26.25 lb	$ 3.67/lb	$ 96.34
Mulch (3 in.; hardwood)	38.4 yd^3	$27.00/$yd^3$	$1,036.80
Total Materials Price			**$1,190.57**

FINAL SOLUTION

The annual maintenance contract for the Owen property requires 34.97 hours. The total selling price of the contract is calculated by adding together the labor and materials prices for a total of $1,984.39.

Labor Price	$ 793.82
Materials Price	$1,190.57
Total	$1,984.39

✐ Practice Problem 9-10 ✐

Calculate the price of the annual maintenance contract for the Newberry property.

References

Cohan, S.M. (2006). *Business Principles of Landscape Contracting*. Upper Saddle River, NJ: Pearson Prentice Hall.

Engles, J.E. (2004). *Landscaping Principles and Practices*, 6th edition. Clifton Park, NY: Thomson Delmar Learning.

Landphair, H.C. and F. Klatt, Jr. (1999). *Landscape Architecture Construction*, 3rd edition. Upper Saddle River, NJ: Prentice Hall.

Spencer, E.R., and RS Means Engineering. (2007). *Means Site Work & Landscape Cost Data*, Kingston, MA: RS Means.

National Landscape Association. (1996). *Landscape Designer and Estimator's Guide*, 3rd edition. National Landscape Association, Washington, DC.

Professional Grounds Management Society. (1995). *Grounds Maintenance Estimating Guidelines*, 7th edition. Professional Grounds Management Society, Hunt Valley, MD.

Professional Landcare Network. (2004). *Guide to Growing a Successful Landscape Maintenance Business*. Professional Landcare Network, Herndon, VA.

Rubenstein, H.M. (1996). *A Guide to Site Planning and Landscape Construction.* New York, NY: John Wiley and Sons, Inc.

VanDerZanden, A.M. and S. N. Rodie. (2008). *Landscape Design Theory and Application*. Clifton Park, NY: Thomson Delmar Learning

Chapter 10: Mathematical Applications for the Greenhouse, Nursery, and Interior Landscape Industries

Introduction

There is overlap in cultural practice and calculations in the greenhouse, nursery, and interior landscape industries so they are presented in a single chapter. An effort has been made to include problems relevant to each of these industries where appropriate.

Calculations Involving Greenhouse Structures

Calculating Greenhouse Bench (Growing Space) Efficiency

The profitability of a greenhouse operation is dependent on many variables. One factor that affects profitability is the efficient use of greenhouse space. The entire volume of the greenhouse has to be heated, ventilated, cooled, covered, and maintained. The more space within a greenhouse structure that is used for growing, the lower the fixed costs will be for each plant grown in that space. Reducing fixed costs for a crop increases the profit derived from the sale of the crop. Efficient use of greenhouse space is dependent on the physical arrangement of the growing space. Growing space may consist of a series of raised benches or beds, ground beds, overhead growing space such as suspended baskets, or floor space. Bench or growing space efficiency is the ratio of the area of growing space to total greenhouse area. Modern, wholesale operations aim for a minimum of 75% growing space and 25% aisle space.

Greenhouse benches or beds are the most common form of growing space. Historically, when cut flower production dominated the industry, benches or beds extended the length of the greenhouse space, parallel to the longest side of the greenhouse (Figure 10-1). Long, narrow aisles separated the benches. Greenhouse employees maintained and harvested crops from these narrow aisles. This parallel or longitudinal bench arrangement, found in cut flower production operations, results in approximately 60 to 70% bench efficiency.

An alternate benching system that was developed as the pot-plant production industry flourished is the peninsular bench arrangement (Figure 10-2). The peninsular bench arrangement employs a wide central aisle that allows for the use of carts, racks, and even monorail systems for the efficient movement of plants in and out

Figure 10-1

Parallel greenhouse bench arrangement

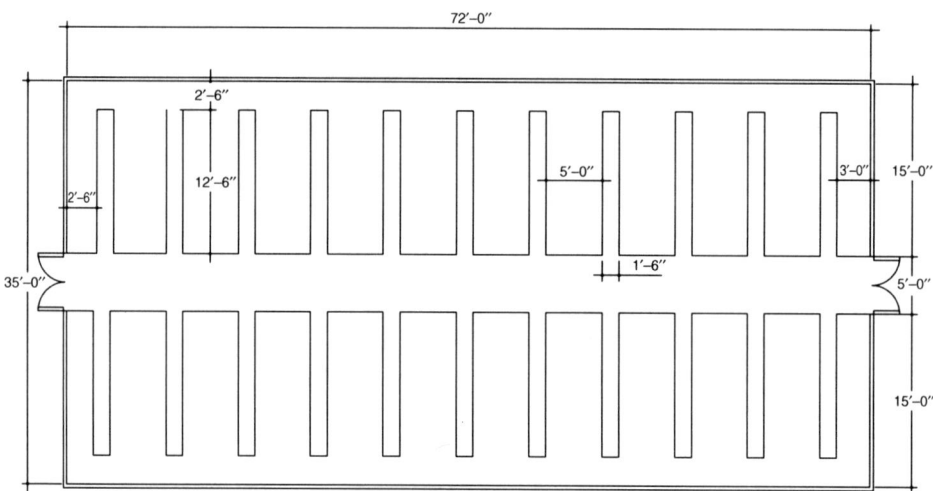

Figure 10-2

Peninsular bench arrangement

of the space. In the peninsular system, benches project out from and perpendicular to the long side of the greenhouse with narrow side aisles separating the benches. This benching system, found in early pot-plant production operations, results in approximately 65 to 75% bench efficiency.

Rolling benches, or the aisle-eliminator system, (Figure 10-3) evolved from the old peninsular benching system. It reduces the space required by the side aisles by providing for side-to-side movement of each bench. One 18- to 24-inch aisle space is allocated for approximately every six benches. An increase in bench or growing space efficiency of approximately 20% can be achieved with the use of rolling benches.

Porous concrete greenhouse floors or floors that can be flooded are the most flexible ways to use greenhouse space. It lacks the rigidity of beds or benches. This

Calculations Involving Greenhouse Structures

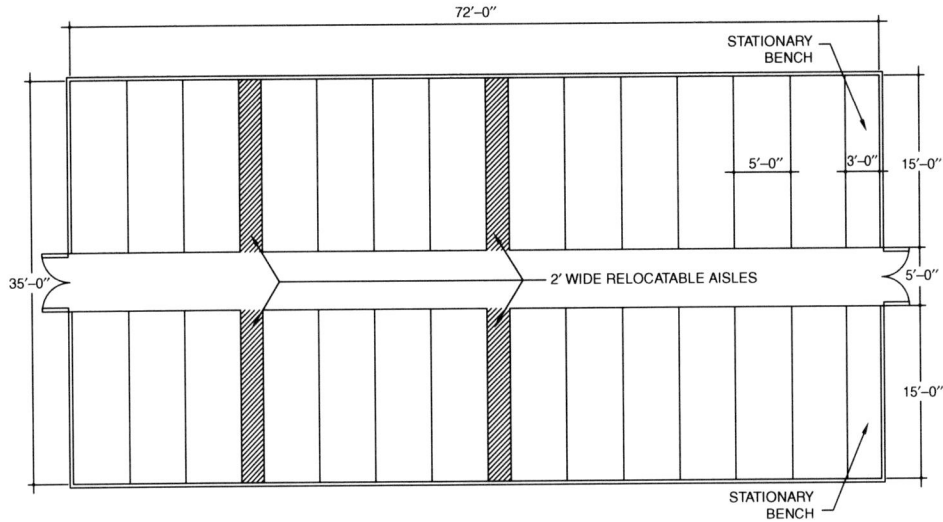

Figure 10-3
Rolling benches or aisle-eliminator system

growing system is more common in the bedding plant production industry. Bench or growing space efficiencies vary with the plant arrangement on greenhouse floors.

Maximum efficiency can be achieved by making use of overhead space above the aisles. Hanging baskets may be suspended over the aisles to make use of this space. Growers must avoid overloading the greenhouse structure when hanging baskets supported by greenhouse structural components. Suspending hanging baskets over the greenhouse aisles can result in a 10 to 20% increase in growing space efficiency.

Calculating bench or growing space efficiency is a simple mathematical task. It is expressed as a percentage and is the ratio of the area of greenhouse space used for growing to the total area of greenhouse space.

> ### ◆ Definition ◆ Bench or Growing Space Efficiency
> **Bench or Growing Space Efficiency** is the ratio of the area in a greenhouse used as growing space to the total greenhouse floor area. It is expressed as a percentage by multiplying the ratio times one hundred.
>
> $$\frac{\text{Total Growing Area (ft}^2)}{\text{Total Greenhouse Area (ft}^2)} \times 100 = \text{Percent (\%) Bench Efficiency}$$

EXAMPLE 10-1

An even-span greenhouse measures 30 ft wide by 108 ft long. There is a 6 ft wide central aisle and three 18 in. side aisles. It is equipped with rolling benches and there are 36 benches that measure 5.5 ft by 12 ft and 2 benches that measure 4.5 by 12 ft. Calculate the bench efficiency of this space.

Step 1
Determine the total greenhouse area by multiplying the length of the greenhouse in feet by the width of the greenhouse in feet.
$$108 \text{ ft} \times 30 \text{ ft} = 3{,}240 \text{ ft}^2$$

Step 2
Determine the total growing area by summing the surface area of all of the benches.
$$36 \text{ benches} \times 5.5 \text{ ft} \times 12 \text{ ft} = 2{,}376 \text{ ft}^2$$
$$2 \text{ benches} \times 4.5 \text{ ft} \times 12 \text{ ft} = 108 \text{ ft}^2$$
$$2{,}376 \text{ ft}^2 + 108 \text{ ft}^2 = 2{,}484 \text{ ft}^2$$

Step 3
Calculate the bench efficiency.
$$\frac{\text{Total Growing Area (ft}^2\text{)}}{\text{Total Greenhouse Area (ft}^2\text{)}} \times 100 = \text{Percent (\%) Bench Efficiency}$$
$$\frac{2{,}484 \text{ ft}^2}{3{,}240 \text{ ft}^2} \times 100 = 76.7\% \text{ Bench Efficiency}$$

Solution
This greenhouse has 76.7% bench efficiency.

EXAMPLE 10-2

Using the greenhouse described in Example 10-1, recalculate bench efficiency when hanging baskets are suspended over the 6 ft wide central aisle.

Step 1
Calculate the growing space gained by suspending baskets over the central aisle by multiplying the width of the central aisle by the length of the greenhouse.
$$6 \text{ ft} \times 108 \text{ ft} = 648 \text{ ft}^2$$

Step 2
Add the space gained by suspending baskets over the central aisle to the existing growing space.
$$648 \text{ ft}^2 + 2{,}484 \text{ ft}^2 = 3{,}132 \text{ ft}^2$$

Step 3
Calculate the bench efficiency.
$$\frac{\text{Total Growing Area (ft}^2\text{)}}{\text{Total Greenhouse Area (ft}^2\text{)}} \times 100 = \text{Percent (\%) Bench Efficiency}$$
$$\frac{3{,}132 \text{ ft}^2}{3{,}240 \text{ ft}^2} \times 100 = 96.7\% \text{ Bench Efficiency}$$

SOLUTION
By adding hanging baskets over the central aisle, bench efficiency is increased from 76.7% to 96.7%.

> ### ✏ Practice Problem Set 10-1 ✏ *Calculating Greenhouse Bench Efficiency*
>
> 1. Calculate the bench efficiency of an older greenhouse that measures 24 ft wide by 100 ft long. It contains four ground beds designed for cut flower production. Each ground bed measures 4 ft wide by 95 ft long and they oriented parallel to one another and to the long side of the greenhouse.
> 2. Calculate the bench efficiency of a greenhouse that measures 41.5 ft wide and 144 ft long. This greenhouse contains 48 rolling benches that measure 5.5 ft by 18 ft each.
> 3. Recalculate the bench efficiency of the greenhouse in question number two after adding hanging baskets over the 5.5 ft wide central aisle.

Calculating Greenhouse Surface Area

Greenhouse managers may be required to determine the surface area of a greenhouse structure. Knowing the area of glazing (greenhouse covering) is important when pricing or comparing prices of replacement glazing material. Determining the surface area of glazing versus the surface area of curtain wall is used in determining the heating needs of the space. Glazed and nonglazed surfaces have different rates of heat loss. In fact, different glazing materials and glazing systems have different rates of heat loss as well. It is critical that these areas be calculated accurately so that appropriately sized heating and cooling systems can be designed for the structure and that replacement glazing materials can be accurately priced and ordered.

Calculating the surface area of a greenhouse involves dividing the exterior surfaces of the greenhouse into simple geometric shapes, calculating the area of each, and then summing each of the areas to determine total surface area.

Even-Span Greenhouse

When calculating the surface area of an even-span greenhouse, separate the greenhouse into two zones. The first zone is the glazed (clear) surface and the second zone is the curtain wall. The glazed surfaces include the roof, gable, gable-end walls, and side walls. The curtain wall surfaces include a portion of the gable-end walls and a portion of the side walls. These spaces represent simple geometric forms so the area of each can be calculated easily (Figure 10-4).

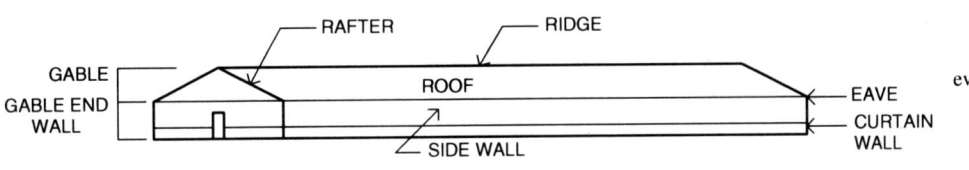

Figure 10-4

Diagram of an even-span greenhouse

EXAMPLE 10-3

Calculate the glazed surface area and the curtain wall surface area of the even-span greenhouse in Figure 10-5. This greenhouse is 36 ft wide and 144 ft long. It has a gable height of 9 ft, rafter length of 20 ft, eave height of 10 ft, and a curtain wall height of 3 ft.

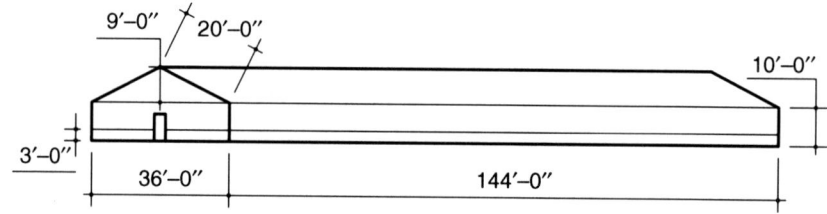

Figure 10-5
Calculating the surface area of an even-span greenhouse

PART ONE: CALCULATE THE GLAZED SURFACE AREA

Step 1

Determine the surface area of the roof.

Roof Surface Area = 2 Roof Sides × Rafter Length × Side Wall Length

Notice that the formula for the area of a rectangle (Area = Length × Width) is used because the roof sides are rectangular in shape. Also notice that it is multiplied by two because there are two sides of the roof.

Roof Surface Area = 2 Roof Sides × Rafter Length × Side Wall Length

Roof Surface Area = 2 × 20 ft × 144 ft

Roof Surface Area = 5,760 ft^2

Step 2

Determine the surface area of the gable.

Gable Surface Area = 2 Gables × $\frac{1}{2}$ (Gable End Width)(Gable Height)

Notice that the formula for the area of a triangle (Area = $\frac{1}{2}$ Base × Height) is used because the gable is triangular in shape. Also notice that it is multiplied by two because there are two gables, one at each end of the greenhouse.

Gable Surface Area = 2 Gables × $\frac{1}{2}$ (Gable End Width)(Gable Height)

Gable Surface Area = 2 × $\frac{1}{2}$ (36 ft)(9 ft)

Gable Surface Area = 324 ft^2

Step 3

Determine the glazed surface area of the gable end walls.

Gable End Wall Area = 2 Gable Ends [Gable End Width

× (Eave Height − Curtain Wall Height)]

The glazed area height is the difference between the heights of the wall at the eave and curtain wall. The area of a rectangle (Area = Length × Width) is used because the gable end wall is rectangular in shape. The area is multiplied by two because there are two gable end walls.

Gable End Wall Area = 2 Gable Ends [Gable End Width

× (Eave Height − Curtain Wall Height)]

Gable End Wall Area = 2 [36 ft (10 ft − 3 ft)]

Gable End Wall Area = 2 × 36 ft × 7 ft

Gable End Wall Area = 504 ft^2

Step 4

Determine the glazed surface area of the side walls.

Side Wall Surface Area = 2 Side Walls [Side Wall Length

× (Eave Height − Curtain Wall Height)]

The glazed area height is the difference between the heights of the wall at the eave and curtain wall. The area of a rectangle (Area = Length × Width) is used because the side wall is rectangular in shape. The area is multiplied by two because there are two side walls.

Side Wall Surface Area = 2 Side Walls [Side Wall Length

× (Eave Height − Curtain Wall Height)]

Side Wall Surface Area = 2 [144 ft × (10 ft − 3 ft)]

Side Wall Surface Area = 2 × 144 ft × 7 ft

Side Wall Surface Area = 2,016 ft^2

Step 5

Sum the surface areas of all the glazed portions of the greenhouse to determine the total glazed surface area.

Total Glazed Surface Area = Roof Area + Gable Area

+ Glazed Gable End Wall Area

+ Glazed Side Wall Area

Total Glazed Surface Area = 5,760 ft^2 + 324 ft^2 + 504 ft^2 + 2,016 ft^2

Total Glazed Surface Area = 8,604 ft^2

SOLUTION FOR PART ONE

The total glazed surface area of this greenhouse is 8,604 ft².

PART TWO: CALCULATE THE SURFACE AREA OF THE CURTAIN WALL

Step 1

Calculate the curtain wall area on the gable ends.

Gable End Curtain Wall Area = 2 Gable Ends × Gable End Width × Height of Curtain Wall

Gable End Curtain Wall Area = 2 × 36 ft × 3 ft

Gable End Curtain Wall Area = 216 ft²

Step 2

Calculate the curtain wall area on the side walls.

Side Wall Curtain Wall Area = 2 Greenhouse Sides × Side Wall Length × Curtain Wall Height

Side Wall Curtain Wall Area = 2 × 144 ft × 3 ft

Side Wall Curtain Wall Area = 864 ft²

Step 3

Find the sum of the surface areas of all the curtain walls.

Total Curtain Wall Area = Gable End Curtain Wall Area + Side Wall Curtain Wall Area

Total Curtain Wall Area = 216 ft² + 864 ft²

Total Curtain Wall Area = 1,080 ft²

SOLUTION FOR PART TWO

Total curtain wall surface area of this greenhouse is 1,080 ft².

Practice Problem 10-2   *Calculating the Surface Area of an Even-Span Greenhouse*

Calculate the glazed surface area and the curtain wall surface area of an even-span greenhouse that measures 30 ft wide and 96 ft long. The eave height is 8 ft, the curtain wall height is 2 ft, the rafter length is 18.75 ft, and the gable height is 11.25 ft.

Arch-Top Greenhouse

Arch-top greenhouses (Figure 10-6) have semicircular or semi-elliptical gables, rather than the triangular gable of the even-span greenhouse. The side walls are the same

Calculations Involving Greenhouse Structures

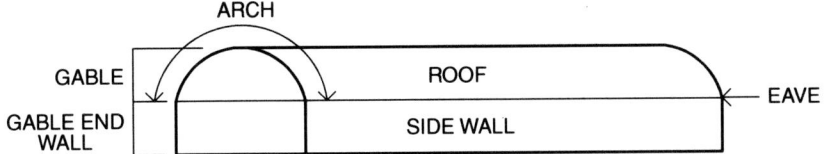

Figure 10-6
Diagram of an arch-top greenhouse

rectangular shape as the side walls of an even-span greenhouse. Although curtain walls are sometimes found on arch-top greenhouses, more often arch-top greenhouses are glazed from the eave to the ground.

EXAMPLE 10-4

Calculate the glazed surface area of the arch-top greenhouse shown in Figure 10-7. This arch-top greenhouse is 30 ft wide by 96 ft long. The eave height is 12 ft, the gable height is 15 ft and the arch length is 47 ft.

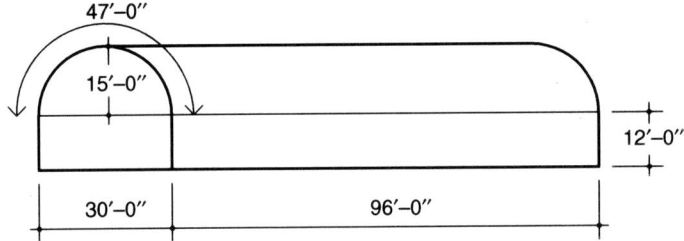

Figure 10-7
Calculating the surface area of an arch-top greenhouse

Step 1

Calculate the surface area of the roof.

Roof Surface Area = Arch Length × Greenhouse Length

Notice that the roof of an arch-top greenhouse is a rectangle. Imagine flattening out the arched roof. Once the arch is flat, it can be viewed as a rectangle. The formula for the area of a rectangle (Area = Length × Width) is applied to this problem.

Roof Surface Area = Arch Length × Greenhouse Length

Roof Surface Area = 47 ft × 96 ft

Roof Surface Area = 4,512 ft²

Step 2

Calculate the surface area of the gable.

$$\text{Surface Area of the Gables} = 2\text{ Gables} \times \left[\frac{\pi \times \text{Gable Height} \times \frac{1}{2}\text{ Greenhouse Width}}{2}\right]$$

Notice that within the brackets, the formula for the area of an ellipse (Area = Π × Minor Radius × Major Radius) is adapted for this space (see Chapter 3, page

60). The bracketed equation is divided by two because each gable is one-half of a circle or ellipse. Finally, the whole equation is multiplied by two because there are two ends of the greenhouse with gables.

The equation can be simplified by eliminating the steps of multiplying and dividing by two because the result is to multiply by one.

Surface Area of the Gables = 2 Gables
$$\times \left[\frac{\pi \times \text{Gable Height} \times \frac{1}{2} \text{Greenhouse Width}}{2} \right]$$

Surface Area of the Gables = $\pi \times$ Gable Height $\times \frac{1}{2}$ Greenhouse Width

Surface Area of the Gables = $\pi \times 15\,\text{ft} \times 15\,\text{ft}$

For this arch-top greenhouse, the gable is a semi-circle. The gable height is equivalent to $\frac{1}{2}$ the greenhouse width. Equal radii indicate a semi-circle. Each arch-top greenhouse is different. Some are more semi-circular and others are semi-elliptical. By using the equation for the area of an ellipse, both shapes are accommodated.

Surface Area of the Gables = $\pi \times 15\,\text{ft} \times 15\,\text{ft}$

Surface Area of the Gables = $706.86\,\text{ft}^2$

Step 3

Calculate the surface area of the gable end walls.

Surface Area of the Gable End Walls = 2 Gable Ends
$\times$ Eave Height $\times$ Greenhouse Width

Surface Area of the Gable End Walls = 2 Gable Ends $\times$ 12 ft $\times$ 30 feet

Surface Area of the Gable End Walls = $720\,\text{ft}^2$

Step 4

Calculate the surface area of the side walls.

Surface Area of the Side Walls = 2 Greenhouse Side Walls
$\times$ Eave Height $\times$ Greenhouse Length

Surface Area of the Side Walls = 2 Greenhouse Side Walls $\times$ 12 ft $\times$ 96 ft

Surface Area of the Side Walls = $2{,}304\,\text{ft}^2$

Step 5

Find the sum of the surface areas of the greenhouse components.

Total Surface Area of the Greenhouse = Roof Surface Area + Gable Surface Area
+ Gable End Wall Surface Area
+ Side Wall Surface Area

Total Surface Area of the Greenhouse $= 4{,}512\,\text{ft}^2 + 706.86\,\text{ft}^2 + 720\,\text{ft}^2 + 2{,}304\,\text{ft}^2$

Total Surface Area of the Greenhouse $= 8{,}242.86\,\text{ft}^2$

SOLUTION

The surface area of this arch-top greenhouse is $8{,}242.86\,\text{ft}^2$.

⌕ Practice Problem 10-3 ⌕ Calculating the Surface Area of an Arch-Top Greenhouse

Calculate the surface area of an arch-top greenhouse that measures 24 ft wide and 48 ft long. The eave height is 12 ft, the gable height is 12 ft, and the arch length is 37.7 ft long.

Gutter-Connected Greenhouses

When calculating the surface area of a gutter-connected greenhouse, the process is similar to the steps used to calculate the surface area of an even-span or arch-top greenhouse. The calculations simply need to be adapted to include the increased number of roof panels, gables, and gable ends.

EXAMPLE 10-5

Calculate the glazed surface area and the curtain wall surface area of the gutter-connected greenhouse shown in Figure 10-8. Each of three greenhouse bays is an even-span structure and is 36 ft wide and 144 ft long. Each bay has a gable height of 9 ft, rafter length of 20 ft, eave height of 10 ft, and a curtain wall height of 3 ft.

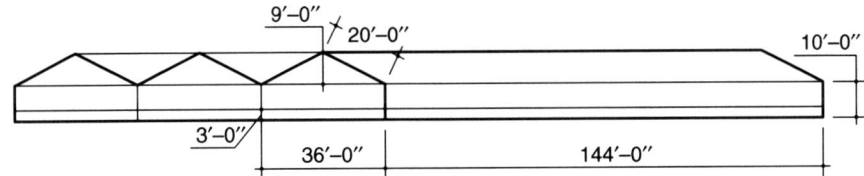

Figure 10-8

Calculating the surface area of a gutter-connected even-span greenhouse

PART ONE: CALCULATE THE GLAZED SURFACE AREA

Step 1

Determine the surface area of the roof.

Roof Surface Area $= 2 \times$ Rafter Length $\times$ Side Wall Length
$\times$ Number of Greenhouse Bays

Roof Surface Area $= 2 \times 20\,\text{ft} \times 144\,\text{ft} \times 3\,\text{Greenhouse Bays}$

Roof Surface Area $= 17{,}280\,\text{ft}^2$

Step 2

Determine the surface area of the gable.

$$\text{Gable Surface Area} = 2 \times \tfrac{1}{2} \text{ (Gable End Width)(Gable Height)}$$
$$\times \text{(Number of Greenhouse Bays)}$$
$$\text{Gable Surface Area} = 2 \times \tfrac{1}{2} \text{ (36 ft)(9 ft)(3)}$$
$$\text{Gable Surface Area} = 972 \text{ ft}^2$$

Step 3

Determine the glazed surface area of the gable end walls.

$$\text{Gable End Wall Area} = 2 \times \text{[Gable End Width(Eave Height}$$
$$-\text{Curtain Wall Height)](Number of Greenhouse Bays)}$$
$$\text{Gable End Wall Area} = 2 \times [36 \text{ ft}(10 \text{ ft} - 3 \text{ ft})](3)$$
$$\text{Gable End Wall Area} = 2 \times 36 \text{ ft} \times 7 \text{ ft} \times 3$$
$$\text{Gable End Wall Area} = 1{,}512 \text{ ft}^2$$

Step 4

Determine the glazed surface area of the side walls.

$$\text{Side Wall Surface Area} = 2 \times \text{Side Wall Length}$$
$$\times \text{(Eave Height} - \text{Curtain Wall Height)}$$
$$\text{Side Wall Surface Area} = 2 \times 144 \text{ ft}(10 \text{ ft} - 3 \text{ ft})$$
$$\text{Side Wall Surface Area} = 2 \times 144 \text{ ft } (7 \text{ ft})$$
$$\text{Side Wall Surface Area} = 2{,}016 \text{ ft}^2$$

Step 5

Find the sum of the surface areas of all the glazed portions of the greenhouse to determine the total glazed surface area.

$$\text{Total Glazed Surface Area} = \text{Roof Area} + \text{Gable Area}$$
$$+ \text{Glazed Gable End Wall Area}$$
$$+ \text{Glazed Side Wall Area}$$
$$\text{Total Glazed Surface Area} = 17{,}280 \text{ ft}^2 + 972 \text{ ft}^2 + 1{,}512 \text{ ft}^2 + 2{,}016 \text{ ft}^2$$
$$\text{Total Glazed Surface Area} = 21{,}780 \text{ ft}^2$$

SOLUTION FOR PART ONE

The total glazed surface area of this greenhouse is 21,780,ft².

Calculations Involving Greenhouse Structures 235

PART TWO: CALCULATE THE SURFACE AREA OF THE CURTAIN WALL

Step 1

Calculate the curtain wall area on the gable ends.

Gable End Curtain Wall Area = 2 × Gable End Width × Height of Curtain Wall
× Number of Greenhouse Bays

Gable End Curtain Wall Area = 2 × 36 ft × 3 ft × 3 Greenhouse Bays

Gable End Curtain Wall Area = 648 ft^2

Step 2

Calculate the curtain wall area on the side walls.

Side Wall Curtain Wall Area = 2 × Side Wall Length × Curtain Wall Height

Side Wall Curtain Wall Area = 2 × 144 ft × 3 ft

Side Wall Curtain Wall Area = 864 ft^2

Step 3

Find the sum of the surface areas of all the curtain walls.

Total Curtain Wall Area = Gable End Curtain Wall Area
+ Side Wall Curtain Wall Area

Total Curtain Wall Area = 648 ft^2 + 864 ft^2

Total Curtain Wall Area = 1,512 ft^2

SOLUTION FOR PART TWO

The total curtain wall surface area of this greenhouse is 1,512 ft^2.

EXAMPLE 10-6

Calculate the glazed surface area of the gutter-connected arch-top greenhouse shown in Figure 10-9. The greenhouse has three greenhouse bays that each measures 30 ft wide by 96 ft long. For each greenhouse bay, the eave height is 12 ft, the gable height is 12 ft, and the arch length is 45 ft.

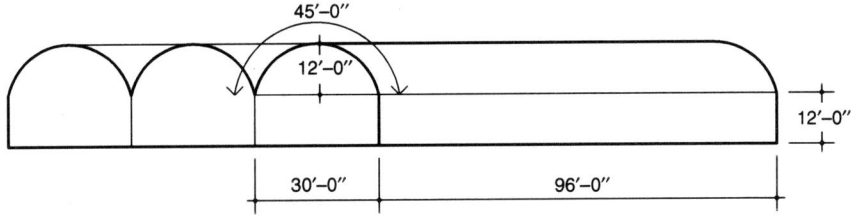

Figure 10-9

Calculating the surface area of a gutter-connected arch-top greenhouse

Step 1

Calculate the surface area of the roof.

$$\text{Roof Surface Area} = \text{Number of Greenhouse Bays} \times \text{Arch Length}$$
$$\times \text{Greenhouse Length}$$
$$\text{Roof Surface Area} = 3 \text{ Greenhouse Bays} \times 45 \text{ ft} \times 96 \text{ ft}$$
$$\text{Roof Surface Area} = 12,960 \text{ ft}^2$$

Step 2

Calculate the surface area of the gables.

$$\text{Surface Area of the Gables} = 3 \text{ Greenhouse Bays} \times 2 \text{ Gables}$$
$$\times \left[\frac{\pi \times \text{Gable Height} \times \frac{1}{2} \text{ Greenhouse Width}}{2} \right]$$
$$\text{Surface Area of the Gables} = 3 \text{ Greenhouse Bays}$$
$$\times \left[\pi \times \text{Gable Height} \times \frac{1}{2} \text{ Greenhouse Width} \right]$$
$$\text{Surface Area of the Gables} = 3 \times \pi \times 12 \text{ ft} \times 15 \text{ ft}$$
$$\text{Surface Area of the Gables} = 1696.46 \text{ ft}^2$$

Step 3

Calculate the surface area of the gable end walls.

$$\text{Surface Area of the Gable End Walls} = 3 \text{ Greenhouse Bays} \times 2 \text{ Gable Ends}$$
$$\times \text{Eave Height} \times \text{Greenhouse Width}$$
$$\text{Surface Area of the Gable End Walls} = 3 \text{ Greenhouse Bays}$$
$$\times 2 \text{ Gable Ends} \times 12 \text{ ft} \times 30 \text{ ft}$$
$$\text{Surface Area of the Gable End Walls} = 2,160 \text{ ft}^2$$

Step 4

Calculate the surface area of the side walls.

$$\text{Surface Area of the Side Walls} = 2 \text{ Greenhouse Side Walls}$$
$$\times \text{Eave Height} \times \text{Greenhouse Length}$$
$$\text{Surface Area of the Side Walls} = 2 \text{ Greenhouse Side Walls} \times 12 \text{ ft} \times 96 \text{ ft}$$
$$\text{Surface Area of the Side Walls} = 2,304 \text{ ft}^2$$

Step 5

Find the sum of the surface areas of the greenhouse components.

Total Surface Area of the Greenhouse = Roof Surface Area + Gable Surface Area

+ Gable End Wall Surface Area

+ Side Wall Surface Area

Total Surface Area of the Greenhouse = $12,960\,\text{ft}^2 + 1,696.46\,\text{ft}^2$

$+ 2,160\,\text{ft}^2 + 2,304\,\text{ft}^2$

Total Surface Area of the Greenhouse = $19,120.46\,\text{ft}^2$

SOLUTION

The surface area of this gutter-connected arch-top greenhouse is $19,120.46\,\text{ft}^2$

Practice Problem Set 10-4 — Calculating the Surface Area of Gutter-Connected Greenhouses

1. Calculate the glazed surface area and the curtain wall surface area of an even-span gutter-connected greenhouse with four bays that each measures 30 ft wide and 96 ft long. The eave height is 8 ft, the curtain wall height is 2 ft, the rafter length is 18.75 ft, and the gable height is 11.25 ft.
2. Calculate the glazed surface area of a gutter-connected arch-top greenhouse with six bays that each measures 24 ft wide and 48 ft long. For each bay, the arch length is 37.7 ft, the gable height is 12 ft, and the eave height is 12 ft. The glazing extends from the eave to the ground so there is no curtain wall.

Quonset Greenhouse

Quonset greenhouses (Figure 10-10) are similar to arch-top greenhouses except that there are no straight side walls. This type of greenhouse has no curtain walls and is glazed from the roof to the ground. In fact, the roof calculation encompasses both the roof and the sides of the greenhouse.

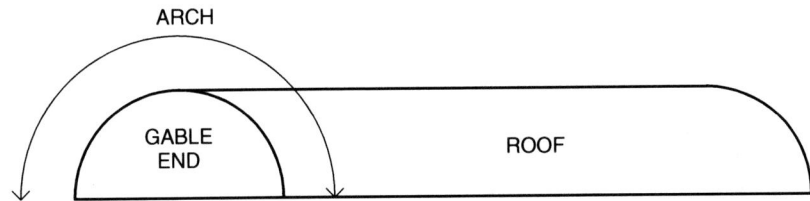

Figure 10-10

Diagram of a Quonset greenhouse

EXAMPLE 10-7

Calculate the surface area of a Quonset greenhouse (Figure 10-11) that measures 24 ft wide and 60 ft long. The gable end is 12 ft high and the arch length is 37.7 ft.

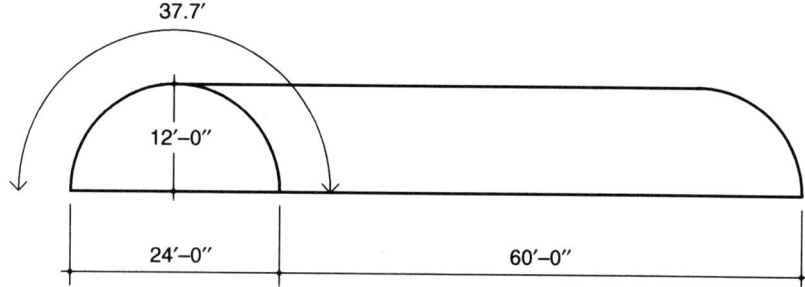

Figure 10-11

Calculating the surface area of a Quonset greenhouse

Step 1

Calculate the surface area of the roof.

$$\text{Surface Area of Roof} = \text{Arch Length} \times \text{Greenhouse Length}$$
$$\text{Surface Area of Roof} = 37.7 \, \text{ft} \times 60 \, \text{ft}$$
$$\text{Surface Area of Roof} = 2{,}262 \, \text{ft}^2$$

Step 2

Calculate the surface area of the gable ends.

$$\text{Surface Area of the Gables} = 2 \text{ Gables} \times \left[\frac{\pi \times \text{Gable Height} \times \frac{1}{2} \text{ Greenhouse Width}}{2} \right]$$

Simplify this equation.

$$\text{Surface Area of the Gables} = \pi \times \text{Gable Height} \times \tfrac{1}{2} \text{ Greenhouse Width}$$

Insert measurements into the equation.

$$\text{Surface Area of the Gables} = \pi \times 12 \, \text{ft} \times \tfrac{1}{2} \, 24 \, \text{ft}$$
$$\text{Surface Area of the Gables} = \pi \times 12 \, \text{ft} \times 12 \, \text{ft}$$
$$\text{Surface Area of the Gables} = 452.39 \, \text{ft}^2$$

Step 3

Find the sum of the surface area of the gables and the roof.

$$\text{Surface Area of a Quonset Greenhouse} = \text{Roof Surface Area} + \text{Gable Surface Area}$$
$$\text{Surface Area of a Quonset Greenhouse} = 2{,}262 \, \text{ft}^2 + 452.39 \, \text{ft}^2$$
$$\text{Surface Area of a Quonset Greenhouse} = 2{,}714.39 \, \text{ft}^2$$

> **SOLUTION**

The total surface area of this Quonset greenhouse is 2,714.39 ft².

✐ Practice Problem 10-5 ✐ *Calculating the Surface Area of a Quonset Greenhouse*

Calculate the surface area of a Quonset greenhouse that is 22 ft wide and 72 ft long. The gable height is 11 ft and the arch length is 34.6 ft long.

Calculating Greenhouse Volume

Calculating the volume of greenhouse space is an important part of the process of designing cooling and ventilating systems for greenhouses. The volume of space, in addition to other factors, helps determine the number and size of ventilating and exhaust fans and the surface area of evaporative cooling pads.

Most greenhouse structures take the form of simple geometric shapes and the formula for calculating volume is straightforward. In simple terms, the volume of a geometric form is equal to the product of the base and the height.

$$\text{Volume} = \text{Base} \times \text{Height}$$

When translating a greenhouse to a geometric form, the volume of the greenhouse space is equal to the product of the area of one of the gable ends and the length of the long side of the greenhouse (Figure 10-12).

> **❦ Definition ❧** *Greenhouse Volume*
>
> *Greenhouse volume* is the product of the area of one of the gable ends and the length of the long side of a greenhouse.
>
> $$\text{Greenhouse Volume} = \text{Gable End Area} \times \text{Length of Long Side}$$

Each greenhouse gable end is composed of geometric forms. An even-span greenhouse gable end is composed of a triangular gable and a rectangular side wall. An arch-top greenhouse has a rectangular side wall and a semicircular or semi-elliptical gable. Several examples demonstrating the proper way to calculate the volume of a greenhouse follow:

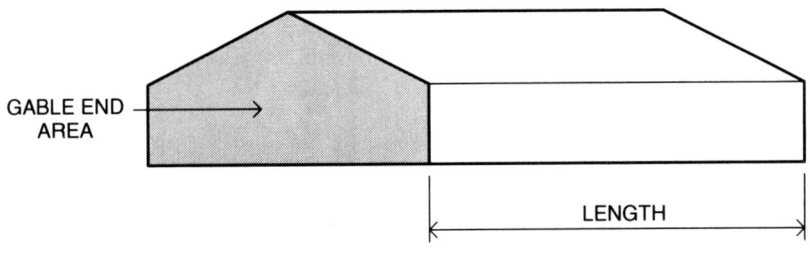

Figure 10-12
Calculating greenhouse volume by multiplying the area of the gable end by the greenhouse length

Even-Span Greenhouse

EXAMPLE 10-8

Calculate the volume of an even-span greenhouse that is 36 ft wide and 144 ft long (Figure 10-13). It has a gable height of 9 ft, rafter length of 20 ft, and an eave height of 10 ft.

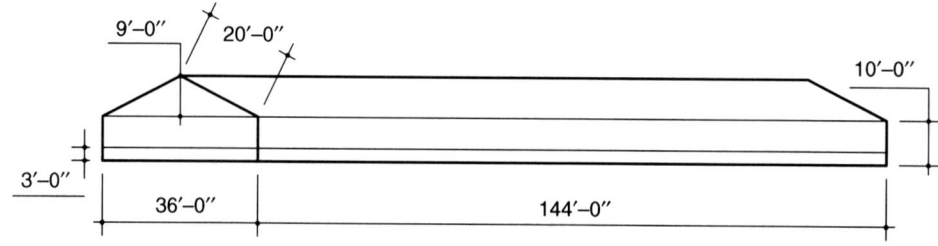

Figure 10-13
Calculating the volume of an even-span greenhouse

Step 1

Calculate the area of the gable. The gable is triangular in shape so the basic formula for the area of a triangle is adapted to the greenhouse.

$$\text{Gable Area} = \tfrac{1}{2} \,(\text{Gable End Width}) \times (\text{Gable Height})$$

$$\text{Gable Area} = \tfrac{1}{2} \,(36\,\text{ft}) \times (9\,\text{ft})$$

$$\text{Gable Area} = 18\,\text{ft} \times 9\,\text{ft}$$

$$\text{Gable Area} = 162\,\text{ft}^2$$

Step 2

Calculate the area of the gable end wall.

$$\text{Gable End Wall Area} = \text{Gable End Width} \times \text{Eave Height}$$

$$\text{Gable End Wall Area} = 36\,\text{ft} \times 10\,\text{ft}$$

$$\text{Gable End Wall Area} = 360\,\text{ft}^2$$

Step 3

Calculate the area of the gable end of the greenhouse.

$$\text{Gable End Area} = \text{Gable Area} + \text{Gable End Wall Area}$$

$$\text{Gable End Area} = 162\,\text{ft}^2 + 360\,\text{ft}^2$$

$$\text{Gable End Area} = 522\,\text{ft}^2$$

Step 4

Multiply the area of the gable end by the length of the greenhouse to determine greenhouse volume.

Volume of Greenhouse = Gable End Area × Greenhouse Length

Volume of Greenhouse = 522 ft² × 144 ft

Volume of Greenhouse = 75,168 ft³

SOLUTION

The volume of this even-span greenhouse is 75,168 ft³.

Practice Problem 10-6 — Calculating the Volume of an Even-Span Greenhouse

Calculate the volume of an even-span greenhouse that measures 30 ft wide and 96 ft long. The eave height is 8 ft and the gable height is 11.25 ft.

Arch-Top Greenhouse

EXAMPLE 10-9

Calculate the volume of an arch-top greenhouse that measures 30 ft wide by 96 ft long (Figure 10-14). The eave height is 12 ft and the gable height is 15 ft.

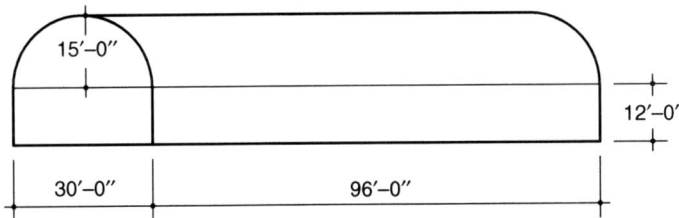

Figure 10-14
Calculating the volume of an arch-top greenhouse

Step 1

Calculate the area of the gable. The gable is semicircular or semi-elliptical in shape so the basic formula for the area of an ellipse is adapted to the greenhouse.

Gable Area = (π × Gable Height × ½ Greenhouse Width) ÷ 2

Gable Area = (π × 15 ft × 15 ft) ÷ 2

Gable Area = 353.43 ft²

Step 2

Calculate the area of the gable end wall.

Gable End Wall Area = Gable End Width × Eave Height

Gable End Wall Area = 30 ft × 12 ft

Gable End Wall Area = 360 ft²

Step 3

Calculate the area of the gable end of the greenhouse.

$$\text{Gable End Area} = \text{Gable Area} + \text{Gable End Wall Area}$$
$$\text{Gable End Area} = 353.43 \text{ ft}^2 + 360 \text{ ft}^2$$
$$\text{Gable End Area} = 713.43 \text{ ft}^2$$

Step 4

Multiply the area of the gable end by the length of the greenhouse to determine greenhouse volume.

$$\text{Volume of Greenhouse} = \text{Gable End Area} \times \text{Greenhouse Length}$$
$$\text{Volume of Greenhouse} = 713.43 \text{ ft}^2 \times 96 \text{ ft}$$
$$\text{Volume of Greenhouse} = 68{,}489.28 \text{ ft}^3$$

Solution

The volume of this arch-top greenhouse is $68{,}489.28 \text{ ft}^3$.

Practice Problem 10-7 *Calculating the Volume of an Arch-Top Greenhouse*

Calculate the volume of an arch-top greenhouse that measures 24 ft wide and 48 ft long. The eave height is 12 ft and the gable height is 12 ft.

Gutter-Connected Greenhouse

Calculating the volume of a gutter-connected greenhouse is a simple process, once the volume of one of the individual bays is calculated. To determine the volume of the greenhouse, the volume of a single bay is multiplied by the number of bays that make up the greenhouse.

EXAMPLE 10-10

Calculate the volume of the gutter-connected greenhouse shown in Figure 10-15. Each of three greenhouse bays is an even-span structure and is 36 ft wide and 144 ft long. Each bay has a gable height of 9 ft and an eave height of 10 ft.

Figure 10-15
Calculating the volume of a gutter-connected even-span greenhouse

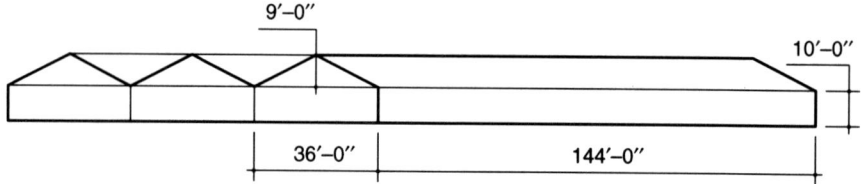

Step 1
Calculate the area of the gable of one of the greenhouse bays. The gable is a triangle.

$$\text{Gable Area} = \tfrac{1}{2} \text{ (Gable End Width)} \times \text{(Gable Height)}$$

$$\text{Gable Area} = \tfrac{1}{2} \text{ (36 ft)} \times \text{(9 ft)}$$

$$\text{Gable Area} = 18 \text{ ft} \times 9 \text{ ft}$$

$$\text{Gable Area} = 162 \text{ ft}^2$$

Step 2
Calculate the area of the gable end wall of one of the greenhouse bays.

$$\text{Gable End Wall Area} = \text{Gable End Width} \times \text{Eave Height}$$

$$\text{Gable End Wall Area} = 36 \text{ ft} \times 10 \text{ ft}$$

$$\text{Gable End Wall Area} = 360 \text{ ft}^2$$

Step 3
Calculate the area of the gable end of one of the greenhouse bays.

$$\text{Gable End Area} = \text{Gable Area} + \text{Gable End Wall Area}$$

$$\text{Gable End Area} = 162 \text{ ft}^2 + 360 \text{ ft}^2$$

$$\text{Gable End Area} = 522 \text{ ft}^2$$

Step 4
Calculate the volume of one of the greenhouse bays.

$$\text{Volume of Greenhouse Bay} = \text{Gable End Area} \times \text{Greenhouse Length}$$

$$\text{Volume of Greenhouse Bay} = 522 \text{ ft}^2 \times 144 \text{ ft}$$

$$\text{Volume of Greenhouse Bay} = 75{,}168 \text{ ft}^3$$

Step 5
Calculate the volume of the greenhouse by multiplying the number of bays by the volume of a single bay.

$$\text{Volume of Greenhouse} = \text{Number of Bays} \times \text{Volume of Greenhouse Bay}$$

$$\text{Volume of Greenhouse} = 3 \text{ Bays} \times 75{,}168 \text{ ft}^3$$

$$\text{Volume of Greenhouse} = 225{,}504 \text{ ft}^3$$

SOLUTION

The volume of this gutter-connected greenhouse is 225,504 ft^3.

EXAMPLE 10-11

Calculate the volume of the gutter-connected arch-top greenhouse shown in Figure 10-16. The greenhouse has four greenhouse bays, each measuring 30 ft wide by 96 ft long. For each greenhouse bay, the eave height is 12 ft, and the gable height is 12 ft.

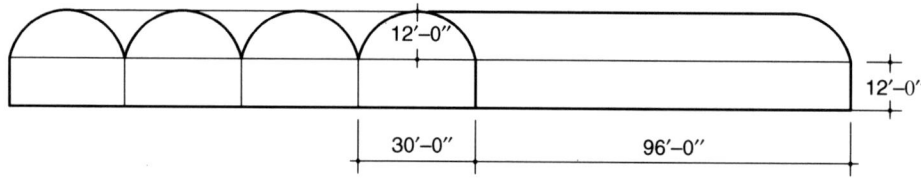

Figure 10-16

Calculating the volume of a gutter-connected arch-top greenhouse

Step 1

Calculate the area of the gable of one of the greenhouse bays.

Gable Area = $(\pi \times \text{Gable Height} \times \frac{1}{2} \text{Greenhouse Width}) \div 2$

Gable Area = $(\pi \times 12\,\text{ft} \times 15\,\text{ft}) \div 2$

Gable Area = $282.74\,\text{ft}^2$

Step 2

Calculate the area of the gable end wall of one of the greenhouse bays.

Gable End Wall Area = Gable End Width × Eave Height

Gable End Wall Area = 30 ft × 12 ft

Gable End Wall Area = $360\,\text{ft}^2$

Step 3

Calculate the area of the gable end of one of the greenhouse bays.

Gable End Area = Gable Area + Gable End Wall Area

Gable End Area = $282.74\,\text{ft}^2 + 360\,\text{ft}^2$

Gable End Area = $642.74\,\text{ft}^2$

Step 4

Calculate the volume of one of the greenhouse bays.

Volume of Greenhouse Bay = Gable End Area × Greenhouse Length

Volume of Greenhouse Bay = $642.74\,\text{ft}^2 \times 96\,\text{ft}$

Volume of Greenhouse Bay = $61{,}703.04\,\text{ft}^3$

Step 5

Calculate the volume of the greenhouse.

Volume of Greenhouse = Number of Bays × Volume of Greenhouse Bay

Volume of Greenhouse = 4 Bays × 61,703.04 ft³

Volume of Greenhouse = 246,812.16 ft³

SOLUTION

The volume of this greenhouse is 246,812.16 ft³.

Practice Problem Set 10-8 — Calculating the Volume of Gutter-Connected Greenhouses

1. Calculate the volume of a gutter-connected even-span greenhouse with four bays that each measures 30 ft wide and 96 ft long. The eave height is 8 ft and the gable height is 11.25 ft.
2. Calculate the volume of a gutter-connected arch-top greenhouse with six bays that each measures 24 ft wide and 48 ft long. For each bay, the gable height is 12 ft and the eave height is 12 ft.

Quonset Greenhouse

Quonset greenhouse gable end areas are either semicircular or semi-elliptical. There are no straight-sided gable end walls so the area of the gable end is a single calculation. The volume is the product of the greenhouse length and the gable end area.

EXAMPLE 10-12

Calculate the volume of a Quonset greenhouse that measures 24 ft wide and 60 ft long (Figure 10-17). The gable end is 12 ft high.

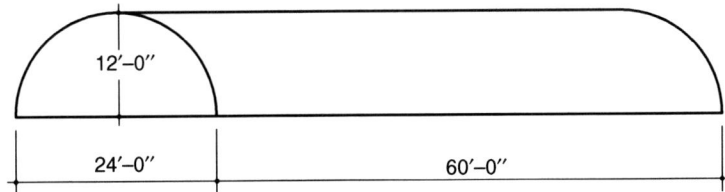

Figure 10-17
Calculating the volume of a Quonset greenhouse

Step 1

Calculate the area of the gable end.

Gable End Area = $\left(\pi \times \text{Gable Height} \times \frac{1}{2} \text{Greenhouse Width}\right) \div 2$

Gable End Area = $\left(\pi \times 12\,\text{ft} \times \frac{1}{2} \times 24\,\text{ft}\right) \div 2$

$$\text{Gable End Area} = (\pi \times 12\,\text{ft} \times 12\,\text{ft}) \div 2$$

$$\text{Gable End Area} = 226.19\,\text{ft}^2$$

Step 2

Calculate the volume of the greenhouse.

$$\text{Greenhouse Volume} = \text{Gable End Area} \times \text{Greenhouse Length}$$

$$\text{Greenhouse Volume} = 226.19\,\text{ft}^2 \times 60\,\text{ft}$$

$$\text{Greenhouse Volume} = 13{,}571.40\,\text{ft}^3$$

SOLUTION

The volume of this greenhouse is $13{,}571.40\,\text{ft}^3$.

Practice Problem 10-9 — Calculating the Volume of a Quonset Greenhouse

Calculate the volume of a Quonset greenhouse that is 22 ft wide and 72 ft long. The gable height is 11 ft.

Calculations Involving Fertilizers and Proportioning Equipment

Proportioning equipment or injectors are designed to introduce, or inject, a small amount of concentrated fertilizer solution (stock) into irrigation water used on plants in a nursery or greenhouse. The final solution delivered to the plant will contain a desired amount of nutrition, dependent on the proportioning equipment, equipment settings, and the type and concentration of fertilizer introduced.

Proportioning Equipment

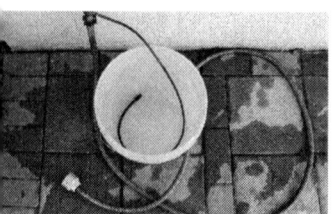

Figure 10-18
A brass-siphon proportioning device

Figure 10-19
A proportioner for a small commercial greenhouse

A brass-siphon proportioning device is the most simple and inexpensive proportioning device available (Figure 10-18). The proportioning ratio often is preset to a 1:16 ratio. That means that 1 part of fertilizer concentrate is mixed with 16 parts of water. More sophisticated and expensive proportioning devices have adjustable ratios that range from 1:20 to 1:1,900 (Figure 10-19). The most common ratios used for greenhouse and nursery crop nutrition are 1:100 and 1:200. Higher ratios require more concentrated stock solutions, and it becomes impossible to dissolve fertilizer salts in the volume of water required for the stock tank solution. The higher ratio proportioners are useful for injecting concentrated acids for the adjustment of water pH. These concentrated acids are injected in very small amounts and the proportioning device needs to be corrosion resistant. As a group, all of these more sophisticated devices are not affected by changes in water pressure and are more reliable in delivering consistent amounts of nutrition.

Brass-Siphon Proportioning Devices

The performance of a brass-siphon proportioning device is dependent on water pressure. As water pressure varies, so does the proportioning ratio. Water pressure can vary from the source and output can vary because of hose length and hose diameter. It is advisable to check the proportioning ratio for each situation through calibration to ensure accurate fertilizer delivery.

Calibration of Brass-Siphon Proportioning Devices

Begin by filling a jar with one cup of water (test stock tank) (Figure 10-20a), insert the siphon hose into the jar of water, open the faucet to its widest opening, and collect water from the hose into a five-gallon bucket until all of the water in the test stock tank is gone (Figure 10-20b). Immediately turn off the faucet when the entire test stock tank has been used up. Measure the amount of water collected in the five-gallon bucket in cups. If there are 17 cups in the bucket, then the proportioning ratio is 1:16, and if there are 15 cups in the bucket, then the proportioning ratio is 1:14, and so on.

(a)

(b)
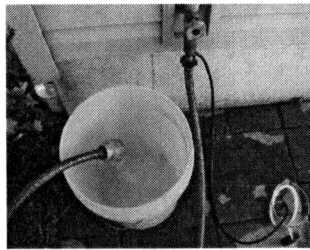

Figure 10-20
To calibrate a brass-siphon proportioning device, use one-cup measure as a test stock tank (a) and collect water from the hose-end into a five-gallon bucket until the one-cup stock tank is empty (b)

EXAMPLE 10-13

This example involves a brass-siphon proportioning device and fertilizer used on a small scale, such as a homeowner situation. The fertilizer is a product that has an analysis of 20-20-20, and it is recommended that it be used at a rate of 1 tablespoon (Tbps) per gal of solution. The homeowner has a brass-siphon proportioning device with a confirmed ratio of 1:16 and a 2 gal volume stock tank. The brass-siphon proportioning device has a ratio of 1:16, which means that for every 16 gal of water, there is 1 gal of stock solution added. This means that a total of 17 gal of dilute solution is delivered.

PART ONE: DETERMINE THE NUMBER OF GALLONS OF SOLUTION DELIVERED WHEN ALL OF THE STOCK IS USED UP

Determine the number of gallons delivered when the entire 2 gal of stock solution is used up. If 1 gal of stock is used when a total of 17 gal is delivered, then how many gallons will be delivered when 2 gal of stock solution is used?

Step 1

Set up a proportion.

$$\frac{1 \text{ gal of stock solution}}{17 \text{ gal of dilute solution delivered}} = \frac{2 \text{ gal of stock solution}}{x \text{ gal of dilute solution delivered}}$$

Step 2

Isolate and solve for x.

$$x = 2 \times 17$$

$$x = 34 \text{ gal}$$

> **SOLUTION**

Thirty-four gallons of dilute solution are delivered when two gallons of stock solution is used.

PART TWO: DETERMINE THE AMOUNT OF FERTILIZER THAT IS ADDED TO THE STOCK TANK

The amount of fertilizer to be added to the stock tank is equivalent to the amount of fertilizer required for 34 gal of solution. If 1 Tbsp of fertilizer is added to 1 gal of dilute solution, then how many tablespoons need to be added to produce 34 gal of dilute solution?

Step 1

Set up a proportion.

$$\frac{1 \text{ Tbsp of fertilizer}}{1 \text{ gal of dilute solution}} = \frac{x \text{ Tbsp of fertilizer}}{34 \text{ gal of dilute solution}}$$

Step 2

Isolate and solve for x.

$$x = 34 \times 1$$
$$x = 34$$

> **SOLUTION**

Thirty-four tablespoons of fertilizer are needed to formulate two gallons of stock solution.

Commercial-Scale Proportioning Devices

In commercial situations, fertilizer recommendations are made in parts per million of a particular nutritional element. One part per million (ppm) is equivalent to one milligram per liter of water or one milligram per kilogram of a dry substance. The U.S. Customary equivalent is 0.013 ounces by weight in 100 gallons.

A conversion factor has been developed that is used as a multiplier to convert parts per million to a more usable form. The conversion factor is 75 parts per million for 1 ounce (by weight) of fertilizer product per 100 gallons of water. In other words, when one ounce of fertilizer product is dissolved in 100 gallons of water, the result is a 75 parts per million solution of that fertilizer product in water. If 1 ounce of 20-10-20 is dissolved in 100 gallons of water, then the resulting solution is 75 ppm of 20-10-20.

This relationship is used in the examples that follow in this section. Example 10-14 shows a step-wise approach to calculating the amount of fertilizer needed for the preparation of stock solution for a commercial-scale proportioning device. Example 10-15 employs a set of short-cut equations to obtain the same result.

Calculations Involving Fertilizers and Proportioning Equipment **249**

EXAMPLE 10-14

How many pounds plus ounces of 20-10-20 is required to prepare stock solution in a 35 gal stock tank when a grower wishes to deliver 250 ppm N using a 1:100 proportioner?

PART ONE

If a 75 ppm fertilizer solution is obtained by having 1 oz of dry fertilizer product in 100 gal of dilute solution, then how many parts per million of nitrogen is contained in this solution if we use a 20-10-20 product? A 20-10-20 fertilizer product contains only 20% N. How many parts per million of nitrogen are contained in a 75 ppm fertilizer product solution? There would be 20% of 75 ppm. To find 20% of 75 ppm, do the following:

Step 1

Convert 20% to a decimal.
$$20\% = \frac{20}{100} = 0.20$$

Step 2

To find 20% of 75 ppm, translate this sentence to a mathematical equation. The word "of" implies multiplication; therefore, multiply 20% (converted to a decimal) by 75 ppm.
$$0.20 \times 75 \text{ ppm} = 15 \text{ ppm}$$

SOLUTION

If 1 oz of 20-10-20 is dissolved in 100 gal of water, it will produce a solution containing 15 ppm N.

PART TWO

A solution containing 15 ppm of nitrogen is not useful in horticulture. Common recommended rates range from 150 ppm of nitrogen to 350 ppm. Taking the calculations one step further, determine how much 20-10-20 needs to be dissolved in 100 gallons of water to produce a solution that contains 250 ppm of nitrogen.

If 1 oz of 20-10-20 dissolved in 100 gal of water produces a 15 ppm N solution, then how many ounces of 20-10-20 need to be dissolved in 100 gal to produce a 250 ppm N solution?

Step 1

Set up a proportion.
$$\frac{1 \text{ oz } 20\text{-}10\text{-}20}{15 \text{ ppm N}} = \frac{x \text{ oz } 20\text{-}10\text{-}20}{250 \text{ ppm N}}$$

Step 2

Isolate and solve for x.

$$15x = 250$$
$$x = \frac{250}{15}$$
$$x = 16.6\overline{6} \text{ oz of 20-10-20}$$

SOLUTION

If 16.67 oz of 20-10-20 are dissolved in 100 gal of water, it will produce a solution containing 250 ppm N.

PART THREE

Expand this problem to include the preparation of stock solution for a 1:100 proportioner. This proportioner delivers 101 gallons of dilute solution for each 1 gallon of stock concentrate used. For each gallon of stock concentrate, enough fertilizer to provide 250 ppm of nitrogen in 101 gallons of dilute solution is needed.

If 16.67 oz of 20-10-20 dissolved in 100 gal of water delivers 250 ppm N, then how many ounces will be needed for 101 gal of water?

Step 1

Set up a proportion.

$$\frac{16.67 \text{ oz of 20-10-20}}{100 \text{ gal}} = \frac{x \text{ oz 20-10-20}}{101 \text{ gal}}$$

Step 2

Isolate and solve for x.

$$100x = 16.67 \times 101$$
$$x = \frac{16.67 \times 101}{100}$$
$$x = 16.8367 \text{ oz of 20-10-20}$$

SOLUTION

If 16.8 oz of 20-10-20 are dissolved in 101 gal of water, a 250 ppm N solution will result.

PART FOUR

This is also the amount of 20-10-20 that will be required for every gallon of stock solution prepared when using a 1:100 proportioner. Stock solution tanks in commercial greenhouse and nursery operations usually hold more than one gallon of stock solution.

If a commercial operation has a 35 gal stock tank that is used with the 1:100 proportioner, then how many ounces of 20-10-20 are needed to prepare 35 gal of stock? The desired concentration of nitrogen to be delivered is still 250 ppm.

Furthermore, if 16.8367 ounces of 20-10-20 are required for every 1 gal of stock, then how many ounces are required for 35 gal of stock solution?

Step 1

Set up a proportion.

$$\frac{16.8367 \text{ oz of 20-10-20}}{1 \text{ gal of stock solution}} = \frac{x \text{ oz of 20-10-20}}{35 \text{ gal of stock}}$$

Step 2

Isolate and solve for x.

$$x = 16.8367 \times 35$$
$$x = 589.2845 \text{ oz of 20-10-20}$$

Step 3

For ease of measuring, convert ounces to pounds plus ounces by setting up a proportion.

$$\frac{16 \text{ oz}}{1 \text{ lb}} = \frac{589.2845 \text{ oz}}{x \text{ lb}}$$

Step 4

Isolate and solve for x.

$$589.2845 = 16x$$
$$x = \frac{589.2845}{16}$$
$$x = 36.83 \text{ lb}$$

Step 5

Convert 0.83 lb to ounces by setting up a proportion and solving for x.

$$\frac{0.83 \text{ lb}}{x \text{ oz}} = \frac{1 \text{ lb}}{16 \text{ oz}}$$
$$x = 0.83 \times 16$$
$$x = 13.3 \text{ oz}$$

SOLUTION

Dissolve 36 lb plus 13.3 ounces of 20-10-20 in 35 gal of water to produce a stock solution that delivers 250 ppm N when using a 1:100 proportioner.

Calculations For Commercial-Scale Proportioning Devices Using Short-Cut Equations

The previous discussion provides the step-by-step process for determining the amount of dry fertilizer needed to prepare a stock solution when all the parameters are described. Those parameters are: desired parts per million of nitrogen, fertilizer product analysis, volume of the stock tank, and injector ratio. Understanding all of these steps is important but speedy derivation of the solution is also important. To that end, short-cut equations have been developed. Those equations follow and then the previous example is recalculated using the short-cut equations.

SHORT-CUT EQUATIONS

Step 1
Determine the number of ounces of dry fertilizer product required to make a desired parts per million of solution in 100 gallons of water.

$$\frac{\text{Desired ppm} \div 75}{\text{Decimal Fraction of the Desired Fertilizer Element}}$$

Step 2
Determine the number of ounces of dry fertilizer to place in a stock tank.

$$\frac{\text{Volume of Stock Tank (gallons)} \times \text{Total Gallons Delivered When 1 Gallon of Stock Is Injected} \times \text{Step 1 Solution}}{100}$$

EXAMPLE 10-15

How many pounds plus ounces of 20-10-20 are required to prepare stock solution for a 35 gal stock tank when a grower wishes to deliver 250 ppm N using a 1:100 proportioner?

Step 1
Determine how many ounces of dry fertilizer product are required to make a desired parts per million of solution in 100 gallons of water by inserting values into the first short-cut equation.

$$\frac{\text{Desired ppm} \div 75}{\text{Decimal Fraction of the Desired Fertilizer Element}}$$

$$\frac{250 \text{ ppm N} \div 75}{0.20} = 16.67 \text{ oz}$$

SOLUTION
Dissolve 16.67 oz of 20-10-20 in 100 gal of water to produce a 250 ppm N solution.

Step 2
Determine how many ounces of dry fertilizer to place in a stock tank by inserting values into the second short-cut equation.

$$\frac{\text{Volume of Stock Tank (gallons)} \times \text{Total Gallons Delivered When 1 Gallon of Stock Is Injected} \times \text{Step 1 Solution}}{100}$$

$$\frac{35 \times 101 \times 16.67}{100} = 589.2845 \text{ oz}$$

For ease of measuring, convert ounces to pounds plus ounces by setting up a proportion.

$$\frac{16\,oz}{1\,lb} = \frac{589.2845\,oz}{x\,lb}$$

Isolate and solve for x.

$$589.2845 = 16x$$
$$x = \frac{589.2845}{16}$$
$$x = 36.83\,lb$$

Convert 0.83 to ounces by setting up a proportion and solving for x.

$$\frac{0.83\,lb}{x\,oz} = \frac{1\,lb}{16\,oz}$$
$$x = 0.83 \times 16$$
$$x = 13.3\,oz$$

SOLUTION

Dissolve 36 lb plus 13.3 oz of 20-10-20 in 35 gal of water to produce a stock solution that delivers 250 ppm N when using a 1:100 proportioner.

How to Determine the Amounts of Elements Other Than Nitrogen That Are Applied When Using Proportioning Devices

The focus of most calculations for fertilizer applications using proportioning equipment is based on the amount of nitrogen being applied. Fertilizer recommendations are commonly described in parts per million of nitrogen. Good plant health is dependent on the amount of nitrogen relative to the amount of phosphorus or potassium, and the proper N:K or N:K$_2$O ratio is often delineated for floriculture crops. The calculations that follow allow the grower to determine the amount of phosphorus and potassium also being applied with a nitrogen-based fertilizer recommendation.

EXAMPLE 10-16

When 250 ppm N is delivered using 20-10-20, how much P is delivered with this solution?

DETERMINING HOW MUCH PHOSPHORUS IS DELIVERED

Step 1
Begin with determining the number of ounces of 20-10-20 needed to be dissolved in 100 gal of water to produce a 250 ppm N solution. (This is the first step in the injector short-cut equations.)

$$\frac{\text{Desired ppm} \div 75}{\text{Decimal Fraction of the Desired Fertilizer Element}}$$

$$\frac{250 \div 75}{0.20} = 16.67 \text{ oz}$$

Step 2

Use the following equation to find how many ppm P_2O_5 are delivered in this solution.

Step 1 Solution × 75 × decimal fraction of P_2O_5 in the fertilizer product
= ppm of P_2O_5
16.67 × 75 × 0.10
= 125 ppm of P_2O_5

Step 3

How many ppm of P are found in 125 ppm of P_2O_5? Chapter 5 indicates that there is 44% P found in the P_2O_5 molecule; therefore, 44% of 125 ppm of P_2O_5 is P or:

$$0.44 \times 125 \text{ ppm } P_2O_5 = 55 \text{ ppm P}$$

SOLUTION

There is 55 ppm P found in a solution containing 250 ppm N made using a 20-10-20 fertilizer product.

DETERMINING HOW MUCH POTASSIUM IS DELIVERED

Step 1

Begin with determining the number of ounces of 20-10-20 needed to be dissolved in 100 gal of water to produce a 250 ppm N solution. (This is the first step in the injector short-cut equations.)

$$\frac{\text{Desired ppm} \div 75}{\text{Decimal Fraction of the Desired Fertilizer Element}}$$

$$\frac{250 \div 75}{0.20} = 16.67 \text{ oz}$$

Step 2

Use the following equation to find how many parts per million of K_2O are delivered in this solution.

Step 1 Solution × 75 × decimal fraction of K$_2$O in the fertilizer product
= ppm of K$_2$O

16.67 × 75 × 0.20

= 250 ppm of K$_2$O

Step 2
How many parts per million of K are found in 250 ppm of K$_2$O? Chapter 5 indicates that there is 83% K found in the K$_2$O molecule; therefore, 83% of 250 ppm of K$_2$O is K or:

$$0.83 \times 250 \text{ ppm K}_2\text{O} = 207.5 \text{ ppm K}$$

SOLUTION

There is 207.5 ppm K found in a solution containing 250 ppm N made using a 20-10-20 fertilizer product.

Practice Problem Set 10-10   Fertilizer Applications Using Proportioning Equipment

1. A homeowner plans to apply a water-soluble fertilizer with a brass-siphon proportioner at the rate of 2.5 Tbsp/gal. The brass-siphon proportioner has been calibrated and delivers at a 1:15 proportion. The homeowner would like to prepare 4 gal of concentrated fertilizer stock solution. How many tablespoons of fertilizer are needed to prepare the 4 gal of stock solution? The dilute fertilizer solution is to be applied at the rate of 1 gal per 25 ft^2. How many square feet can be covered if all of the stock solution is used?

2. A nursery manager would like to use a brass-siphon device to apply fertilizer solution to a small crop of potted nursery stock. The manager set up the delivery system and calibrated the device. A total of 17 cups of solution were collected when 1 cup of test stock solution was used. What is the proportioner ratio for this brass-siphon device?

3. A nursery manager would like to fertilize a hydrangea crop with 13-0-22 at a rate of 150 ppm N. The injector ratio is 1:200 and the stock tank volume is 100 gallons. How much 13-0-22 is needed to prepare stock solution in the 100 gal tank? How many parts per million of potassium are delivered with the nitrogen?

4. A greenhouse manager would like to fertilize a poinsettia crop using a combination of potassium nitrate (13-0-44) at 100 ppm N and calcium nitrate (15.5-0-0) at 100 ppm N. The injector ratio is 1:100 and the stock tank volume is 50 gal. How much 13-0-44 and 15.5-0-0 are needed to prepare the stock solution in the 50 gal tank? How many parts per million of potassium are delivered with both sources of nitrogen?

Calculations Involving Containers and Root Medium

Calculating the Volume of a Growing Container

Knowing the volume of a growing container is the first step in calculating the volume of root media needed to fill containers for growing a containerized crop. Recent government regulations, designed to protect the consumer, have required retailers to label plant products with the actual volume of the growing container. Container manufacturers now identify their products with the volume of the container in addition to the traditional designations of pot diameter, pot style, or nursery container numbering system. Some containers may not need to conform to these labeling requirements, and these may include decorative planters and urns. If a horticulturist needs to calculate the volume of a growing container, the first step in this calculation is to determine if the container has parallel or tapered sides.

Containers with Parallel Sides

Containers with parallel sides include: nursery containers, window boxes, free-standing planters, built-in planters boxes, and ground and raised beds found in greenhouses. These containers may be shaped like cubes, cylinders, or rectangular prisms. Keep in mind the basic formula for calculating volume is volume equals the product of the base and the height. See Chapter 3, page 72.

$$\text{Volume} = \text{Base} \times \text{Height}$$

where the Base = Area of cross-section.

Square Container with Parallel Sides

For a square container, the base is a square and the height is the third dimension. Each dimension equals the other to form a cube.

> **◆ Definition ◆** *Volume of a Square Container with Parallel Sides (Cube) is the product of the length of the three sides or the length of a side cubed.*
>
> $$\text{Volume} = \text{Side Length}^3$$

EXAMPLE 10-17

Calculate the volume in cubic feet of a large, square decorative container that measures 3 ft high, 3 ft deep, and 3 ft wide.

Step 1

Insert the dimensions of the container into the formula and solve.

$$\text{Volume} = \text{Side Length}^3$$
$$\text{Volume} = (3\,\text{ft})^3$$
$$\text{Volume} = 27\,\text{ft}^3$$

SOLUTION

The decorative container has a volume of $27\,\text{ft}^3$.

Rectangular Container with Parallel Sides

For a rectangular container, the base could be a square or a rectangle and the height is the long dimension of the container.

> **◆ Definition ◆** *Volume of a Rectangular Container with Parallel Sides (Rectangular Prism)*
>
> *Volume of a rectangular container with parallel sides (rectangular prism)* is the product of the area of the base and the height or the product of the length and the width and the height.
>
> $$\text{Volume} = \text{Base} \times \text{Height}$$
>
> where $\text{Base} = \text{Length} \times \text{Width}$
>
> or
>
> $$\text{Volume} = \text{Length} \times \text{Width} \times \text{Height}$$

EXAMPLE 10-18

Calculate the volume of a rectangular planter box that measures 5 ft long by 1 ft tall by 1.5 ft wide.

Step 1

Insert the dimensions into the equation.

$$\text{Volume} = \text{Length} \times \text{Width} \times \text{Height}$$
$$\text{Volume} = 5.0\,\text{ft} \times 1.5\,\text{ft} \times 1.0\,\text{ft}$$

Step 2

Solve the equation.

$$\text{Volume} = 5.0\,\text{ft} \times 1.5\,\text{ft} \times 1.0\,\text{ft}$$
$$\text{Volume} = 7.5\,\text{ft}^3$$

SOLUTION

The volume of the rectangular planter box is 7.5 ft³.

Cylindrical Container

For a cylindrical container, the base is a circle and the height is the tall dimension of the container.

> **◄ Definition ►** *Volume of a Cylindrical Container*
>
> *Volume of a cylindrical container* is the product of the area of the circular base and the height of the container.
>
> $$\text{Volume} = \text{Base} \times \text{Height}$$
>
> where Base $= \pi r^2$ and $r =$ radius of container base
>
> or
>
> $$\text{Volume} = \pi r^2 \times \text{Height}$$

EXAMPLE 10-19

Calculate the volume of a large, round interior planter that measures 5 ft in diameter and is 3.5 ft tall.

Step 1

Calculate the radius of the planter.

$$\text{Radius} = \tfrac{1}{2} \text{ Diameter}$$
$$\text{Radius} = \tfrac{1}{2} \text{ (5 ft)}$$
$$\text{Radius} = 2.5 \text{ ft}$$

Step 2

Insert the dimensions into the equation.

$$\text{Volume} = \pi r^2 \times \text{Height}$$
$$\text{Volume} = \pi \ (2.5 \text{ ft})^2 \times 3.5 \text{ ft}$$

Step 3

Solve the equation.

$$\text{Volume} = \pi (2.5 \text{ ft})^2 \times 3.5 \text{ ft}$$
$$\text{Volume} = 68.72 \text{ ft}^3$$

SOLUTION

The volume of the large, cylindrical interior container is 68.72 ft³.

Containers with Tapered Sides

Containers with tapered sides include traditional standard pots, azalea pots, and bulb pans used in greenhouse operations. Nursery containers also may have tapered sides and vary in size from containers used to grow herbaceous perennials to those used to grow larger nursery stock. Occasionally, large planters have tapered sides as well. Containers with tapered sides may be round or square in cross-section. The width of the cross-section of containers with tapered sides changes from the widest dimension at the top to the narrowest dimension at the bottom. The formula to calculate the volume of these containers uses the average of the area of the top and the bottom.

Round Container with Tapered Sides

Round containers with tapered sides have a round top and bottom, and the diameter decreases from top to bottom.

> **◆ Definition ◆** *Volume of a Round, Tapered Container*
> **Volume of a round, tapered container** is the product of the average of the area of the top and the area of the bottom and the height of the container.
>
> $$\text{Volume} = \text{Base} \times \text{Height}$$
>
> $$\text{where Base} = \frac{\text{Area}_{Top} + \text{Area}_{Bottom}}{2}$$
>
> or
>
> $$\text{Volume} = \left(\frac{\pi r^2_{Top} + \pi r^2_{Bottom}}{2}\right) \times \text{Height}$$

EXAMPLE 10-20

Calculate the volume of a flower pot that measures 7.5 cm tall, 7.5 cm diameter at the top, and 5 cm diameter at the bottom.

Step 1

Calculate the radius of the top of the pot.

$$\text{Radius} = \tfrac{1}{2} \text{ Diameter}$$

$$\text{Radius} = \tfrac{1}{2}(7.5\,\text{cm})$$

$$\text{Radius} = 3.75\,\text{cm}$$

Calculate the radius of the bottom of the pot.

$$\text{Radius} = \tfrac{1}{2} \text{ Diameter}$$

$$\text{Radius} = \tfrac{1}{2}(5\,\text{cm})$$

$$\text{Radius} = 2.5\,\text{cm}$$

Step 2

Insert the dimensions into the equation.

$$\text{Volume} = \left(\frac{\pi r_{\text{Top}}^2 + \pi r_{\text{Bottom}}^2}{2}\right) \times \text{Height}$$

$$\text{Volume} = \left(\frac{\pi(3.75\,\text{cm})_{\text{Top}}^2 + \pi(2.50\,\text{cm})_{\text{Bottom}}^2}{2}\right) \times 7.5\,\text{cm}$$

Step 3

Solve the equation.

$$\text{Volume} = \left(\frac{\pi(3.75\,\text{cm})_{\text{Top}}^2 + \pi(2.50\,\text{cm})_{\text{Bottom}}^2}{2}\right) \times 7.5\,\text{cm}$$

$$\text{Volume} = \left(\frac{44.18\,\text{cm}^2 + 19.63\,\text{cm}^2}{2}\right) \times 7.5\,\text{cm}$$

$$\text{Volume} = 31.905\,\text{cm}^2 \times 7.5\,\text{cm}$$

$$\text{Volume} = 239.29\,\text{cm}^3$$

SOLUTION

The volume of the round, tapered flower pot is 239.29 cm^3.

Square Container with Tapered Sides

Square containers with tapered sides first came to the industry after the development of the round plastic pot. Square pots are more space efficient when placed pot to pot and fit nicely into flats. Today, many configurations and uses can be found for square, tapered pots.

> **◀ Definition ▶** ***Volume of a Square, Tapered Container***
>
> ***Volume of a square, tapered container*** is the product of the average of the area of the top and the area of the bottom and the height of the container.
>
> $$\text{Volume} = \text{Base} \times \text{Height}$$
>
> $$\text{where Base} = \frac{\text{Area}_{\text{Top}} + \text{Area}_{\text{Bottom}}}{2}$$
>
> or
>
> $$\text{Volume} = \left(\frac{\text{Side}_{\text{Top}}^2 + \text{Side}_{\text{Bottom}}^2}{2}\right) \times \text{Height}$$

EXAMPLE 10-21

Calculate the volume of a container used to produce herbaceous perennials that measure 4.375 in. by 4.375 in. on the top, 3 in. by 3 in. on the bottom, and stands 4.75 in. tall.

Step 1
Insert the dimensions into the equation.

$$\text{Volume} = \left(\frac{\text{Side}^2_{\text{Top}} + \text{Side}^2_{\text{Bottom}}}{2}\right) \times \text{Height}$$

$$\text{Volume} = \left(\frac{4.375^2 \text{ in.}_{\text{Top}} + 3^2 \text{ in.}_{\text{Bottom}}}{2}\right) \times 4.75 \text{ in.}$$

Step 2
Solve the equation.

$$\text{Volume} = \left(\frac{4.375^2 \text{ in.}_{\text{Top}} + 3^2 \text{ in.}_{\text{Bottom}}}{2}\right) \times 4.75 \text{ in.}$$

$$\text{Volume} = \left(\frac{19.14 \text{ in.}^2 + 9 \text{ in.}^2}{2}\right) \times 4.75 \text{ in.}$$

$$\text{Volume} = 14.07 \text{ in.}^2 \times 4.75 \text{ in.}$$

$$\text{Volume} = 66.83 \text{ in.}^3$$

SOLUTION

The volume of the square, tapered container is 66.83 in.^{3}.

Practice Problem Set 10-11   Calculating the Volume of a Growing Container

1. Calculate the volume of a cube-shaped free-standing planter that measures 2 ft^2.
2. Calculate the volume of an L-shaped planter in a shopping mall interior that measures 1.5 ft deep and 4 ft wide. One arm of the L measures 15 ft long and the other arm of the L measures 25 ft long.
3. Cylindrical free-standing containers that measure 8 ft in diameter and 3.5 ft deep are found in an indoor courtyard. What is the volume of each of these containers?
4. Calculate the volume of a round, tapered container that measures 10 in. in diameter at the top, 7 in. in diameter at the bottom, and 5.5 in. tall.
5. Calculate the volume of a square, tapered container that measures 6.5 in. square on the top, 5 in. square on the bottom, and 6.5 in. tall.

Calculating Root Medium Volume Requirements

Once the volume of a single container is determined, the volume of root medium required to fill a given number of containers can be determined. The volume of common growing containers produced by a manufacturer given in U.S. customary units and in metric units can be found in most greenhouse supply catalogs. If volume is not listed in a supply catalog, most container manufacturers maintain web sites that provide specifications for each of their products. Note that each container is unique and that the volume of a six-inch azalea pot from Manufacturer A can be different than the volume of a six-inch azalea pot from Manufacturer B.

Root media manufacturers distribute their product in several forms. Loose-filled bags, loose-filled bulk bags, compressed bales, and bulk-compressed bales are the most common forms. The volume of root media found in loose-filled bags and some bulk bags is printed on the bag, and this information can be used to determine how much media needs to be ordered for a given crop. Compressed bales are described in two ways. The volume of the bale in the compressed state is often printed on the bale and is used in its description. When loosened, a compressed bale provides a greater volume of media than the volume that is printed on the bag. Each manufacturer will state in product specifications the expected volume of media resulting from loosening a compressed bale either as a compression ratio (2:1 or 2.25:1) or as a bale yield. This number should be used to determine the number of bales that need to be ordered for a project.

Calculating How Much Root Media Is Needed and How Many Bags or Bales to Order

To calculate the amount of root media required for a crop, the parameters first need to be defined. Define the volume of the container, the number of containers to be filled, and the source/form of root media used.

EXAMPLE 10-22

A greenhouse grower is planning a crop of 5,000 pot mums that are to be grown in 6.5 in. azalea pots. If each pot has a volume of 1.71 L, how much root media is required to fill 5,000 pots? The grower purchases root medium in 70 ft^3 bulk bags. How many bulk bags are needed for this crop?

Step 1

Determine the total volume of root media needed in liters by multiplying the number of pots by the volume of each pot.

$$5{,}000 \text{ pots} \times 1.71 \frac{\text{L}}{\text{pot}} = 8{,}550 \text{ L}$$

SOLUTION

A volume of 8,550 L of root medium is needed to fill five thousand 6.5 in. azalea pots.

Step 2

Convert liters of root medium to cubic feet of root medium.

Set up a proportion.

$$\frac{28.316846592 \text{ L}}{1 \text{ ft}^3} = \frac{8{,}550 \text{ L}}{x \text{ ft}^3}$$

Isolate and solve for x.

$$28.316846592x = 8{,}550$$

$$x = \frac{8{,}550}{28.316846592}$$

$$x = 301.94 \text{ ft}^3$$

SOLUTION

A volume of 301.94 ft³ of root medium is needed to fill five thousand 6.5 in. azalea pots.

Step 3
Determine how many 70 ft³ bulk bags are equivalent to 301.94 ft³.

Set up a proportion.

$$\frac{1 \text{ bulk bag}}{70 \text{ ft}^3} = \frac{x \text{ bulk bags}}{301.94 \text{ ft}^3}$$

Isolate and solve for x.

$$70x = 301.94$$

$$x = \frac{301.94}{70}$$

$$x = 4.3 \text{ bulk bags}$$

SOLUTION

A total of 4.3 bulk bags (70 ft³ each) are required for this crop.

EXAMPLE 10-23

In the situation above, the grower needs 301.94 ft³ of root medium for the crop. Instead of using bulk bagged mix, the grower chose to use a 55 ft³ compressed bale that has a compression ratio of 2.25:1. How many 55 ft³ bulk bales are required for this crop?

Step 1
Determine the bale yield for a 55 ft³ bulk bale with a compression ratio of 2.25:1.

Set up a proportion.

$$\frac{1}{2.25} = \frac{55 \text{ ft}^3}{x \text{ ft}^3}$$

Isolate and solve for x.

$$x = 55 \times 2.25$$
$$x = 123.75 \, \text{ft}^3$$

SOLUTION

The bale yield is 123.75 ft³ of root medium from a 55 ft³ compressed bale with a compression ratio of 2.25:1.

Step 2

Determine the number of bales equivalent to 301.94 ft³.
 Set up a proportion.

$$\frac{1 \, \text{bale}}{123.75 \, \text{ft}^3} = \frac{x \, \text{bales}}{301.94 \, \text{ft}^3}$$

Isolate and solve for x.

$$123.75x = 301.94$$
$$x = \frac{301.94}{123.75}$$
$$x = 2.4 \, \text{bales}$$

SOLUTION

A total of 2.4 bales is equivalent to 301.94 ft³.

Practice Problem Set 10-12 — Calculating Root Medium Volume Requirements

1. A nursery manager is planning a crop of 3,000 garden mums grown in 2.9 L mum pans. How many 3.8 ft³ compressed bales that yield 7.5 ft³ each need to be ordered for this crop?
2. A greenhouse manager is planning a crop of 17,500 flats of petunias grown in flat inserts that hold 0.1667 ft³ of root media each. How many 55 ft³ loose bulk bags are needed for this crop?
3. An interior landscape manager needs to fill 17 round planters measuring 6 ft in diameter and 3 ft tall with root media. How many 60 ft³ bulk bags of root media are needed for this project?
4. A club house manager at a country club needs to fill 25 planter boxes on a patio deck that measures 18 in. by 18 in. by 6 ft long. How many 3 ft³ bags of root medium are required to fill all of the planter boxes?

Formulating Root Media

Although premixed root media are available in a variety of formulations, occasionally a greenhouse or nursery manager may need to prepare customized root media for crops with special needs. All root media are formulated and prepared by measuring the components on a volume basis. A listing of example formulations is found in Table 10-1.

TABLE 10-1 • SEVERAL ROOT MEDIUM FORMULAE AND THEIR USE

Medium Components					Use
1 vermiculite	1 peat moss				Propagation Mix
	1 peat moss		1 perlite		Propagation Mix
	1 peat moss			1 sand	Propagation Mix
1 vermiculite	2 peat moss		1 perlite		Container Mix
	3 peat moss		1 perlite		Container Mix
1 vermiculite		1 pine bark			Container Mix
2 vermiculite		2 pine bark	1 perlite		Container Mix
2 vermiculite	1 peat moss	1 pine bark	1 perlite		Container Mix
	1 peat moss	3 pine bark		1 sand	Container Mix
	1 peat moss	3 hardwood bark		1 sand	Container Mix
1 rock wool	1 peat moss				Container Mix
3 rock wool	7 peat moss				Container Mix
1 coir	2 peat moss				Container Mix

Adapted from Nelson, P.V. (2003). *Greenhouse Operation and Management, 6th edition*. Pearson Education, Inc., Upper Saddle River: New Jersey.

Various proportions of medium components are combined to create a root medium formulation. For example, a common germination mix is composed of one part vermiculite and one part peat moss. To produce two cubic feet of this formula, a grower would combine one cubic foot of vermiculite with one cubic foot of peat moss. If a grower needed 3 cubic feet of this 1 to 1 (1:1) vermiculite to peat moss mix, then 1.5 cubic feet of peat moss would be combined with 1.5 cubic feet of vermiculite. As the complexity of the root medium formula increases, so does the complexity of the mathematical problem.

EXAMPLE 10-24

A greenhouse manager needs to mix 2 yd³ of a root medium that contains 2 parts vermiculite, 1 part peat moss, 1 part pine bark, and 1 part perlite. How many cubic feet of each component are required for 2 yd³ of this mix?

Step 1

Convert 2 yd³ to the equivalent in cubic feet.

Set up a proportion.

$$\frac{27 \text{ ft}^3}{1 \text{ yd}^3} = \frac{x \text{ ft}^3}{2 \text{ yd}^3}$$

Isolate and solve for x.

$$x = 27 \times 2$$
$$x = 54 \text{ ft}^3$$

> **SOLUTION**

Two cubic yards are equivalent to 54 ft^3.

Step 2

Determine how many units are found in the formulation by adding all the parts of the formula.

2 vermiculite + 1 peat moss + 1 pine bark + 1 perlite = 5 units in the formula

Step 3

Determine the volume of each unit by dividing the total volume of mix by the number of units.

$$54 \text{ ft}^3 \text{ of mix} \div 5 \text{ units} = 10.8 \frac{\text{ft}^3}{\text{unit}}$$

Step 4

Calculate the volume of each component required for this mix.

$$2 \text{ units vermiculite} \times 10.8 \text{ ft}^3/\text{unit} = 21.6 \text{ ft}^3 \text{ vermiculite}$$
$$1 \text{ unit peat moss} \times 10.8 \text{ ft}^3/\text{unit} = 10.8 \text{ ft}^3 \text{ peat moss}$$
$$1 \text{ unit pine bark} \times 10.8 \text{ ft}^3/\text{unit} = 10.8 \text{ ft}^3 \text{ pine bark}$$
$$1 \text{ unit perlite} \times 10.8 \text{ ft}^3/\text{unit} = 10.8 \text{ ft}^3 \text{ perlite}$$

Step 5

Find the sum of the volumes of each component as a check.

$$21.6 + 10.8 + 10.8 + 10.8 = 54$$

> **SOLUTION**

The total is 54, and this agrees with the total volume desired of 54 ft^3 of medium.

> **EXAMPLE 10-25**

How much will the 40 ft^3 of a 1:2:1 mix of vermiculite:peat moss:perlite cost based on the values in Table 10-2?

TABLE 10-2 • PRICING AND SIZING OF VARIOUS ROOT MEDIUM COMPONENTS

Component	Size	Price
Sphagnum peat moss	3.8 ft^3 compressed bale 2:1 compression ratio	$9.35/bale
Perlite	6 ft^3 bag	$16.49/bag
Vermiculite	6 ft^3 bag	$17.39/bag

Step 1
Determine how many units are found in the formulation by adding all the parts of the formula.

$$1 \text{ vermiculite} + 2 \text{ peat moss} + 1 \text{ perlite} = 4 \text{ units in the formula}$$

Step 2
Determine the volume of each unit by dividing the total volume of mix desired by the number of units.

$$40 \text{ ft}^3 \text{ of mix} \div 4 \text{ units} = 10 \frac{\text{ft}^3}{\text{unit}}$$

Step 3
Calculate the volume of each component required for this mix.

$$1 \text{ unit vermiculite} \times 10 \text{ ft}^3/\text{unit} = 10 \text{ ft}^3 \text{ vermiculite}$$

$$2 \text{ units peat moss} \times 10 \text{ ft}^3/\text{unit} = 20 \text{ ft}^3 \text{ peat moss}$$

$$1 \text{ unit perlite} \times 10 \text{ ft}^3/\text{unit} = 10 \text{ ft}^3 \text{ perlite}$$

Step 4
Calculate the cost of each component by first determining the cost per cubic foot of each component.

Step 4a: Vermiculite
Vermiculite is sold in 6 ft^3 bags and costs $17.39/bag.

How much does each cubic foot of vermiculite cost?

Set up a proportion.

$$\frac{6 \text{ ft}^3}{\$17.39} = \frac{1 \text{ ft}^3}{\$x}$$

Isolate and solve for x.

$$6x = 17.39$$
$$x = \frac{17.39}{6}$$
$$x = 2.90$$

SOLUTION

Each cubic foot of vermiculite costs $2.90.

Step 4b: Peat Moss
Peat moss is sold in 3.8 ft^3 compressed bales with a compression ratio of 2:1. That means each bale yields 7.6 ft^3. Confirm this by setting up a proportion, as follows:

$$\frac{1}{2} = \frac{3.8 \text{ ft}^3}{x \text{ ft}^3}$$

Isolate and solve for x.

$$x = 3.8 \times 2$$
$$x = 7.6$$

If 7.6 ft^3 of peat moss costs \$9.35, then how much does 1 ft^3 of peat moss cost? Set up a proportion.

$$\frac{7.6 \text{ ft}^3}{\$9.35} = \frac{1 \text{ ft}^3}{\$x}$$

Isolate and solve for x.

$$7.6x = 9.35$$
$$x = \frac{9.35}{7.6}$$
$$x = 1.23$$

SOLUTION

Each cubic foot of peat moss costs \$1.23.

Step 4c: Perlite

Perlite is sold in 6 ft^3 bags and costs \$16.49/bag.
How much does each cubic foot of perlite cost? Set up a proportion.

$$\frac{6 \text{ ft}^3}{\$16.49} = \frac{1 \text{ ft}^3}{\$x}$$

Isolate and solve for x.

$$6x = 16.49$$
$$x = \frac{16.49}{6}$$
$$x = 2.75$$

SOLUTION

Each cubic foot of perlite costs \$2.75.

Step 5

Calculate the cost of each component required for this mix.

$$10 \text{ ft}^3 \text{ vermiculite} \times \$2.90/\text{ft}^3 = \$29.00$$
$$20 \text{ ft}^3 \text{ peat moss} \times \$1.23/\text{ft}^3 = \$24.60$$
$$10 \text{ ft}^3 \text{ perlite} \times \$2.75/\text{ft}^3 = \$27.50$$

To calculate the total cost of the mix, find the sum of the cost of each component.

$$\$29.00 + \$24.60 + \$27.50 = \$81.10$$

SOLUTION

The cost to formulate 40 ft^3 of a 1:2:1 mix of vermiculite:peat moss:perlite is \$81.10.

Practice Problem Set 10-13 — *Formulating Root Media*

A greenhouse manager is planning a crop of 450 pots of hydrangeas. They are to be grown in 8 in. azalea pots (0.1429 ft^3/pot) filled with a root medium containing 2 parts vermiculite, 2 parts peat moss, and 1 part perlite.

1. What volume of root medium is needed for this crop?
2. How many cubic feet of each root medium component are required?
3. How much will each component cost (use Table 10-2)?
4. What is the total cost of materials for the root medium for this crop?

Incorporation of Fertilizers into Root Medium During Formulation

When a greenhouse or nursery manager elects to produce root media in-house, rather than buy prepared root media, the incorporation of fertilizers into the medium is a common practice. This allows for a pH adjustment of media that contain acid-reacting components, such as *Sphagnum* peat moss and pine bark; allows for the incorporation of an initial nutrient charge; and allows for the incorporation of micronutrients necessary for the production of a quality crop grown in a soil-less mix. Tables 10-3 and 10-4 provide common fertilizer additions and rates used in the preparation of root media.

TABLE 10-3 • APPLICATION RATES OF FERTILIZERS CONTAINING MACRONUTRIENTS THAT ARE ADDED TO ROOT MEDIA DURING FORMULATION

Product	Rate lb/yd^3	N	P	K	Ca	Mg	S
Calcium nitrate (15.5-0-0-20Ca)	1	X			X		
Potassium nitrate (13-0-44)	1	X		X			
Superphosphate (0-45-0)	2.25		X				X
Dolomitic Limestone CaMg(CO$_3$)$_2$ (To increase pH, add calcium and magnesium)	10				X	X	
Gypsum CaSO$_4$ •2(H$_2$O) (To provide calcium and is often combined with Epsom salt—see below.)	0-5				X		X
Epsom Salt MgSO$_4$ •7H$_2$O (To provide magnesium and is often combined with Gypsum at the rate of 0-5 lb/yd^3.)	0-1					X	X
Gypsum CaSO$_4$ •2(H$_2$O) (To provide sulfur)	1.5				X		X

Adapted from Nelson, P.V. (2003). *Greenhouse Operation and Management, 6th edition*. Pearson Education, Inc., Upper Saddle River: New Jersey.

TABLE 10-4 • APPLICATION RATES OF FERTILIZERS CONTAINING MICRONUTRIENTS THAT ARE ADDED TO ROOT MEDIA DURING FORMULATION

Product*	Rate lb/yd^3	N	P	K	Ca	Mg	S	Fe	Mn	Zn	Cu	Bo	Mo
Uni-Mix® 11-5-11	2	X	X	X	X	X	X	X	X	X	X	X	X
Micromax® Plus 0-4-0	10		X		X	X	X	X	X	X	X	X	X
Micromax® Granular	1–2				X	X	X	X	X	X	X	X	X
Micronutrients LG (Large Granular)	1–1.5				X	X	X	X	X	X	X	X	X
Step® HiMag	1.5–3					X	X	X	X	X	X		

*Reference to commercial products or trade names is made with the understanding that no discrimination is intended and no endorsement by the authors or publisher is implied.
Products of The Scotts Company LLC, Marysville, OH.

EXAMPLE 10-26

A grower needs to prepare 3.5 yd^3 of a soilless mix. The grower plans to add dolomitic limestone, superphosphate, Micromax® (1.5 lb/yd^3), calcium nitrate, and potassium nitrate to the mix. How many pounds of each of the fertilizer products should be incorporated into 3.5 yd^3 of this mix?

Step 1a

Refer to Table 10-3 and find the recommended rate of required dolomitic limestone.
 10 lb of dolomitic limestone are required for every cubic yard of root medium.

Step 2a

Set up a proportion to find out how many pounds of dolomitic limestone are required for 3.5 yd^3 of root medium.

$$\frac{10 \text{ lb dolomitic limestone}}{1 \text{ yd}^3 \text{ of medium}} = \frac{x \text{ lb of dolomitic limestone}}{3.5 \text{ yd}^3 \text{ of medium}}$$

Step 3a

Isolate and solve for x.

$$x = 35$$

SOLUTION A

Thirty-five pounds of dolomitic limestone are required for 3.5 yd^3 of root medium.

Step 1b

Refer to Table 10-3 and find the recommended rate of required superphosphate.
 2.25 lb of superphosphate are required for every cubic yard of root medium.

Step 2b

Set up a proportion to find out how many pounds of superphosphate are required for 3.5 yd³ of root medium.

$$\frac{2.25 \text{ lb superphosphate}}{1 \text{ yd}^3 \text{ of medium}} = \frac{x \text{ lb of superphosphate}}{3.5 \text{ yd}^3 \text{ of medium}}$$

Step 3b

Isolate and solve for x.

$$x = 2.25 \times 3.5$$
$$x = 7.88$$

SOLUTION B

7.88 lb of superphosphate are required for 3.5 yd³ of root medium.

Step 1c

Review the question to determine how many pounds of Micromax® are required per cubic yard of root medium.

1.5 lb of Micromax® are required for every cubic yard of root medium.

Step 2c

Set up a proportion to find out how many pounds of Micromax® are required for 3.5 yd³ of root medium.

$$\frac{1.5 \text{ lb Micromax}^{\circledR}}{1 \text{ yd}^3 \text{ of medium}} = \frac{x \text{ lb Micromax}^{\circledR}}{3.5 \text{ yd}^3 \text{ of medium}}$$

Step 3c

Isolate and solve for x.

$$x = 1.5 \times 3.5$$
$$x = 5.25$$

SOLUTION C

5.25 lb of Micromax® are required for 3.5 yd³ of root medium.

Step 1d

Refer to Table 10-3 and find the recommended rate of required calcium nitrate and potassium nitrate.

One pound each of calcium nitrate and potassium nitrate are required for every cubic yard of root medium.

STEP 2D

Set up a proportion to find out how many pounds of calcium nitrate or potassium nitrate are required for 3.5 yd³ of root medium.

$$\frac{1 \text{ lb calcium nitrate or potassium nitrate}}{1 \text{ yd}^3 \text{ of medium}} = \frac{x \text{ lb calcium nitrate or potassium nitrate}}{3.5 \text{ yd}^3 \text{ of medium}}$$

STEP 3D
Isolate and solve for x.
$$x = 3.5$$

SOLUTION D
A total of 3.5 lb each of calcium nitrate and potassium nitrate are required for 3.5 yd^3 of root medium.

Practice Problem Set 10-14 — Incorporation of Fertilizers into Root Media during Formulation

1. A greenhouse manager plans to produce 500 flats of bedding plants using a 1206 flat insert. Each 1206 flat insert holds 0.1087 ft^3 of mix. How much soilless root medium is required? How much dolomitic limestone and Uni-Mix® are required to add to this volume of root medium?

2. A nursery manager is preparing 8 yd^3 of root medium. How much dolomitic limestone, superphosphate, micronutrients LG (1.5 lb/yd^3 rate), calcium nitrate, and potassium nitrate need to be added to this volume of root medium?

Monitoring Crop Nutritional Status: Electrical Conductivity

Electrical conductivity is the measure of total dissolved salts in a solution. Growers use this measure to monitor the nutritional status of a crop through root medium analysis. The method used to extract solution from root medium varies and is not within the scope of this publication. When units for values (reported by testing labs), grower references, and conductivity meters differ, it is helpful to understand the relationship between those units. This publication can bring clarity to the different types of units that are used to report electrical conductivity.

> **Definition — Electrical Conductivity (EC)**
> *Electrical conductivity* is quality or power of a substance that conducts an electrical current. Electrical conductivity is the inverse or reciprocal ($\frac{1}{x}$) of electrical resistance.

Electrical conductivity meters measure the ability of a solution to conduct a harmless electrical current between two poles. The two poles are spaced one centimeter apart and are found within a probe that is submerged in the solution. The higher the level of soluble salts in the solution, the faster the electrical current moves from one pole to the other and the higher the EC reading. More soluble salts in solution equals

Monitoring Crop Nutritional Status: Electrical Conductivity 273

a higher EC level. The most common unit used for reporting EC is millisiemens per centimeter (mS/cm). An alternate unit of measure of EC is millimhos per centimeter (mmhos/cm) and is often the unit of measure for older electrical conductivity meters. A third unit of measure that is used in scientific publications and by some testing labs is decisiemens per meter (dS/m). All three of these units of measure are equivalent. The equations below describe these relationships. Also see Table A-1 in the Appendix A for a refresher on metric prefixes.

$$1.00 \text{ mS/cm} = 1.00 \text{ mmhos/cm}$$

and

$$\frac{\text{mS}}{\text{cm}} = \frac{10^{-3}\,\text{S}}{\text{cm}}, \frac{10^{-3}\,\text{S}}{\text{cm}} \times \frac{100}{100} = \frac{10^{-1}\,\text{S}}{\text{m}} = \frac{\text{dS}}{\text{m}}$$

$$1.00 \text{ mS/cm} = 1.00 \text{ dS/m}$$

Occasionally a laboratory will report results in either mhos $\times 10^{-5}$/cm or micromhos per centimeter (μmhos/cm). These labs prefer to report results in whole numbers rather than in decimal fractions. The following equivalents demonstrate how this is done:

A result of 1.85 mmhos/cm can be reported as: 185 mhos $\times 10^{-5}$/cm. It is derived this way:

$$\frac{1.85 \text{ mmhos}}{\text{cm}} = \frac{[1.85 \text{ mhos}(\times 100)] \times [10^{-3}(\div 100)]}{\text{cm}}$$

$$= \frac{185 \text{ mhos} \times 10^{-5}}{\text{cm}}$$

A result of $\frac{1.85 \text{ mmhos}}{\text{cm}}$ can be reported as: $\frac{1{,}850\,\mu\text{mhos}}{\text{cm}}$.
It is derived this way:

$$\frac{1.85 \text{ mmhos}}{\text{cm}} = \frac{[1.85 \text{ mhos}(\times 1{,}000)] \times [10^{-3}(\div 1{,}000)]}{\text{cm}}$$

$$= \frac{1{,}850 \text{ mhos} \times 10^{-6}}{\text{cm}} = \frac{1{,}850\,\mu\text{mhos}}{\text{cm}}$$

EXAMPLE 10-27

A laboratory report indicates the average electrical conductivity of a poinsettia crop is 285 mhos $\times 10^{-5}$/cm. The grower's reference material indicates that the target electrical conductivity range is 2.8 to 4.1 mS/cm. Is the crop's electrical conductivity within the target range?

Step 1
Convert 285 mhos $\times 10^{-5}$/cm to millisiemens per centimeter.

$$\text{mS/cm} = \text{mmhos/cm}$$

Therefore, converting to millimhos per centimeter is equivalent to converting to millisiemens per centimeter.

$$\frac{[285 \text{ mhos}(\div 100)] \times [10^{-5}(\times 100)]}{\text{cm}} = \frac{2.85 \text{ mhos} \times 10^{-3}}{\text{cm}}$$

$$= \frac{2.85 \text{ mmhos}}{\text{cm}} = \frac{2.85 \text{ mS}}{\text{cm}}$$

SOLUTION

The crop electrical conductivity is 2.85 mS/cm, which is within the target range of 2.8 to 4.1 mS/cm.

Finally, electrical conductivity results may also be reported as total dissolved solids (TDS). The units for reporting TDS are parts per million (ppm). The relationship between parts per million and millisiemens per centimeter is dependent on the nature of the dissolved fertilizer salts and is given as a range of 640 to 700 ppm to every millisiemens per centimeter. Using the average of the range, an equivalent of 670 ppm to every 1.0 mS/cm can be used. This conversion is an estimate and reporting electrical conductivity in millisiemens per centimeter or millimhos per centimeter is preferred.

EXAMPLE 10-28

A grower has a target electrical conductivity range of 2.8 to 4.1 mS/cm for a poinsettia crop. An EC meter that reports total dissolved solids (TDS) in parts per million is available for use until the grower can obtain an EC meter that reports in millisiemens per centimeter. If the grower uses 670 ppm equivalent to 1.0 mS/cm, what is the target range in parts per million?

Step 1

Convert the range 2.8 to 4.1 mS/cm to parts per million.

Set up a proportion for 2.8 mS/cm.

$$\frac{1.0 \text{ mS/cm}}{670 \text{ ppm}} = \frac{2.8 \text{ mS/cm}}{x \text{ ppm}}$$

Isolate and solve for x.

$$x = 2.8 \times 670$$

$$x = 1,876 \text{ ppm}$$

Set up a proportion for 4.1 mS/cm.

$$\frac{1.0 \text{ mS/cm}}{670 \text{ ppm}} = \frac{4.1 \text{ mS/cm}}{x \text{ ppm}}$$

Isolate and solve for x.

$$x = 4.1 \times 670$$

$$x = 2,747 \text{ ppm}$$

SOLUTION

The range of 2.8 to 4.1 mS/cm is equivalent to 1,876 to 2,747 ppm using an EC meter that reads the total dissolved solids in parts per million.

✏ Practice Problem 10-15 ✏ Monitoring Crop Nutritional Status: Electrical Conductivity

Fill in Table 10-5 by converting the electrical conductivity measurements.

TABLE 10-5 •

mmhos/cm	mhos × 10^{-5}/cm	mS/cm	µmhos/cm	dS/m
4.12 mmhos/cm				
	183 mhos × 10^{-5}/cm			
		2.35 mS/cm		
			2500 µmhos/cm	
				1.95 dS/m

Monitoring Light Intensity

Light intensity in either modified or controlled environments can be affected by many factors. Glazing materials, supplemental lighting, shading compound, shade cloth, and shade screens all affect the light levels in a greenhouse. The greenhouse manager often manipulates the last three to modify light intensity so that it is appropriate for the type and age of a crop. In addition, overhead obstructions such as greenhouse structural members and hanging baskets could impact greenhouse light levels. In modified environments, such as cold frames and shade houses, the selection of the covering material will impact light levels within. A nursery manager will select a shade-house shading fabric with a specific percentage shade rating to create an environment appropriate for the crop. Cold frames and unheated greenhouse structures used for over-wintering plants often are covered with white, rather than clear polyethylene, to limit light transmittance. Interior environment such as offices or commercial shopping malls are also types of controlled environments where plants are maintained. To select the proper plants for interior environments, interior landscapers match interior light intensity readings to light intensity recommendations for each plant species.

Light intensity recommendations are usually made in footcandle (fc) units. Some international references provide recommendations in lux, which is a metric unit. Light meters report light intensity levels in footcandles, lux (lx), or kilolux (klx). These are photometric units or measures of illuminance.

❧ Definition ❧ *Footcandle (fc)*

$$\text{Footcandle} = \text{one lumen per square foot}$$

❧ Definition ❧

LUX (lx)

$$\text{Lux} = \text{one lumen per square meter}$$

KILOLUX (klx)

$$\text{Kilolux} = 1{,}000 \text{ lux}$$

Light meters, plant production references, and indoor plant references may report light intensity in any of the above units, and a manager must be able to easily convert between them.

❧ Definition ❧ *Equivalents for Units of Irradiance*

$$1 \text{ footcandle} = 10.764 \text{ lux}$$
$$1 \text{ lux} = 0.0929 \text{ footcandle}$$
$$1 \text{ kilolux} = 1{,}000 \text{ lux}$$
$$1 \text{ kilolux} = 92.9 \text{ footcandles}$$
$$1{,}000 \text{ footcandles} = 10.764 \text{ kilolux}$$

EXAMPLE 10-29

An initial light level of 4,500 fc is recommended for rooted poinsettia cuttings that are transplanted into pots. The greenhouse manager has a light meter that reports light levels in kilolux units. Convert 4,500 fc to klx.

Step 1

Set up a proportion using the equivalents above.

$$\frac{10.764 \text{ klux}}{1{,}000 \text{ fc}} = \frac{x \text{ klux}}{4{,}500 \text{ fc}}$$

Step 2

Isolate and solve for x.

$$1{,}000x = (4{,}500)(10.764)$$
$$x = \frac{(4{,}500)(10.764)}{1{,}000}$$
$$x = 48.44 \text{ klx}$$

SOLUTION

A light level of 4,500 fc is equivalent to 48 klx.

⌾ Practice Problem Set 10-16   ⌾ Monitoring Light Intensity

Convert each of the following light intensity values:

1. Convert 54 klx to footcandles.
2. Convert 6,000 fc to kilolux.
3. Convert 48 klx to lux.
4. Convert 50 fc to lux.
5. Convert 3,000 lx to footcandles.

Calculations Involving Pesticides and Plant Growth Regulators

Floriculture and nursery crop pesticide and plant growth regulator treatments are often made via spray and drench applications. Specific spray or drench volumes are recommended in addition to a prescribed product rate. As a result, additional mathematical work is needed to correctly apply these materials. Some examples of these calculations follow.

Foliar Spray Applications with Recommended Spray Volumes

The efficacy of pesticide and plant growth regulators varies with the volume of spray solution used per unit area. Usually plants with well-developed canopies require higher spray volumes. In addition, spray volume recommendations vary due to the targeted site of absorption. Lower spray volumes, in the form of a light overhead spray application, target the foliage in foliar-absorbed products. Higher spray volumes are needed to penetrate foliage, if the stems are the sites of absorption. Products that are applied as a foliar spray but also have the root or crown as a site of absorption are carefully metered. Excess spray volume, intended as a foliar application, can move into the crown or the root system and may result in phytotoxicity.

EXAMPLE 10-30

A greenhouse manager needs to spray 14,000 ft^2 of bedding plants with a plant growth regulator at the rate of 1,000 ppm using 2.5 quarts of spray solution for every 100 ft^2 of crop. It takes 32 ml of product to prepare every gallon of 1,000 ppm concentration spray solution. How many gallons of spray solution are required for this crop? How many milliliters of product are required to prepare this amount of spray solution?

Step 1

Determine the volume of spray solution needed for this crop if the rate is 2.5 qt per 100 ft² and there are 14,000 ft² of the crop.

Set up a proportion.

$$\frac{2.5 \text{ qt spray solution}}{100 \text{ ft}^2} = \frac{x \text{ qt spray solution}}{14,000 \text{ ft}^2}$$

Isolate and solve for x.

$$x \times 100 = 14,000 \times 2.5$$

$$\frac{x \times 100}{100} = \frac{14,000 \times 2.5}{100}$$

$$x = \frac{14,000 \times 2.5}{100}$$

$$x = 350 \text{ qt}$$

A total of 350 quarts of spray solution are needed for this crop.
Convert the volume from quarts to gallons by setting up a proportion.

$$\frac{1 \text{ gal}}{4 \text{ qt}} = \frac{x \text{ gal}}{350 \text{ qt}}$$

Isolate and solve for x.

$$x \times 4 = 350$$

$$\frac{x \times 4}{4} = \frac{350}{4}$$

$$x = \frac{350}{4}$$

$$x = 87.5 \text{ gal}$$

A total of 87.5 gallons of spray solution are needed for this crop.

Step 2

Determine how much product is required to prepare 87.5 gal of spray solution if 32 mL of product are needed for each gallon of spray solution.

Set up a proportion. (Note: Use milliliters because of the small quantity of product.)

$$\frac{32 \text{ ml of product}}{1 \text{ gal of spray solution}} = \frac{x \text{ ml of product}}{87.5 \text{ gal of spray solution}}$$

Isolate and solve for x.

$$x = 87.5 \times 32$$

$$x = 2,800 \text{ ml}$$

Calculations Involving Pesticides and Plant Growth Regulators 279

SOLUTION

A total of 2,800 mL of growth regulator product are needed for this application.

Banded Spray Applications

Banded spray applications are made near the root system of field-grown crops. Product rate is based on the crop, the pest, and the distance between the rows. Rate charts on the label help the nursery manager determine which rate is right for the situation. In addition, the spray application rate in gallons per linear foot of row is also described.

EXAMPLE 10-31

A Christmas tree grower needs to make a banded spray application of an insecticide to a crop of Frazier fir trees. The rate of insecticide product recommended is 1.47 oz/1,000 linear foot of row, and the minimum spray volume is 2 gal of water/1,000 linear foot of row. The grower has approximately 1 ac in production, and it is arranged as fifty-four 100 ft long rows. How many gallons of water should be used to make this application? How many ounces of product are needed for this application?

Step 1

Determine the total number of linear feet in production by multiplying the length of the row by the number of rows.

$$100 \text{ feet} \times 54 \text{ rows} = 5,400 \text{ linear feet of row}$$

Step 2

Determine the volume of water required to make the application.

Set up a proportion.

$$\frac{2 \text{ gallons}}{1,000 \text{ linear feet of row}} = \frac{x \text{ gallons}}{5,400 \text{ linear feet of row}}$$

Isolate and solve for x.

$$x \times 1,000 = 5,400 \times 2$$

$$\frac{x \times 1,000}{1,000} = \frac{5,400 \times 2}{1,000}$$

$$x = \frac{5,400 \times 2}{1,000}$$

$$x = 10.8 \text{ gal}$$

A total of 10.8 gal of water are required for this application, and the Christmas tree grower will prepare 11 gal of solution.

Step 3

Determine the number of ounces of insecticide required for this application.

Set up a proportion.

$$\frac{1.47 \text{ ounces of product}}{1,000 \text{ linear feet of row}} = \frac{x \text{ ounces of product}}{5,400 \text{ linear feet of row}}$$

Isolate and solve for x.

$$x \times 1,000 = 5,400 \times 1.47$$

$$\frac{x \times 1,000}{1,000} = \frac{5,400 \times 1.47}{1,000}$$

$$x = \frac{5,400 \times 1.47}{1,000}$$

$$x = 7.938 \text{ oz}$$

SOLUTION

A total of 7.9 oz of insecticide are required for this application. This amount of product needs to be added to 11 gal of water and should be applied at the rate of 2 gal of solution per every 1,000 linear feet of row.

Drench Applications

Drench applications are used when a pesticide or plant growth regulator must come in contact with the plant crown, plant root system, soil, or root medium. Either the crown or root is the site of absorption, or the soil or root medium is the site of the pest infestation. When applied to plants in containers, the volume of drench is often given as an estimate of the number of ounces of solution that are required per pot. The volume of drench is also described as the volume sufficient to wet root media with 0% run-through, or it is described as limiting the product run-through to 10%.

EXAMPLE 10-32

A greenhouse manager is preparing plant growth regulator solution for a crop of 1,500 poinsettias grown in 6 in. azalea pots. If 4 fl oz of drench solution are required for each pot, how many gallons of drench are required for this application? If a 1.0 ppm concentration solution is desired and 0.032 fl oz of plant growth regulator product/gal produces a 1.0 ppm solution, then how many fluid ounces of product are required for this application?

Step 1a

Determine how many fluid ounces of drench are required to treat 1,500 pots at a rate of 4 fl oz/pot.

Set up a proportion.

$$\frac{4 \text{ fl oz of drench}}{1 \text{ pot}} = \frac{x \text{ fl oz of drench}}{1,500 \text{ pots}} =$$

Isolate and solve for x.

$$x = 1{,}500 \times 4$$
$$x = 6{,}000 \text{ fl oz of drench}$$

Step 2a

Convert fluid ounces to gallons by setting up a proportion.

$$\frac{1 \text{ gal}}{128 \text{ fl oz}} = \frac{x \text{ gal}}{6{,}000 \text{ fl oz}}$$

Isolate and solve for x.

$$x \times 128 = 6{,}000$$
$$\frac{x \times 128}{128} = \frac{6{,}000}{128}$$
$$x = 46.875 \text{ gal}$$

SOLUTION A

A total of 46.875 gal of drench solution are required to treat 1,500 pots of poinsettia. A greenhouse manager will prepare 47 gal of solution.

Step 1b

Determine how many ounces of growth regulator product are required to prepare 47 gal of solution when 0.032 fl oz are needed to prepare a gallon of solution.

Set up a proportion.

$$\frac{0.032 \text{ fl oz of product}}{1 \text{ gal of solution}} = \frac{x \text{ fl oz of product}}{47 \text{ gal of solution}}$$

Isolate and solve for x.

$$x = 47 \times 0.032$$
$$x = 1.5 \text{ fl oz of product}$$

SOLUTION B

To prepare 47 gal of dilute plant growth regulator solution for a drench, 1.5 fl oz of plant growth regulator product is needed.

Calculation of the Amount of Active Ingredient Applied in a Given Pesticide or Growth Regulator Application

In an effort to reduce risk to the environment, the Environmental Protection Agency limits the amount of active ingredient of many pesticide and growth regulator products that may be applied to a given area per year or per cropping cycle. Each nursery or greenhouse manager needs to keep records of each application and then add the total amount of each active ingredient used. Once the total of each active ingredient meets the legal limit, use of products that contain that active ingredient must cease. Note that many, but not all, pesticides and growth regulators have established limits on the label. It is also important to note that many pesticide products on the market today are a combination of products that contain more than one (often two to three) active ingredient. Using two of the hypothetical examples above, the total amount of active ingredient applied will be calculated for each.

EXAMPLE 10-33

In Example 10-34, a foliar spray of a growth regulator is applied to bedding plants and a total of 2,800 mL is used. This product contains 1.0 lb a.i./gal of product. How many pounds of active ingredient are used in this application?

Step 1

If 1 lb of active ingredient is found in 1 gal of product, then 1 lb of product is found in 3,785.41184 ml of product (the milliliter equivalent to 1 gal is found in Table B-8).

$$1 \text{ gallon is equivalent to } 3{,}785.41184 \text{ milliliters}$$

$$\frac{1 \text{ lb a.i.}}{1 \text{ gal}} = \frac{1 \text{ lb a.i.}}{3{,}785.41184 \text{ ml of product}}$$

Step 2

Use a proportion to find out how many pounds of active ingredient are in 2,800 mL of product.

$$\frac{1 \text{ lb a.i.}}{3{,}785.41184 \text{ ml}} = \frac{x \text{ lb a.i.}}{2{,}800 \text{ ml}} =$$

Isolate and solve for x.

$$x \times 3{,}785.41184 = 2{,}800$$

$$\frac{x \times 3{,}785.41184}{3{,}785.41184} = \frac{2{,}800}{3{,}785.41184}$$

$$x = 0.74 \text{ lb a.i.}$$

SOLUTION

A total 0.74 lb of active ingredient can be found in 2,800 mL of this growth regulator product.

EXAMPLE 10-34

In Example 10-31, a banded spray application is made to Christmas trees and 7.9 oz of an insecticide product is used. The product label indicates that it is a dry product, and it contains 25% by weight of active ingredient. How many ounces of active ingredient are used in this application?

Step 1

Determine the number of ounces of active ingredient in 7.9 oz of product by using a proportion. If 25% or 25 oz in 100 oz are active ingredient, then the proportion is:

$$\frac{25 \text{ oz a.i.}}{100 \text{ oz of product}} = \frac{x \text{ oz a.i.}}{7.9 \text{ oz of product}}$$

Isolate and solve for x.

$$25 \times 7.9 = 100 \times x$$

$$\frac{25 \times 7.9}{100} = \frac{100 \times x}{100}$$

$$x = \frac{25 \times 7.9}{100}$$

$$x = 1.975 \text{ oz a.i.}$$

SOLUTION

There are 1.975 oz of active ingredient in 7.9 oz of this insecticide product.

An alternate method for calculating the amount of active ingredient in this product would be to say that 25% of the product is active ingredient.

Translated to a mathematical equation:

$$0.25 \text{ a.i.} \times 7.9 \text{ oz} = 1.975 \text{ oz a.i.}$$

The 25% is written in a decimal form (0.25), and the word "of" in mathematics translates to the process of multiplication.

Practice Problem Set 10-17   Calculations Involving Pesticides and Growth Regulators

1. A greenhouse manager needs to make a drench application of an insecticide to a crop of 1,800 poinsettias grown in 8 in. pots. A drench volume of 8 fl oz is required, and 12 oz of dry insecticide product are dissolved in every 100 gal of drench solution. How many gallons of drench solution are required for this application? How many ounces of dry insecticide product are needed for this application?

2. If the dry insecticide product used in question 1 is a 25% active ingredient formulation, how many ounces of active ingredient are used in this application?

3. A greenhouse manager needs to make a foliar application of a growth regulator to 25,000 ft² of a young bedding plant crop. The label indicates that 1 qt of spray volume for every 100 ft² of crop is appropriate, and a rate of 4.7 mL of product in every gallon of spray is recommended for this crop. How many gallons of spray need to be prepared for this application? How many fluid ounces of product are required to prepare this spray solution?

4. How many grams of active ingredient are used in this application if 0.12 g of active ingredient are found in 1 fl oz of product?

APPENDIX A
METRIC SYSTEM PREFIXES

TABLE A-1 • METRIC SYSTEM PREFIXES

Prefix	Abbreviation	Magnitude	Example (in grams)	Example Abbreviation
Vocto-	y	10^{-24}	Yoctogram	yg
Zepto-	z	10^{-21}	Zeptogram	zg
Atto-	a	10^{-18}	Attogram	ag
Femto-	f	10^{-15}	Femtogram	fg
Pico-	p	10^{-12}	Pictogram	pg
Nano-	n	10^{-9}	Nanogram	ng
Micro-	μ	10^{-6}	Microgram	μg
Milli-	m	10^{-3}	Milligram	mg
Centi-	c	10^{-2}	Centigram	cg
Deci-	d	10^{-1}	Decigram	dg
		10^{0}	Gram	g
Deka-	da	10	Dekagram	dag
Hecto-	h	10^{2}	Hectogram	hg
Kilo-	k	10^{3}	Kilogram	kg
Mega-	M	10^{6}	Megagram	Mg
Giga-	G	10^{9}	Gigagram	Gg
Tera-	T	10^{12}	Teragram	Tg
Peta-	P	10^{15}	Petagram	Pg
Exa-	E	10^{18}	Exagram	Eg
Zeta-	Z	10^{21}	Zetagram	Zg
Yotta-	Y	10^{24}	Yottagram	Yg

APPENDIX B
TABLES OF EQUIVALENTS

**TABLE B-1 • UNITS OF LENGTH: U.S. CUSTOMARY SYSTEM
(UNDERLINED VALUES ARE EXACT NUMBERS)**

\multicolumn{4}{l}{U.S. Customary Units of Length}
Unit	Abbreviation	U.S. Customary Equivalents	Metric Equivalents
Mile	mi	1,760 yards 5,280 feet 63,360 inches	1.609344 kilometers 1,609.344 meters 160,934.4 centimeters
Yard	yd	0.0005681818 miles 3 feet 36 inches	0.9144 meters 91.44 centimeters
Foot	ft	0.0001893939 miles 0.3333333 yards 12 inches	0.3048 meters 30.48 centimeters
Inch	in.	0.00001578283 miles 0.02777778 yards 0.08333333 feet	0.0254 meters 2.54 centimeters 25.4 millimeters

TABLE B-2 • UNITS OF LENGTH: METRIC SYSTEM
(UNDERLINED VALUES ARE EXACT NUMBERS)

Metric Units of Length

Unit	Abbreviation	Metric Equivalents	U.S. Customary Equivalents
Kilometer	km	1,000 meters 10 hectometers	0.6213712 miles 1,093.613 yards 3,280.840 feet
Hectometer	hm	100 meters 10 dekameters	328.0840 feet
Dekameter	dam	10 meters	32.80840 feet
Meter	m	10 decimeters 100 centimeters 1,000 millimeters 1,000,000 micrometers	0.0006213712 miles 1.093613 yards 3.280840 feet 39.37008 inches
Decimeter	dm	0.1 meters 10 centimeters 100 millimeters	3.937008 inches
Centimeter	cm	0.01 meters 10 millimeters	0.03280840 feet 0.3937008 inches
Millimeter	mm	0.001 meters	0.03937008 inches
Micrometer	μm	0.000001 meters	0.00003937008 inches

TABLE B-3 • UNITS OF AREA: U.S. CUSTOMARY SYSTEM
(UNDERLINED VALUES ARE EXACT NUMBERS)

U.S. Customary Units of Area

Unit	Abbreviation	U.S. Customary Equivalents	Metric Equivalents
Square mile	mi^2 or sq mi	640 acres 3,097,600 square yards 27,878,400 square feet 4,014,489,600 square inches	2.589988110336 square kilometers 258.9988110336 hectares 2589988.110336 square meters
Acre	ac	0.0015625 square miles 4,840 square yards 43,560 square feet	0.004046873 square kilometers 0.4046873 hectare 4,046.873 square meters
Square yard	yd^2 or sq yd	0.0000003228306 square miles 9 square feet 1,296 square inches	0.83612736 square meters 8,361.2736 square centimeters
Square foot	ft^2 or sq ft	0.00000003587006 square miles 0.00002295684 acres 0.1111111 square yards 144 square inches	0.092903041 square meters 929.03041 square centimeters
Square inch	in^2 or sq in	0.0000000002490977 square miles 0.0007716049 square yard 0.00694444 square feet	0.00064516 square meters 6.4516 square centimeters

Appendix B Tables of Equivalents **289**

TABLE B-4 • UNITS OF AREA: METRIC SYSTEM
(UNDERLINED VALUES ARE EXACT NUMBERS)

Metric Units of Area

Unit	Abbreviation	Metric Equivalents	U.S. Customary Equivalents
Square kilometer	km^2 or sq km	100 hectares 1,000,000 square meters	0.3861022 square mile 247.1044 acres
Hectare	ha	0.01 square kilometers 10,000 square meters	0.003861022 square mile 2.471044 acres 107,639.1 square feet
Are	a	100 square meters	0.02471044 acre 119.5990 square yards
Square meter	m^2 or sq m	10,000 square centimeters 1,000,000 square millimeters	0.0000003861022 square miles 0.0002471044 acres 1.195990 square yards 10.76391 square feet 1,550.003 square inches
Square centimeter	cm^2 or sq cm	0.0001 square meter	0.0001195990 square yards 0.001076397 square feet 0.1550003 square inches

TABLE B-5 • UNITS OF VOLUME: U.S. CUSTOMARY SYSTEM
(UNDERLINED VALUES ARE EXACT NUMBERS)

U.S. Customary Units of Volume

Unit	Abbreviation	U.S. Customary Equivalents	Metric Equivalents
Cubic yard	yd^3 or cu yd	27 cubic feet 46,656 cubic inches 201.974 gallons	0.764554857984 cubic meters 764.554857984 liters 764,554.857984 cubic centimeters
Cubic foot	ft^3 or cu ft	0.03703704 cubic yards 1,728 cubic inches 957.5065 fluid ounces 59.84416 liquid pints 51.42809 dry pints 29.92208 liquid quarts 25.71405 dry quarts 7.480519 gallons 3.214256 pecks 0.80356395 bushels	0.028316846592 cubic meters 28.316846592 liters 28,316.846592 cubic centimeters
Cubic inch	in.^{3} or cu in.	0.00002143347 cubic yard 0.0005787037 cubic feet 0.5541126 fluid ounces 0.03463203 liquid pints 0.0297616 dry pints 0.01731602 liquid quarts 0.0148808 dry quarts 0.004329004 gallons 0.00186010 pecks 0.000465025 bushels	0.000016387064 cubic meters 0.016387064 liters 16.387064 cubic centimeters

TABLE B-6 • UNITS OF VOLUME: METRIC SYSTEM
(UNDERLINED VALUES ARE EXACT NUMBERS)

Metric Units of Volume

Unit	Abbreviation	Metric Equivalents	U.S. Customary Equivalents
Cubic meter	m^3	1,000,000 cubic centimeters 1,000 cubic decimeters 1,000 liters	1.307951 cubic yards 35.31467 cubic feet 61,023.74 cubic inches 264.1721 gallons 1,816.166 dry pints 908.0830 dry quarts 113.5104 pecks 28.37759 bushels
Cubic decimeter	dm^3	0.001 cubic meters 1 liter 1,000 cubic centimeters	61.02374 cubic inches 0.2641721 gallons 2.113376 liquid pints 1.816166 dry pints 1.056688 liquid quarts 0.9080830 dry quarts 0.1135104 pecks 0.02837759 bushels
Cubic centimeter	cm^3 or cc	0.000001 cubic meters 0.001 liters 1 milliliter	0.061023 cubic inches 0.002113376 liquid pints 0.001056688 liquid quarts 0.0002641721 gallons

TABLE B-7 • UNITS OF DRY VOLUME: U.S. CUSTOMARY SYSTEM
(UNDERLINED VALUES ARE EXACT NUMBERS)

U.S. Customary Units of Dry Volume

Unit	Abbreviation	U.S. Customary Equivalents	Metric Equivalents
Dry pint	dry pt	0.50 dry quarts 0.0625 pecks 0.015625 bushels 0.01944463 cubic feet 33.6003125 cubic inches	0.0005506105 cubic meters 0.5506105 liters
Dry quart	dry qt	2 dry pints 0.125 pecks 0.03125 bushels 0.03888925 cubic feet 67.200625 cubic inches	0.001101221 cubic meters 1.101221 liters
Peck	pk	16 dry pints 8 dry quarts 0.25 bushels 0.311114 cubic feet 537.605 cubic inches	0.008809768 cubic meters 8.809768 liters
Bushel	bu	64 dry pints 32 dry quarts 4 pecks 1.244456 cubic feet 2,150.42 cubic inches	0.03523907 cubic meters 35.23907 liters

TABLE B-8 • UNITS OF CAPACITY OR LIQUID VOLUME: U.S. CUSTOMARY SYSTEM (UNDERLINED VALUES ARE EXACT NUMBERS)

U.S. Customary Units of Capacity or Liquid Volume			
Unit	Abbreviation	U.S. Customary Equivalents	Metric Equivalents
Gallon	gal	4 liquid quarts 8 liquid pints 16 cups 128 fluid ounces 0.1337 cubic feet 231 cubic inches	3.78541184 liters 3,785.41184 milliliters
Quart	qt or liq qt	0.25 gallons 2 liquid pints 4 cups 32 fluid ounces 0.3342 cubic feet 57.749 cubic inches	0.94635296 liter 946.35296 milliliters
Pint	pt or liq pt	0.125 gallons 0.50 liquid quarts 2 cups 16 fluid ounces 0.0167 cubic feet 28.875 cubic inches	0.47317648 liters 473.17648 milliliters
Cup	C	0.625 gallons 0.25 liquid quarts 0.50 liquid pints 8 fluid ounces 14.438 cubic inches 16 tablespoons	0.23658824 liters 236.58824 milliliters
Fluid ounce	fl oz	0.0625 liquid pints 0.03125 liquid quarts 0.0078125 gallons 1.8046875 cubic inches 2 tablespoons 6 teaspoons	0.02957353 liters 29.57353 milliliters
Tablespoon	Tbsp	0.50 fluid ounce 3 teaspoons	14.786765 milliliters
Teaspoon	tsp	0.3333333 tablespoon 0.1666666 fluid ounce 60 drops	4.928921667 milliliters

TABLE B-9 • UNITS OF CAPACITY OR LIQUID VOLUME: METRIC SYSTEM
(UNDERLINED VALUES ARE EXACT NUMBERS)

Metric Units of Capacity or Liquid Volume			
Unit	**Abbreviation**	**Metric Equivalents**	**U.S. Customary Equivalents**
Kiloliter	kL	1,000 liters	264.1721 gallons
Hectoliter	hL	100 liters	26.41721 gallons
Dekaliter	daL	10 liters	2.641721 gallons
Liter[1]	L	1 cubic decimeter 1,000 milliliters 1,000 cubic centimeters	0.2641721 gallons 1.056688 liquid quarts 2.113376 liquid pints 33.81402 fluid ounces 61.02374 cubic inches
Deciliter	dL	0.1 liters	3.381402 fluid ounces
Centiliter	cL	0.01 liters	0.3381402 fluid ounces
Milliliter	mL	0.001 liters	0.03381402 fluid ounces
Microliter	μL	0.000001 liters	0.000003381402 fluid ounces

[1] The 12th General Conference on Weights and Measures announced in 1964 the name liter as a special name for the cubic decimeter.

TABLE B-10 • UNITS OF WEIGHT: U.S. CUSTOMARY SYSTEM
(UNDERLINED VALUES ARE EXACT NUMBERS)

U.S. Customary Units of Weight			
Unit	**Abbreviation**	**U.S. Customary Equivalents**	**Metric Equivalents**
Long ton	--	2,240 pounds	1.0160469088 metric ton 1016.0469088 kilograms
Short ton (ton)	--	0.8928571 long tons 2,000 pounds 32,000 ounces	0.90718474 metric ton 907.18474 kilograms
Pound	lb	0.0004464286 long tons 0.0005 short tons 16 ounces	0.00045359237 metric tons 0.45359237 kilograms 453.59237 grams
Ounce	oz	0.00002790179 long tons 0.00003125 short tons 0.0625 pound	0.000028349523125 metric tons 0.028349523125 kilograms 28.349523125 grams

TABLE B-11 • UNITS OF WEIGHT: METRIC SYSTEM
(UNDERLINED VALUES ARE EXACT NUMBERS)

Metric Units of Weight

Unit	Abbreviation	Metric Equivalents	U.S. Customary Equivalents
Metric ton	t	1,000,000 grams	0.9842065 long tons 1.102311 short tons 2,204.623 pounds 35,273.96 ounces
Kilogram	kg	1,000 grams	2.204623 pounds 35.27396 ounces
Hectogram	hg	100 grams	3.527396 ounces
Dekagram	dag	10 grams	0.3527396 ounces
Gram	g	0.001 kilograms 1,000 milligrams	0.002204623 pounds 0.03527396 ounces
Decigram	dg	0.1 grams	0.003527396 ounces
Centigram	cg	0.01 grams	0.0003527396 ounces
Milligram	mg	0.001 grams	0.00003527396 ounces
Microgram	μg	0.000001 grams	0.00000003527396 ounces

TABLE B-12 • UNITS OF CONCENTRATION

Unit	Volume/Volume	Weight/Volume	Weight/Weight
1 percent 1% 0.01 1 in 100 1 in 10^2 10,000 ppm	1 milliliters/100 milliliters 10 milliliters/liter	1 gram/100 milliliters 10 grams/liter 8.336 pounds/ 100 gallons 1.28 ounces/gallon	1 gram/100 grams
1 part per million 1 in 1,000,000 1 in 10^6 0.000001 0.0001%	1 microliter/liter nanoliter/milliliter	1 milligram/liter 1 microgram/milliliter 0.379 gram/ 100 gallons	1 milligram/kilogram 1 microgram/gram 1 pound/500 tons
1 part per billion 1 in 1,000,000,000 1 in 10^9 0.000000001 0.0000001%	1 nanoliter/liter 1 picoliter/milliliter	1 microgram/liter 1 nanogram/milliliter	1 microgram/kilogram 1 nanogram/gram

TABLE B-13 • UNITS OF TIME

Unit	Abbreviation	Equivalents
Second	sec	1/60 minute
Minute	min	60 seconds 1/60 hour
Hour	hr	3,600 seconds 60 minutes 1/24 day
Day	day	86,400 seconds 1,440 minutes 24 hours

TABLE B-14 • UNITS OF ELECTRICAL CONDUCTIVITY

Unit	Abbreviation	Equivalents
$\dfrac{\text{Decisiemens}}{\text{meter}}$	$\dfrac{dS}{m}$	$1.0\ \dfrac{mmhos}{cm}$ $1.0\ \dfrac{mS}{cm}$ $1.0 \times 10^3\ \dfrac{\mu mhos}{cm}$
$\dfrac{\text{Millisiemens}}{\text{centimeter}}$	$\dfrac{mS}{cm}$	$1.0\ \dfrac{dS}{m}$ $1.0\ \dfrac{mmhos}{cm}$ $1.0 \times 10^3\ \dfrac{\mu mhos}{cm}$
$\dfrac{\text{Millimhos}}{\text{centimeter}}$	$\dfrac{mmhos}{cm}$	$1.0\ \dfrac{dS}{m}$ $1.0\ \dfrac{mS}{cm}$ $1.0 \times 10^3\ \dfrac{\mu mhos}{cm}$
$\dfrac{\text{Micromhos}}{\text{centimeter}}$	$\dfrac{\mu mhos}{cm}$	$1.0 \times 10^{-3}\ \dfrac{dS}{m}$ $1.0 \times 10^{-3}\ \dfrac{mmhos}{cm}$ $1.0 \times 10^{-3}\ \dfrac{mS}{cm}$

TABLE B-15 • UNITS OF IRRADIANCE

Unit	Abbreviation	Equivalents
Footcandle	fc	0.010764 klx
		10.764 lux
Lux	lx	0.0929 fc
		1×10^{-3} klx
Kilolux	klx	9.29 fc
		1,000 lx

APPENDIX C

TABLE OF CONVERSION FACTORS

TABLE C-1 • CONVERSION FACTORS

To Convert From:	To:	Multiply by:
Acre	Hectare	0.4046873
Acre	Square Foot	43,560
Acre	Square Kilometer	0.004046873
Acre	Square Meter	4,046.873
Acre	Square Mile	0.0015625
Acre	Square Yard	4,840
Acre Foot	Cubic Foot	43,560
Acre Foot	Cubic Meter	1,233.489
Acre Foot	Gallon	325,900
Are	Acre	0.02471043692
Are	Square Meter	100
Are	Square Yard	119.599
Bushel	Cubic Feet	1.244456
Bushel	Cubic Inch	2,150.42
Bushel	Cubic Meter	0.03523907
Bushel	Liter	35.23907
Bushel	Peck	4
Bushel	Pint (dry)	64
Bushel	Quart (dry)	32
Centimeter	Foot	0.0328084
Centimeter	Inch	0.3937008
Centimeter	Kilometer	0.00001
Centimeter	Meter	0.01
Centimeter	Mile	0.000006.213712
Centimeter	Millimeter	10

(continues)

TABLE C-1 • CONVERSION FACTORS (CONTINUED)

To Convert From:	To:	Multiply by:
Centimeter	Yard	0.01093613
Cubic Centimeter	Cubic Foot	35,314,670
Cubic Centimeter	Cubic Inch	0.061023
Cubic Centimeter	Cubic Meter	0.000001
Cubic Centimeter	Cubic Yard	0.000001307951
Cubic Centimeter	Gallon	0.0002641721
Cubic Centimeter	Liter	0.001
Cubic Centimeter	Pint (liquid)	0.002113376
Cubic Centimeter	Quart (liquid)	0.001056688
Cubic Foot	Bushel	0.80356395
Cubic Foot	Cubic Centimeter	28,316.846592
Cubic Foot	Cubic Inch	1,728
Cubic Foot	Cubic Meter	0.028316846592
Cubic Foot	Cubic Yard	0.03703704
Cubic Foot	Gallon	7.480519
Cubic Foot	Liter	28.316845692
Cubic Foot	Pint (liquid)	59.84416
Cubic Foot	Quart (liquid)	29.92208
Cubic Inch	Cubic Centimeter	16.387064
Cubic Inch	Cubic Foot	0.0005787037
Cubic Inch	Cubic Meter	0.000016387064
Cubic Inch	Cubic Yard	0.00002143347
Cubic Inch	Gallon	0.004329004
Cubic Inch	Liter	0.016387064
Cubic Inch	Pint (liquid)	0.03463203
Cubic Inch	Quart (liquid)	0.01731602
Cubic Meter	Bushel	28.37759
Cubic Meter	Cubic Centimeter	1,000,000
Cubic Meter	Cubic Foot	35.31467
Cubic Meter	Cubic Inch	61,023.74
Cubic Meter	Cubic Yard	1.307951
Cubic Meter	Gallon	264.1721
Cubic Meter	Liter	1,000

TABLE C-1 • CONVERSION FACTORS (CONTINUED)

To Convert From:	To:	Multiply by:
Cubic Meter	Pint (liquid)	2,113.376
Cubic Meter	Quart (liquid)	1,056.688
Cubic Yard	Cubic Centimeter	764,554.857984
Cubic Yard	Cubic Foot	27
Cubic Yard	Cubic Inches	46,656
Cubic Yard	Cubic Meter	0.764554857984
Cubic Yard	Gallon	201.974
Cubic Yard	Liter	764.554857984
Cubic Yard	Pint (liquid)	1,615.9
Cubic Yard	Quart (liquid)	807.9
Cup	Cubic Inch	14.438
Cup	Cubic Meter	0.00023658824
Cup	Fluid Ounce	8
Cup	Gallon	0.625
Cup	Liter	0.23658824
Cup	Milliliter	236.58824
Cup	Pint	0.50
Cup	Quart (liquid)	0.25
Cup	Tablespoon	16
Day	Second	86,400
Day	Minute	1,440
Day	Hour	24
Fluid Ounce	Cubic Inch	1.8046875
Fluid Ounce	Cubic Meter	0.00002957353
Fluid Ounce	Gallon	0.0078125
Fluid Ounce	Liter	0.02957353
Fluid Ounce	Milliliter	29.57353
Fluid Ounce	Pint (liquid)	0.0625
Fluid Ounce	Quart (liquid)	0.25
Fluid Ounce	Tablespoon	2
Fluid Ounce	Teaspoon	6
Foot	Centimeter	30.48
Foot	Kilometer	0.0003048

(continues)

TABLE C-1 • CONVERSION FACTORS (CONTINUED)

To Convert From:	To:	Multiply by:
Foot	Meter	0.3048
Foot	Mile	0.0001893939
Foot	Millimeter	304.8
Foot	Yard	0.333333
Foot-candle	Lumen/Square Meter	10.76391
Foot-candle	Lux	10.76391
Gallon	Cubic Centimeter	3,785.41184
Gallon	Cubic Feet	0.1337
Gallon	Cubic Inch	231
Gallon	Cubic Meter	0.00378541184
Gallon	Cubic Yard	0.004951
Gallon	Liter	3.78541184
Gram	Kilogram	0.001
Gram	Milligram	1,000
Gram	Ounce	0.03527396
Gram	Pound	0.002204623
Hectare	Acre	2.471043692
Hectare	Square Foot	107,639.1
Hectare	Square Meter	10,000
Hour	Day	0.04167
Hour	Second	3600
Hour	Week	0.005952
Inch	Centimeter	2.54
Inch	Foot	0.08333333
Inch	Meter	0.0254
Inch	Mile	0.00001578283
Inch	Millimeter	25.4
Inch	Yard	0.02777778
Kilogram	Gram	1,000
Kilogram	Ounce	35.27396
Kilogram	Pound	2.204623
Kilometer	Centimeter	100,000
Kilometer	Foot	3,280.84

TABLE C-1 • CONVERSION FACTORS (CONTINUED)

To Convert From:	To:	Multiply by:
Kilometer	Inch	39,370
Kilometer	Meter	1,000
Kilometer	Mile	0.6213712
Kilometer	Millimeter	1,000,000
Kilometer	Yard	1,093.613
Liter	Bushel	0.02838
Liter	Cubic Centimeter	1,000
Liter	Cubic Foot	0.035314664
Liter	Cubic Inch	61.02374
Liter	Cubic Meter	0.001
Liter	Cubic Yard	0.001307951
Liter	Fluid Ounce	33.81402
Liter	Gallon	0.2641721
Liter	Pint (liquid)	2.113376
Liter	Quart (liquid)	1.056688
Lumen per Square Foot	Foot-candle	1
Lumen per Square Foot	Lux	10.76391
Lux	Foot-candle	0.0929
Meter	Centimeter	100
Meter	Foot	3.28084
Meter	Inch	39.3700787
Meter	Kilometer	0.001.0
Meter	Mile	0.0006213712
Meter	Millimeter	1,000
Meter	Yard	1.093613
Mile	Centimeter	0.00001609344
Mile	Foot	5,280
Mile	Inch	63,360
Mile	Kilometer	1.609344
Mile	Meter	1,609.344
Mile	Yard	1,760
Milliliter	Cubic Centimeter	1
Milliliter	Liter	1,000
Millimeter	Centimeter	0.10

(continues)

TABLE C-1 • CONVERSION FACTORS (CONTINUED)

To Convert From:	To:	Multiply by:
Millimeter	Foot	0.00328084
Millimeter	Inch	0.03937008
Millimeter	Kilometer	0.000001
Millimeter	Meter	0.001
Millimeter	Mile	0.0000006213712
Millimeter	Yard	0.001093613
Minute	Day	0.0006.944
Minute	Hour	0.01667
Minute	Second	60
Minute	Week	0.000099206
Ounce	Gram	28.349523125
Ounce	Kilogram	0.028349523125
Ounce	Pound	0.0625
Peck	Cubic Meter	0.008809768
Peck	Liter	8.809768
Pint (dry)	Bushel	0.015625
Pint (dry)	Cubic Inch	33.6003125
Pint (dry)	Liter	0.55056105
Pint (dry)	Quart	0.50
Pint (liquid)	Cubic Centimeters	473.17648
Pint (liquid)	Cubic Foot	0.01671
Pint (liquid)	Cubic Inch	28.875
Pint (liquid)	Cubic Meter	0.00047317648
Pint (liquid)	Cubic Yard	0.0006189
Pint (liquid)	Gallon	0.125
Pint (liquid)	Liter	0.47317648
Pint (liquid)	Quart (liquid)	0.50
Pound	Gram	453.59237
Pound	Kilogram	0.45359237
Pound	Ounce	16
Pound	Long Ton	0.0004464286
Pound	Short Ton	0.0005
Quart (dry)	Cubic Inches	67.200625
Quart (dry)	Cubic Meter	0.001101221
Quart (liquid)	Cubic Centimeter	946.35296

TABLE C-1 • CONVERSION FACTORS (CONTINUED)

To Convert From:	To:	Multiply by:
Quart (liquid)	Cubic Foot	0.03342
Quart (liquid)	Cubic Inch	57.749
Quart (liquid)	Cubic Meter	0.00094635296
Quart (liquid)	Cubic Yard	0.001238
Quart (liquid)	Gallon	0.25
Quart (liquid)	Liter	0.94635296
Square Centimeter	Square Foot	0.001076397
Square Centimeter	Square Inch	0.1550003
Square Centimeter	Square Meter	0.0001
Square Centimeter	Square Mile	0.00000000003861022
Square Centimeter	Square Millimeter	100
Square Centimeter	Square Yard	0.000119599
Square Foot	Acre	0.00002295684
Square Foot	Square Centimeter	929.03041
Square Foot	Square Inch	144
Square Foot	Square Meter	0.0092903041
Square Foot	Square Mile	0.00000003587006
Square Foot	Square Millimeter	92,903.041
Square Foot	Square Yard	0.111111
Square Inch	Square Centimeter	6.4516
Square Inch	Square Foot	0.00694444
Square Inch	Square Meter	0.00064516
Square Inch	Square Millimeter	645.16
Square Inch	Square Yard	0.0007716049
Square Kilometer	Acre	247.1044
Square Kilometer	Square Centimeter	10,000,000,000
Square Kilometer	Square Foot	10,763,910
Square Kilometer	Square Inch	1,550,003,000
Square Kilometer	Square Meter	1,000,000
Square Kilometer	Square Mile	0.3861022
Square Kilometer	Square Yard	1,195,990
Square Meter	Acre	0.0002471044
Square Meter	Square Centimeter	10,000
Square Meter	Square Foot	10.76391
Square Meter	Square Inch	1,550.003

(continues)

TABLE C-1 • CONVERSION FACTORS (CONTINUED)

To Convert From:	To:	Multiply by:
Square Meter	Square Mile	0.0000003861022
Square Meter	Square Millimeter	1,000,000
Square Meter	Square Yard	1.19599
Square Mile	Acre	640
Square Mile	Square Foot	27,878,400
Square Mile	Square Kilometer	2.589988110336
Square Mile	Square Meter	2,589,988.110336
Square Mile	Square Yard	3,097,600
Square Millimeter	Square Centimeter	0.01
Square Millimeter	Square Foot	0.00001076391
Square Millimeter	Square Inch	0.001550003
Square Yard	Acre	0.0002066
Square Yard	Square Centimeter	8,361.2736
Square Yard	Square Foot	9
Square Yard	Square Inch	1,296
Square Yard	Square Meter	0.83612736
Square Yard	Square Mile	0.0000003228306
Square Yard	Square Millimeter	836,100
Tablespoon	Fluid Ounce	0.50
Tablespoon	Milliliter	14.786765
Tablespoon	Teaspoon	3
Teaspoon	Milliliter	4.928921667
Teaspoon	Tablespoon	0.333333
Teaspoon	Drop	60
Ton (long)	Kilogram	1,016.0469088
Ton (long)	Pound	2,240
Ton (long)	Ton (short)	1.12
Ton (metric)	Kilogram	1,000
Ton (metric)	Pound	2,204.623
Ton (short)	Kilogram	907.18474
Ton (short)	Ounce	32,000
Ton (short)	Pound	2,000
Ton (short)	Ton (long)	0.8928571
Ton (short)	Ton (metric)	0.90718474
Week	Hour	168

TABLE C-1 • CONVERSION FACTORS (CONTINUED)

To Convert From:	To:	Multiply by:
Week	Minute	10,080
Week	Second	604,800
Yard	Centimeter	91.44
Yard	Feet	3
Yard	Inch	36
Yard	Kilometer	0.00009144
Yard	Meter	0.9144
Yard	Mile	0.0005681818
Yard	Millimeter	914.4
Year	Day	365.256
Year	Hour	8,766.1

APPENDIX D
SQUARING-UP GARDENS AND GARDEN STRUCTURES

Imagine this situation. You have been asked to prepare an area for a large rectangular garden that measures 25 feet wide by 75 feet long. No one has marked the area yet and you are faced with a sea of green grass.

How will you be able to get the sides of the rectangle parallel with no reference points or objects to work from?

How will you know if the corners are square without the use of a huge carpenter's square?

Pythagoras, the Greek philosopher and mathematician, can help you. His theorem, the Pythagorean Theorem, provides the information you need to produce a perfect right triangle.

Why create a right triangle? A square or rectangle is composed of two right triangles with their longest sides (hypotenuse) oriented back to back.

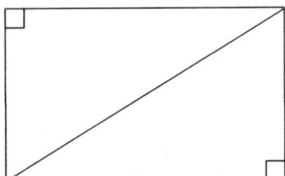

In this example, the garden measures 25 feet wide and 75 feet long. Use the Pythagorean Theorem to solve for the diagonal measure of the garden plot.

$$a^2 + b^2 = c^2$$
$$25^2 + 75^2 = c^2$$
$$625 + 5625 = 6250$$
$$c^2 = 6250$$
$$c = 79.1 \text{ ft} = 79'1'' \, (0.1 \text{ ft} = 1.2 \text{ in})$$

With the aid of three measuring tapes and one assistant, you can create a right triangle. This will result in a perfectly squared-up garden or garden structure foundation.

Begin with the zero marks of two tapes at the upper left-hand corner of your rectangular plot. Put a stake into the ground at this spot and hook the tape ends onto the stake. Extend each tape out to the desired length. For our example, one tape will be extended out to 25 feet and the other extended out to 75 feet. Lay each tape on the ground and work with your assistant to extend the third tape measure out to the length of the diagonal (79 feet 1 inch, for our example).

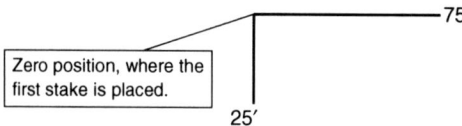

You and your assistant should walk back to your original tapes holding the third tape in your hands. If you are holding the zero position of the third tape and you are responsible for the 25-foot position of tape number two, then your assistant will be holding the 79 foot 1 inch position on tape three and the 75-foot position on tape one.

Adjust your positions until the appropriate measures on the three tapes line up perfectly. This will allow the placement of stakes in the upper right and lower left positions of the rectangular garden plot.

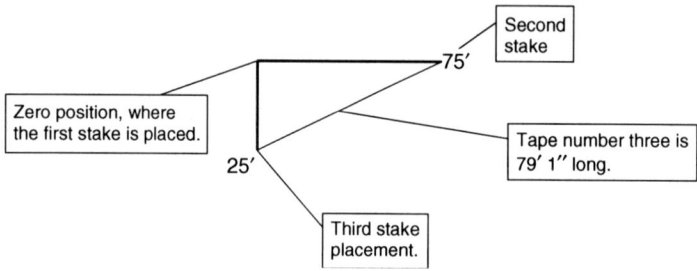

Put tape number three down and remove tapes one and two from the stake at the upper left corner. Hook one tape on the upper right corner stake and one tape on the lower left corner stake. Measure out 25 feet down from the upper right corner and 75 feet across from the lower left corner. Have the 25-foot mark of one tape cross the 75-foot mark of the other tape. Place a fourth stake at this mark. Once the length of all four sides and the diagonal are double-checked, a "squared-up" garden plot is guaranteed.

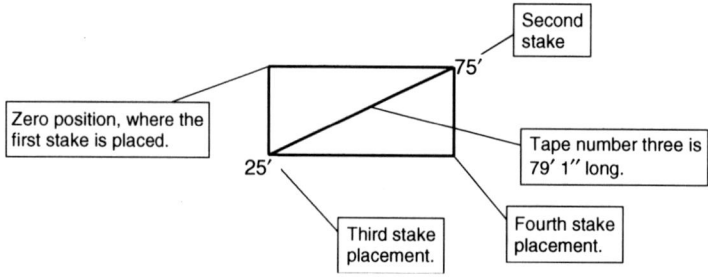

APPENDIX E
SOLUTIONS TO PRACTICE PROBLEMS

Chapter 1

Practice Problem Set 1-1: Whole Numbers, Fractions, and Mixed Numbers

Identify each of the following numbers as a whole number, proper fraction, improper fraction, or mixed number.

1. $\frac{9}{8}$; Improper Fraction
2. $\frac{3}{4}$; Proper Fraction
3. 7; Whole Number
4. $1\frac{5}{8}$; Mixed Number
5. $\frac{25}{25}$; Improper Fraction

Convert each of the improper fractions to a mixed number.

6. $\frac{11}{8} = 1\frac{3}{8}$
7. $\frac{67}{32} = 2\frac{3}{32}$
8. $\frac{128}{5} = 25\frac{3}{5}$
9. $\frac{13}{4} = 3\frac{1}{4}$
10. $\frac{117}{16} = 7\frac{5}{16}$

Practice Problem Set 1-2: Adding, Subtracting, Multiplying, and Dividing Fractions

Solve the following problems involving fractions:

1. $\frac{3}{32} + \frac{15}{16} = \frac{a}{b} + \frac{c}{d} = \frac{ad+bc}{bd} = \frac{(3)(16)+(32)(15)}{(32)(16)} = \frac{48+480}{512} = \frac{528}{512} = 1\frac{16}{512} = 1\frac{1}{32}$
2. $\frac{3}{8} + \frac{1}{2} = \frac{a}{b} + \frac{c}{d} = \frac{ad+bc}{bd} = \frac{(3)(2)+(8)(1)}{(8)(2)} = \frac{6+8}{16} = \frac{14}{16} = \frac{7}{18}$
3. $\frac{10}{32} - \frac{1}{4} = \frac{a}{b} - \frac{c}{d} = \frac{ad-bc}{bd} = \frac{(10)(4)-(32)(1)}{(32)(4)} = \frac{40-32}{128} = \frac{8}{128} = \frac{1}{16}$
4. $\frac{7}{8} - \frac{1}{3} = \frac{a}{b} - \frac{c}{d} = \frac{ad-bc}{bd} = \frac{(7)(3)-(8)(1)}{(8)(3)} = \frac{21-8}{24} = \frac{13}{24}$
5. $\frac{3}{8} \times \frac{1}{4} = \frac{a}{b} \times \frac{c}{d} = \frac{ac}{bd} = \frac{(3)(1)}{(8)(4)} = \frac{3}{32}$

6. $\frac{2}{16} \times \frac{1}{10} = \frac{a}{b} \times \frac{c}{d} = \frac{ac}{bd} = \frac{(2)(1)}{(16)(10)} = \frac{2}{160} = \frac{1}{80}$

7. $\frac{10}{16} \div \frac{3}{8} = \frac{a}{b} \div \frac{c}{d} = \frac{ad}{bc} = \frac{(10)(8)}{(16)(3)} = \frac{80}{48} = \frac{5}{3} = 1\frac{2}{3}$

8. $\frac{1}{2} \div \frac{1}{3} = \frac{a}{b} \div \frac{c}{d} = \frac{ad}{bc} = \frac{(1)(3)}{(2)(1)} = \frac{3}{2} = 1\frac{1}{2}$

Practice Problem Set 1-3: Decimal Numbers and Place Value

Write each of the following numbers in expanded form:

1. $43{,}560 = 4(10{,}000) + 3(1{,}000) + 5(100) + 6(10) + 0(1)$
2. $5{,}280.5 = 5(1{,}000) + 2(100) + 8(10) + 0(1) + 5\left(\frac{1}{10}\right)$
3. $3.14 = 3(1) + 1\left(\frac{1}{10}\right) + 4\left(\frac{1}{100}\right)$
4. $0.05 = 0(1) + 0\left(\frac{1}{10}\right) + 5\left(\frac{1}{100}\right)$
5. $24.175 = 2(10) + 4(1) + 1\left(\frac{1}{10}\right) + 7\left(\frac{1}{100}\right) + 5\left(\frac{1}{1{,}000}\right)$

Rewrite each of the following fractions in the form of a decimal fraction:

6. $\frac{1}{5} = \frac{1}{5} \times \frac{2}{2} = \frac{2}{10}$
7. $\frac{4}{20} = \frac{4}{20} \times \frac{5}{5} = \frac{20}{100} = \frac{2}{10}$
8. $\frac{3}{4} = \frac{3}{4} \times \frac{25}{25} = \frac{75}{100}$
9. $\frac{1}{25} = \frac{1}{25} \times \frac{4}{4} = \frac{4}{100}$
10. $\frac{7}{5} = \frac{7}{5} \times \frac{2}{2} = \frac{14}{10}$

Practice Problem Set 1-4: Converting Decimals into Fractions and Fractions into Decimals

Convert the following decimals into decimal fractions:

1. $0.50 = \frac{5}{10}$
2. $0.88 = \frac{88}{100}$
3. $0.125 = \frac{125}{1{,}000}$
4. $0.3456 = \frac{3{,}456}{10{,}000}$
5. $0.75896 = \frac{75{,}896}{100{,}000}$

Convert the following fractions into decimal form:

6. $\frac{3}{8} = 3 \div 8 = 0.375$
7. $\frac{25}{75} = 25 \div 75 = 0.\overline{33}$
8. $\frac{75}{100} = 75 \div 100 = 0.75$
9. $\frac{13}{16} = 13 \div 16 = 0.8125$
10. $\frac{9}{12} = 9 \div 12 = 0.75$

Practice Problem Set 1-5: Scientific Notation, Exponential Notation, and Square Root

Express the following numbers using scientific notation:
1. $56,200,000 = 5.62 \times 10^7$
2. $0.0000302 = 3.02 \times 10^{-5}$
3. $775,000 = 7.75 \times 10^5$
4. $310,000,000,000 = 3.10 \times 10^{11}$
5. $0.000000000922 = 9.22 \times 10^{-10}$

Provide the numerical equivalent to each of the following numbers expressed as exponents:
6. $12^2 = 12 \times 12 = 144$
7. $4^4 = 4 \times 4 \times 4 \times 4 = 256$
8. $3^{-1} = \frac{1}{3} = 0.\overline{33}$
9. $7^0 = 1$
10. $4^{-2} = \frac{1}{4} \times \frac{1}{4} = \frac{1}{16} = 0.0625$

Complete the following problems involving square roots:
11. $\sqrt{625} = 25$
12. $\sqrt{2500} = 50$
13. $\sqrt{562} = 23.71$
14. What is the length of the hypotenuse of a right triangle when one side is 15 ft and the other side is 20 ft?
$$a^2 + b^2 = c^2$$
$$15^2 + 20^2 = c^2$$
$$225 + 400 = c^2$$
$$c^2 = 625$$
$$c = 25$$

15. What is the length of the hypotenuse of a right triangle when one side is 9 ft and the other side is 12 ft?
$$a^2 + b^2 = c^2$$
$$9^2 + 12^2 = c^2$$
$$81 + 144 = c^2$$
$$c^2 = 225$$
$$c = 15$$

Practice Problem Set 1-6: Order of Mathematical Operations

Complete the following problems using PEMDAS:

1.
$$\frac{5}{3}[(3+4) + 2(8+3+5) + 4(5+5+6+4)] = \frac{5}{3}[(7) + 2(16) + 4(20)]$$
$$= \frac{5}{3}[7 + 32 + 80]$$
$$= \frac{5}{3}[119] = 198.\overline{33}$$

2.
$$\sqrt{6(6-4)(6-3)(6-5)} = \sqrt{6(2)(3)(1)} = \sqrt{36} = 6$$

3.
$$32(6^2 + 5^3) = 32(36 + 125) = 32(161) = 5{,}152$$

4.
$$7(3+4^2) + 8(25+7^3) = 7(3+16) + 8(25+343) = 7(19) + 8(368)$$
$$= 133 + 2{,}944 = 3{,}077$$

5.
$$2(21 \div 3) \div (3+4) = 2(7) \div (7) = 14 \div 7 = 2$$

Practice Problem Set 1-7: Solving for x

Isolate and solve for x.

1.
$$x + 25 = 100$$
$$(x + 25) - 25 = 100 - 25$$
$$x = 75$$

2.
$$\frac{x}{7} = 49$$
$$\frac{x}{7} \times 7 = 49 \times 7$$
$$x = 343$$

3.
$$x^2 + 16 = 25$$
$$(x^2 + 16) - 16 = 25 - 16$$
$$x^2 = 9$$
$$\sqrt{x^2} = \sqrt{9}$$
$$x = 3$$

Appendix E Solutions to Practice Problems **313**

4.
$$2x + (7 - x) = 10$$
$$2x + (7 - x) - 2x = 10 - 2x$$
$$7 - x = 10 - 2x$$
$$7 - x + 2x = 10 - 2x + 2x$$
$$7 + x = 10$$
$$7 + x - 7 = 10 - 7$$
$$x = 3$$

5.
$$x(4 + 20) = 384$$
$$x(24) = 384$$
$$\frac{x(24)}{24} = \frac{384}{24}$$
$$x = 16$$

Practice Problem Set 1-8: Ratios and Proportions

1. Convert 1.75 lb to grams using a proportion. (1 lb = 453.59243 g)

   Set up a proportion and then isolate and solve for x.

   $$\frac{453.59243 \text{ g}}{1.00 \text{ lb}} = \frac{x \text{ g}}{1.75 \text{ lb}}$$
   $$x = 453.59243 \times 1.75$$
   $$x = 793.79 \text{ g}$$

2. How many fluid ounces of oil need to be added to 1.5 gallons of gas if the desired ratio of gas to oil is 32 to 1 (Hint:1 gallon = 128 fluid ounces)?

   Step 1: Convert 1.5 gal to fluid ounces using a proportion.

   $$\frac{128 \text{ fl oz}}{1.0 \text{ gal}} = \frac{x \text{ fl oz}}{1.5 \text{ gal}}$$
   $$x = 1.5 \times 128$$
   $$x = 192 \text{ fl oz}$$

   192 fl oz are equivalent to 1.5 gal.

314 Appendix E Solutions to Practice Problems

Step 2: Set up a proportion and then isolate and solve for x.

$$\frac{1 \text{ fl oz of oil}}{32 \text{ fl oz of gas}} = \frac{x \text{ fl oz of oil}}{192 \text{ fl oz of gas}}$$

$$x \times 32 = 192$$

$$x = \frac{192}{32}$$

$$x = 6 \text{ fl oz of oil}$$

Solution
Six fl oz of oil should be added to 1.5 gal of gas to achieve a 32:1 gas to oil ratio.

3. If a nursery manager would like to prepare 2 yd³ of potting mix with a ratio of 2 parts peat moss to 2 parts aged bark to 1 part sand, how many cubic feet of each component are needed to prepare the mix? (Hint: 1 yd³ = 27 ft³)

Step 1: Convert cubic yards of potting mix to cubic feet by using a proportion.

$$\frac{27 \text{ ft}^3}{1 \text{ yd}^3} = \frac{x \text{ ft}^3}{2 \text{ yd}^3}$$

$$x = 2 \times 27$$

$$x = 54$$

54 ft³ is equivalent to 2 yd³.

Step 2: How many parts total are there in this mix?

$$2 + 2 + 1 = 5$$

There are five total parts in the mix.

Step 3a: Set up a proportion to determine how many cubic feet of peat moss or bark are required for the mix.

$$\frac{2 \text{ parts peat moss or bark}}{5 \text{ parts mix}} = \frac{x \text{ ft}^3 \text{ of peat moss or bark}}{54 \text{ ft}^3 \text{ of mix}}$$

$$x \times 5 = 54 \times 2$$

$$x = \frac{54 \times 2}{5}$$

$$x = 21.6 \text{ ft}^3 \text{ each of peat moss and aged bark}$$

Step 3b: Set up a proportion to determine how many cubic feet of sand are required for the mix.

$$\frac{1 \text{ part sand}}{5 \text{ parts mix}} = \frac{x \text{ ft}^3 \text{ of peat moss or bark}}{54 \text{ ft}^3 \text{ of mix}}$$

$$x \times 5 = 54$$

$$x = \frac{54}{5}$$

$$x = 10.8 \text{ ft}^3 \text{ of sand}$$

Solution

To prepare 54 ft³ of potting mix, 10.8 ft³ of sand, 21.6 ft³ of aged bark, and 21.6 ft³ of peat moss are required.

4. Rate of speed equals the ratio of distance to time (Rate of Speed = $\frac{\text{Distance}}{\text{Time}}$). If a tractor travels at a rate of speed of 3 mph, then how long will it take (in seconds) for the tractor to travel 100 ft? (Hint: 5,280 ft = 1 mile and 1 hour = 3,600 seconds)

Step 1: Convert miles per hour to feet per second.
One hour is equivalent to 3,600 seconds.

$$\frac{3 \text{ miles}}{1 \text{ hour}} = \frac{x \text{ feet}}{3,600 \text{ seconds}}$$

Three miles are equivalent to how many feet? Set up a proportion.

$$\frac{5,280 \text{ feet}}{1 \text{ mile}} = \frac{x \text{ feet}}{3 \text{ miles}}$$

$$x = 5,280 \times 3$$

$$x = 15,840 \text{ feet}$$

$$\frac{3 \text{ miles}}{1 \text{ hour}} = \frac{15,840 \text{ feet}}{3,600 \text{ seconds}}$$

Step 2: Set up a proportion to determine how long it will take for the tractor to travel 100 ft.

$$\frac{15,840 \text{ feet}}{3,600 \text{ seconds}} = \frac{100 \text{ feet}}{x \text{ seconds}}$$

$$x \times 15,840 = 3,600 \times 100$$

$$x = \frac{3,600 \times 100}{15,840}$$

$$x = 22.7 \text{ seconds}$$

Solution

It will take 22.7 sec for the tractor to travel 100 ft.

5. If the recommended rate of fertilizer is $\frac{1.5 \text{ lb fertilizer}}{100 \text{ ft}^2 \text{ of garden}}$, then how many pounds of fertilizer are required for 750 ft² of garden space?

Set up a proportion to determine how many pounds of fertilizer are required for 750 ft² of garden space.

$$\frac{1.5 \text{ lb fertilizer}}{100 \text{ ft}^2 \text{ of garden space}} = \frac{x \text{ lb fertilizer}}{750 \text{ ft}^2 \text{ of garden space}}$$

$$x \times 100 = 1.5 \times 750$$

$$x = \frac{1.5 \times 750}{100}$$

$$x = 11.25 \text{ lb}$$

A total of 11.25 lb of fertilizer are required for 750 ft² of garden space.

6. If 750 pots of a 25,000-pot chrysanthemum crop are not suitable for sale, what is the percentage of loss for this production cycle?

Seven hundred fifty pots represent what percentage shrinkage of the crop? Set up a ratio.

$$\frac{750 \text{ pots lost}}{25,000 \text{ pots total}} = 0.03 \text{ percentage lost}$$

Percentage expressed with a percent sign.

$$0.03 \times 100 = 3\%$$

Solution

This production cycle has 3% shrinkage.

Chapter 2

Practice Problem Set 2-1: Measurement, Accuracy, and Precision

1. Which of the following represent the proper way to report a measurement of 25 g from a gram-scale with a degree of accuracy of ± 0.01?

    a. 25
    b. 25.0
    c. 25.00
    d. 25.000

    Answer: c or 25.00 is correct

2. Which of the following represents the proper way to report a measurement of 5 hundredths of a gram from an analytical balance with a degree of accuracy of ± 0.0001?

Appendix E Solutions to Practice Problems **317**

a. 0.0500
b. 0.5000
c. 0.0005
d. 0.0050

Answer: a or 0.0500 is correct

3. For each of the measurements indicated in Figure 2-4, report the result to the proper degree of accuracy.

Answers:
a. 3.25
b. 3.60
c. 4.00
d. 4.05

Practice Problem Set 2-2: Significant Digits

For each of the numbers below, report the number of significant digits.

1. 4.36   Answer: 3
2. 300   Answer: 1
3. 310   Answer: 2
4. 312   Answer: 3
5. 7.3×10^3 Answer: 2
6. 0.00063   Answer: 2
7. 00.2000   Answer: 4
8. 5,673   Answer: 4
9. 10,213.789   Answer: 8
10. 3.75×10^{12} Answer: 3

Practice Problem Set 2-3: Rounding Numbers

Round the following numbers to one significant digit:

1. 3.75   Answer: 4
2. 5.3   Answer: 5
3. 4.03   Answer: 4
4. 4.5   Answer: 5

Round the following numbers to three significant digits:

5. 365.6   Answer: 366
6. 0.004573   Answer: 0.00457
7. 0.4207   Answer: 0.421
8. 7.655×10^3 Answer: 7.66×10^3

318 Appendix E Solutions to Practice Problems

Round the following numbers to two significant digits:
9. 0.253   Answer: 0.25
10. 75.4   Answer: 75
11. 0.00729   Answer: 0.0073
12. 0.301   Answer: 0.30

Practice Problem Set 2-4: Converting Units of Temperature

Convert the following temperatures in Celsius to Fahrenheit:
1. 30°C

$$°F = \frac{9}{5}(°C) + 32$$
$$°F = \frac{9}{5}(30) + 32$$
$$°F = 54 + 32$$
$$°F = 86$$

2. 27°C

$$°F = \frac{9}{5}(°C) + 32$$
$$°F = \frac{9}{5}(27) + 32$$
$$°F = 48.6 + 32$$
$$°F = 80.6$$

3. 5°C

$$°F = \frac{9}{5}(°C) + 32$$
$$°F = \frac{9}{5}(5) + 32$$
$$°F = 9 + 32$$
$$°F = 41$$

Convert the following temperatures in Fahrenheit to Celsius:
4. 65°F

$$°C = \frac{5}{9}(°F - 32)$$
$$°C = \frac{5}{9}(65 - 32)$$

$$°C = \frac{5}{9}(33)$$

$$°C = 18.3$$

5. 78°F

$$°C = \frac{5}{9}(°F - 32)$$

$$°C = \frac{5}{9}(78 - 32)$$

$$°C = \frac{5}{9}(46)$$

$$°C = 25.6$$

6. 29°F

$$°C = \frac{5}{9}(°F - 32)$$

$$°C = \frac{5}{9}(29 - 32)$$

$$°C = \frac{5}{9}(-3)$$

$$°C = -1.7$$

Practice Problem Set 2-5: Unit Conversions

Perform the following unit conversions using equivalents:

1. 4 inches to centimeters

$$4 \text{ in.} = x \text{ cm}$$

$$\frac{2.54 \text{ cm}}{1 \text{ in.}} = \frac{x \text{ cm}}{4 \text{ in.}}$$

$$x = 2.54 \times 4$$

$$x = 10.16 \text{ cm}$$

$$4 \text{ in.} = 10.16 \text{ cm}$$

2. 7,200 square inches to square feet

$$7,200 \text{ in.}^2 = x \text{ ft}^2$$

$$\frac{144 \text{ in.}^2}{1 \text{ ft}^2} = \frac{7,200 \text{ in.}^2}{x \text{ ft}^2}$$

320 Appendix E Solutions to Practice Problems

$$7{,}200 = 144 \times x$$

$$x = \frac{7{,}200}{144}$$

$$x = 50 \text{ ft}^2$$

$$7{,}200 \text{ in.}^2 = 50 \text{ ft}^2$$

3. 100 gallons to cubic feet

$$100 \text{ gal} = x \text{ ft}^3$$

$$\frac{0.1337 \text{ ft}^3}{1 \text{ gal}} = \frac{x \text{ ft}^3}{100 \text{ gal}}$$

$$x = 0.1337 \times 100$$

$$x = 13.37 \text{ ft}^3$$

$$100 \text{ gal} = 13.37 \text{ ft}^3$$

4. 3,484,800 square feet to acres

$$3{,}484{,}800 \text{ ft}^2 = x \text{ ac}$$

$$\frac{43{,}560 \text{ ft}^2}{1 \text{ ac}} = \frac{3{,}484{,}800 \text{ ft}^2}{x \text{ ac}}$$

$$3{,}484{,}800 = 43{,}560 \times x$$

$$x = \frac{3{,}484{,}800}{43{,}560}$$

$$x = 80 \text{ ac}$$

$$3{,}484{,}800 \text{ ft}^2 = 80 \text{ ac}$$

5. 2.5 gallons to fluid ounces

$$2.5 \text{ gal} = x \text{ fl oz}$$

$$\frac{128 \text{ fl oz}}{1 \text{ gal}} = \frac{x \text{ fl oz}}{2.5 \text{ gal}}$$

$$x = 128 \times 2.5$$

$$x = 320 \text{ fl oz}$$

$$2.5 \text{ gal} = 320 \text{ fl oz}$$

Perform the following unit conversions using conversion factors:

6. 212.5 centimeters to inches

$$212.5 \text{ cm} = x \text{ in.}$$
$$212.5 \text{ cm} \times 0.3937008 \text{ in./cm} = 83.66142 \text{ in.}$$
$$212.5 \text{ cm} = 83.7 \text{ in.}$$

7. 7 liquid pints to liters

$$7 \text{ pt} = x \text{ L}$$
$$7 \text{ pt} \times 0.47317648 \text{ L/pt} = 3.31223536 \text{ L}$$
$$7 \text{ pt} = 3.3 \text{ L}$$

8. 45.72 meters to yards

$$45.72 \text{ m} = x \text{ yd}$$
$$45.72 \text{ m} \times 1.093613 \text{ yd/m} = 49.99998636 \text{ yd}$$
$$45.72 \text{ m} = 50 \text{ yd}$$

9. 27 acres to square meters

$$27 \text{ ac} = x \text{ m}^2$$
$$27 \text{ ac} \times 4{,}046.873 \text{ m}^2/\text{ac} = 109{,}265.571 \text{ m}^2$$
$$27 \text{ ac} = 109{,}265.6 \text{ m}^2$$

10. 91 liters to gallons

$$91 \text{ L} = x \text{ gal}$$
$$91 \text{ L} \times 0.2641721 \text{ gal}/L = 24.0396611$$
$$91 \text{ L} = 24.0 \text{ gal}$$

Chapter 3

Practice Problem Set 3-1: Area of Quadrilaterals

Calculate the area of each rectangle or square.

1. $l = 25$ ft, $w = 15$ ft (unit = ft^2)

$$\text{Area} = (l)(w)$$
$$(25 \text{ ft})(15 \text{ ft}) = 375 \text{ ft}^2$$

322 Appendix E Solutions to Practice Problems

2. $l = 5$ ft, $w = 27$ in. (unit = ft^2)
Convert 27 in. into ft

$$\frac{27 \text{ in.}}{12 \text{ in./ft}} = 2.25 \text{ ft}$$

$$\text{Area} = (l)(w)$$

$$(5 \text{ ft})(2.25 \text{ ft}) = 11.25 \text{ ft}^2$$

3. $s = 11$ ft (unit = ft^2)

$$\text{Area} = s^2$$

$$11^2 \text{ ft} = 121 \text{ ft}^2$$

Calculate the area of each trapezoid.

4. $b_1 = 8$ ft, $b_2 = 13$ ft, $h = 5$ ft (unit = ft^2)

$$\text{Area} = [(b_1 + b_2) \div 2] \times \text{height}$$
$$\text{Area} = [(8 \text{ ft} + 13 \text{ ft}) \div 2] \times 5 \text{ ft}$$
$$\text{Area} = [21 \text{ ft} \div 2] \times 5 \text{ ft}$$
$$\text{Area} = 10.5 \text{ ft} \times 5 \text{ ft}$$
$$\text{Area} = 52.5 \text{ ft}^2$$

5. $b_1 = 18$ in., $b_2 = 16$ in., $h = 0.75$ ft (unit = in.^{2})
Convert 0.75 ft into inches.

$$(0.75 \text{ ft})(12 \text{ in./ft}) = 9 \text{ in.}$$
$$\text{Area} = [(b_1 + b_2) \div 2] \times \text{height}$$
$$\text{Area} = [(18 \text{ in.} + 16 \text{ in.}) \div 2] \times 9 \text{ in.}$$
$$\text{Area} = [34 \text{ in.} \div 2] \times 9 \text{ in.}$$
$$\text{Area} = 17 \text{ in.} \times 9 \text{ in.}$$
$$\text{Area} = 153 \text{ in.}^2$$

6. $b_1 = 8,000$ ft, $b_2 = 10,000$ ft, $h = 250$ ft (unit = ac)

$$\text{Area} = [(b_1 + b_2) \div 2] \times \text{height}$$
$$\text{Area} = [(8,000 \text{ ft} + 10,000 \text{ ft}) \div 2] \times 250 \text{ ft}$$
$$\text{Area} = [18,000 \text{ ft} \div 2] \times 250 \text{ ft}$$
$$\text{Area} = 9,000 \text{ ft} \times 250 \text{ ft}$$
$$\text{Area} = 2,250,000 \text{ ft}^2$$

There are 43,560 ft² in 1 ac.
$$\frac{2{,}250{,}000 \text{ ft}^2}{43{,}560 \text{ ft}^2} = 51.65 \text{ ac}$$

Practice Problem Set 3-2: Area of Triangles

Calculate the area of each triangle.

1. $h = 10$ ft, $b = 15$ ft

$$\text{Area} = (b \times h) \div 2$$
$$\text{Area} = (10 \text{ ft} \times 15 \text{ ft}) \div 2$$
$$\text{Area} = 150 \text{ ft}^2 \div 2$$
$$\text{Area} = 75 \text{ ft}^2$$

2. $h = 3$ ft, $b = 21$ ft

$$\text{Area} = (b \times h) \div 2$$
$$\text{Area} = (3 \text{ ft} \times 21 \text{ ft}) \div 2$$
$$\text{Area} = 63 \text{ ft}^2 \div 2$$
$$\text{Area} = 31.5 \text{ ft}^2$$

3. $a = 5$ ft, $b = 10$ ft, $c = 12$ ft

$$\text{Area} = \sqrt{s(s-a)(s-b)(s-c)}$$
$$s = \tfrac{1}{2}(5 + 10 + 12) = \tfrac{1}{2}(27) = 13.5$$
$$\text{Area} = \sqrt{13.5(13.5 - 5)(13.5 - 10)(13.5 - 12)}$$
$$\text{Area} = \sqrt{13.5(8.5)(3.5)(1.5)}$$
$$\text{Area} = \sqrt{602.4}$$
$$\text{Area} = 24.5 \text{ ft}^2$$

4. Calculate the length of the hypotenuse using the Pythagorean Theorem in Figure 3-12 that has a base of 13 ft and a height of 4 ft.

$a^2 + b^2 = c^2$ is the formula for calculating the lengths of the sides of a right triangle. $a = 13$ ft, $b = 4$ ft and c is unknown.

$$13^2 \text{ ft} + 4^2 \text{ ft} = c^2 \text{ ft}$$
$$169 \text{ ft} + 16 \text{ ft} = 185 \text{ ft}$$
$$c^2 = 185 \text{ ft}$$
$$\text{The length of } c = \sqrt{185} \text{ ft}$$
$$c = 13.6 \text{ ft}$$

Practice Problem Set 3-3: Area of Circles and Ellipses

Calculate the area of each circle.

1. $r = 12$ ft (unit = ft^2)

$$\text{Area} = \pi(r^2)$$
$$A = \pi(12 \text{ ft})^2$$
$$A = \pi(144 \text{ ft}^2)$$
$$A = 452.4 \text{ ft}^2$$

2. $d = 52$ ft (unit = ft^2)

$$\text{Area} = \pi(r^2)$$

First determine the radius (r) of the circle. $r = \frac{1}{2}d$

$$\tfrac{1}{2}(52 \text{ ft}) = 26 \text{ ft}$$
$$A = \pi(26 \text{ ft})^2$$
$$A = \pi(676 \text{ ft}^2)$$
$$A = 2123.7 \text{ ft}^2$$

3. $c = 27$ yd (unit = yd^2)

$$\text{Area} = \frac{c^2}{4(\pi)}$$
$$A = \frac{27^2}{4(\pi)}$$
$$A = \frac{729 \text{ yd}^2}{12.57}$$
$$A = 58.0 \text{ yd}^2$$

Calculate the area of each ellipse.

4. major axis = 29 ft, minor axis = 11 ft (unit = ft^2)

$$\text{Area} = \pi(r_{\text{major}} \times r_{\text{minor}})$$

Determine the major and minor radii by dividing the major and minor axis by 2. The minor axis is 11 ft and the major axis is 29 ft.

$$r_{\text{minor}} = \tfrac{1}{2}(11 \text{ ft}) = 5.5 \text{ ft}$$
$$r_{\text{major}} = \tfrac{1}{2}(29 \text{ ft}) = 14.5 \text{ ft}$$
$$A = \pi[(5.5 \text{ ft})(14.5 \text{ ft})]$$

$$A = \pi(79.75 \text{ ft}^2)$$
$$A = 250.5 \text{ ft}^2$$

5. major axis = 6 ft, minor axis = 2 yd (unit = ft^2)
$$\text{Area} = \pi[(wr)(lr)]$$

First convert yards into feet and then determine the length of the major and minor radii.
There are 3 ft in 1 yd, thus, 2 yd × 3 ft/yd = 6 ft.
$$r_{minor} = \tfrac{1}{2}(6) = 3 \text{ ft}$$
$$r_{major} = \tfrac{1}{2}(6) = 3 \text{ ft}$$
$$A = \pi[(3 \text{ ft})(3 \text{ ft})]$$
$$A = \pi(9 \text{ ft}^2)$$
$$A = 28.27 \text{ ft}^2$$

Practice Problem Set 3-4: Area of Irregular Figures

1. Calculate the area of Figure 3-35 using the Offset Method.

   Step 1: The length has been determined to be 90 ft
   Step 2: The offset lines are spaced 10 ft apart.
   Step 3: Each offset line measures as follows: 23.8 ft, 34.3 ft, 38.8 ft, 38.5 ft, 32.5 ft, 30 ft, 28.5 ft, 18.5 ft.
   Step 4: Add up the length of all the offset lines.
   $$23.8 \text{ ft} + 34.3 \text{ ft} + 38.8 \text{ ft} + 38.5 \text{ ft} + 32.5 \text{ ft}$$
   $$+ 30 \text{ ft} + 28.5 \text{ ft} + 18.5 \text{ ft} = 244.9 \text{ ft}^2$$
   Step 5: Multiply the sum from Step 4 by the distance between offset lines.
   $$(244.9 \text{ ft})(10 \text{ ft}) = 2{,}449 \text{ ft}^2$$

2. Calculate the area of Figure 3-36 using the Modified Offset Method.

   Step 1: The length is determined 33 ft and the width is 15 ft.
   Step 2: The offset lines are spaced every 3 ft.
   Step 3: Measure the length of the offset line segments.
   Each pair of offset lines measures as follows: (E1 = 3 ft and E2 = 1.25 ft), (F1 = 1.0 ft and F2 = 0.75 ft), (G1 = 0.75 ft and G2 = 0.75 ft), (H1 = 1.25 ft and H2 = 1 ft), (I1 = 2.25 ft and I2 = 1.75 ft), (J1 = 3.25 ft and J2 = 1.25 ft), (K1 = 3 ft and K2 = 0.75 ft), (L1 = 2.25 ft and L2 = 1.5 ft), (M1 = 2.0 ft and M2 = 3.0 ft), and (N1 = 2.75 ft and N2 = 4 ft).

Then add up each pair of offset measurements.

$$\Sigma E = E1 + E2 = 3\,\text{ft} + 1.25\,\text{ft} = 4.25\,\text{ft}$$
$$\Sigma F = F1 + F2 = 1\,\text{ft} + 0.75\,\text{ft} = 1.75\,\text{ft}$$
$$\Sigma G = G1 + G2 = 0.75\,\text{ft} + 0.75\,\text{ft} = 1.5\,\text{ft}$$
$$\Sigma H = H1 + H2 = 1.25\,\text{ft} + 1\,\text{ft} = 2.25\,\text{ft}$$
$$\Sigma I = I1 + I2 = 2.25\,\text{ft} + 1.75\,\text{ft} = 4\,\text{ft}$$
$$\Sigma J = J1 + J2 = 3.25\,\text{ft} + 1.25\,\text{ft} = 4.5\,\text{ft}$$
$$\Sigma K = K1 + K2 = 3\,\text{ft} + 0.75\,\text{ft} = 3.75\,\text{ft}$$
$$\Sigma L = L1 + L2 = 2.25\,\text{ft} + 1.5\,\text{ft} = 3.75\,\text{ft}$$
$$\Sigma M = M1 + M2 = 2\,\text{ft} + 3\,\text{ft} = 5\,\text{ft}$$
$$\Sigma N = N1 + N2 = 2.75\,\text{ft} + 4\,\text{ft} = 6.75\,\text{ft}$$

Step 4: Subtract each of the sums of the line segments (E through I) from the width of the rectangle. The width is 15 ft.

$$E = (w - \Sigma E) \text{ or } 15\,\text{ft} - 4.25\,\text{ft} = 10.75\,\text{ft}$$
$$F = (w - \Sigma F) \text{ or } 15\,\text{ft} - 1.75\,\text{ft} = 13.25\,\text{ft}$$
$$G = (w - \Sigma G) \text{ or } 15\,\text{ft} - 1.5\,\text{ft} = 13.5\,\text{ft}$$
$$H = (w - \Sigma H) \text{ or } 15\,\text{ft} - 2.25\,\text{ft} = 12.75\,\text{ft}$$
$$I = (w - \Sigma I) \text{ or } 15\,\text{ft} - 4\,\text{ft} = 11\,\text{ft}$$
$$J = (w - \Sigma J) \text{ or } 15\,\text{ft} - 4.5\,\text{ft} = 10.5$$
$$K = (w - \Sigma K) \text{ or } 15\,\text{ft} - 3.75\,\text{ft} = 11.25\,\text{ft}$$
$$L = (w - \Sigma L) \text{ or } 15\,\text{ft} - 3.75\,\text{ft} = 11.25\,\text{ft}$$
$$M = (w - \Sigma M) \text{ or } 15\,\text{ft} - 5\,\text{ft} = 10\,\text{ft}$$
$$N = (w - \Sigma N) \text{ or } 15\,\text{ft} - 6.75\,\text{ft} = 8.25\,\text{ft}$$

Step 5: Add up the widths of the figure as determined in Step 4 for each of the offsets.

$$10.75\,\text{ft} + 13.25\,\text{ft} + 13.5\,\text{ft} + 12.75\,\text{ft} + 11\,\text{ft} + 10.5\,\text{ft} + 11.25\,\text{ft}$$
$$+ 11.25\,\text{ft} + 10\,\text{ft} + 8.25\,\text{ft} = 112.5\,\text{ft}$$

Step 6: Multiply the summed value found in Step 5 by the distance between the offsets (3 ft).

$$(112.5\,\text{ft})(3\,\text{ft}) = 337.5\,\text{ft}^2$$

3. Calculate the area of Figure 3-37 using Simpson's Rule.

Appendix E Solutions to Practice Problems **327**

Insert your measurements into the following formula where $h = 9$ ft:

$$A = \frac{h}{3}[(L_1 + L_9) + 2(L_3 + L_5 + L_7) + 4(L_2 + L_4 + L_6 + L_8)]$$

Measurement:

$L_1 = 26$ ft $\quad L_4 = 36$ ft $\quad L_7 = 36$ ft
$L_2 = 34$ ft $\quad L_5 = 33$ ft $\quad L_8 = 34$ ft
$L_3 = 38$ ft $\quad L_6 = 35$ ft $\quad L_9 = 30$ ft

$$A = \frac{9}{3}[(26 + 30) + 2(38 + 33 + 36) + 4(34 + 36 + 35 + 34)]$$

$$A = \frac{9}{3}[(56) + 2(107) + 4(139)]$$

$$A = \frac{9}{3}[56 + 214 + 556]$$

$$A = \frac{9}{3}(826)$$

$$A = 2{,}478 \text{ ft}^2$$

Practice Problem Set 3-5: Volume of Geometric Figures

1. Calculate the volume of a rectangular-shaped prism that has a base that measures 6 ft by 10 ft and a height of 14 ft.

$$B = l \times w$$
$$B = 6 \text{ ft} \times 10 \text{ ft}$$
$$B = 60 \text{ ft}^2$$
$$V = \text{base} \times \text{height}$$
$$V = 60 \text{ ft}^2 \times 14 \text{ ft}$$
$$V = 840 \text{ ft}^3$$

2. Calculate the volume of a cylinder that has a base with a diameter of 12 in. and a height of 3 ft.

$$B = \pi r^2$$
$$B = \pi (0.5^2)$$
$$B = \pi \times 0.25 \text{ ft}^2$$
$$B = 0.7854 \text{ ft}^2$$
$$V = \text{base} \times \text{height}$$

328 Appendix E Solutions to Practice Problems

$$V = 0.7854 \text{ ft}^2 \times 3 \text{ ft}$$
$$V = 2.356 \text{ ft}^3$$

3. Calculate the volume of a cone that has a base with a radius of 1 ft 3 in. and a height of 2.5 ft.

Convert 1 ft 3 in. into ft. If 1 ft equals 12 in., how many feet are in 3 in.?

$$\frac{1 \text{ ft}}{12 \text{ in.}} = \frac{x \text{ ft}}{3 \text{ in.}}$$

$$\frac{3 \text{ in.}}{12 \text{ in.}} = 0.25 \text{ ft}$$

1 ft 3 in. equals 1.25 ft

$$B = \pi r^2$$
$$B = \pi (1.25^2)$$
$$B = \pi \times 1.5625 \text{ ft}^2$$
$$B = 4.91 \text{ ft}^2$$
$$V = \frac{1}{3}(B \times h)$$
$$V = \frac{1}{3}(4.91 \text{ ft}^2 \times 2.5 \text{ ft})$$
$$V = \frac{1}{3}(12.27 \text{ ft}^3)$$
$$V = 4.09 \text{ ft}^3$$

4. Calculate the volume of a sphere that has a diameter of 18 in.

$$r = \frac{1}{2} 18 \text{ in.}$$
$$r = 9 \text{ in.}$$
$$V = \frac{4}{3}\pi r^3$$
$$V = \frac{4}{3}[\pi (9^3)]$$
$$V = \frac{4}{3}[\pi (729 \text{ in.}^3)]$$
$$V = \frac{4}{3} \times 2290.2 \text{ in.}^3$$
$$V = 3053.6 \text{ in.}^3$$

Chapter 4

Practice Problem Set 4-1: Area of Geometric Figures

Calculate the area of each of the following:

1. A rugby field that has a width of 225 ft and a length of 330 ft (unit = ft²).

   Area of a rectangle = Length × Width

   Area = 225 ft × 330 ft

   Area of the rugby field = 74,250 ft²

2. A lacrosse field that has a width of 180 ft and a length of 330 ft (unit = ft²).

   Area of a rectangle = Length × Width

   Area = 180 ft × 330 ft

   Area of the lacrosse field = 59,400 ft²

3. An oval-shaped flower bed that has a diameter of 25 ft and a length of 65 ft (unit = ft²).

   First determine half of the total width and half of the total length.

   $wr = \frac{1}{2}$ (25) ft = 12.5 ft

   $lr = \frac{1}{2}$ (65) ft = 32.5 ft

   Area of an ellipse = $\pi[(wr)(lr)]$

   Area = $\pi[(12.5 \text{ ft})(32.5)]$

   Area = $\pi(406.25 \text{ ft}^2)$

   Area of the flower bed = 1,276 ft²

4. There are 12 trees in a landscape that have circular tree rings that measure 8 ft in diameter.

   Radius (r) = $\frac{1}{2}$ diameter (d)

   $r = \frac{1}{2} 8$ ft

   $r = 4$ ft

   a. What is the area of one tree ring?

   Area of a circle = πr^2

   Area = $\pi (4 \text{ ft})^2$

   Area = $\pi (16 \text{ ft}^2)$

   Area of one tree ring = 50.27 ft²

b. What is the total area of all the tree rings? (unit = ft^2)
There are eight tree rings that each have a surface area of 50.27 ft^2.

$$\text{Area} = (8 \text{ tree rings})(50.27 \text{ ft}^2 \text{ per tree ring})$$

$$\text{Area of 8 tree rings} = 402.2 \text{ ft}^2$$

Practice Problem Set 4-2: Area of Irregular Landscape Features

1. Calculate the area of the wildflower planting in Figure 4-15. The total length of the area is 110 ft^2.

  a. Using the Offset Method:
    Step 1: The length line is 110 ft.
    Step 2: Offset lines are on 5 ft spacing along the length line.
    Step 3: The measurements for the offset lines are listed in Figure 4-15.
    Step 4: Add the lengths of the offset lines:

$$24 \text{ ft} + 32 \text{ ft} + 38 \text{ ft} + 41 \text{ ft} + 40 \text{ ft} + 37 \text{ ft} + 34 \text{ ft} + 32 \text{ ft} + 31 \text{ ft}$$

$$+ 30 \text{ ft} + 31 \text{ ft} + 33 \text{ ft} + 35 \text{ ft} + 35 \text{ ft} + 33 \text{ ft} + 32 \text{ ft} + 32 \text{ ft} + 33 \text{ ft}$$

$$+ 31 \text{ ft} + 27 \text{ ft} + 20 \text{ ft} = 681 \text{ ft}$$

    Step 5: Multiply the sum of the offset lines by the distance between the offset lines.

$$681 \text{ ft} \times 5 \text{ ft} = 3,405 \text{ ft}^2$$

Solution

The wildflower bed is 3,405 ft^2 as calculated by the Offset Method.

  b. Using Simpson's Rule.
    Step 1: The length of the line between points A and B is 110 ft.
    Step 2: Divide the line into an even number of equally spaced subsegments. This example uses 22 subsegments. The width of the subsegments is called h. For this example h = 5 ft.
    Step 3: Draw lines perpendicular from the line that connects A to B at each mark. Label each line L_1 to L_{21}.
    Step 4: Measure the length of each line.

$L_1 = 24$ ft	$L_8 = 32$ ft	$L_{15} = 33$ ft
$L_2 = 32$ ft	$L_9 = 31$ ft	$L_{16} = 32$ ft
$L_3 = 38$ ft	$L_{10} = 30$ ft	$L_{17} = 32$ ft
$L_4 = 41$ ft	$L_{11} = 31$ ft	$L_{18} = 33$ ft
$L_5 = 40$ ft	$L_{12} = 33$ ft	$L_{19} = 31$ ft
$L_6 = 37$ ft	$L_{13} = 35$ ft	$L_{20} = 27$ ft
$L_7 = 34$ ft	$L_{14} = 35$ ft	$L_{21} = 20$ ft

Step 5: Insert your measurements into the following formula:

$$\text{Area} = \frac{h}{3}[(L_1 + L_{21}) + 2(L_3 + L_5 + L_7 + L_9 + L_{11} + L_{13}$$
$$+ L_{15} + L_{17} + L_{19}) + 4(L_2 + L_4 + L_6 + L_8 + L_{10}$$
$$+ L_{12} + L_{14} + L_{16} + L_{18} + L_{20})]$$

$$\text{Area} = \frac{5}{3}[(24 + 20) + 2(38 + 40 + 34 + 31 + 31 + 35 + 33 + 32 + 31)$$
$$+ 4(32 + 41 + 37 + 32 + 30 + 33 + 35 + 32 + 33 + 27)]$$

$$\text{Area} = \frac{5}{3}[(44) + 2(305) + 4(332)]$$

$$\text{Area} = \frac{5}{3}[44 + 610 + 1{,}328]$$

$$\text{Area} = \frac{5}{3}(1{,}982)$$

$$\text{Area} = 3{,}303 \text{ ft}^2$$

Solution

The wildflower bed is 3,303 ft^2, as calculated by Simpson's Rule.

2. Calculate the area of a small garden pond in Figure 4-16 using the Modified Offset Method.

   Step 1: Create a rectangle around the area that is to be measured. The lengths of line segments AB or CD are equal to 12 ft. The lengths of line segments AC or BD are equal to 10 ft.
   Step 2: Establish offset lines every 2 ft perpendicular to lines AB and CD.
   Step 3: Measure each offset line from the edge of the rectangle to the perimeter of the landscape feature. The offset lines measure as follows:

   E1 = 2.5 ft   F1 = 1.5 ft   G1 = 1.0 ft   H1 = 1.0 ft   I1 = 1.3 ft
   E2 = 2.7 ft   F2 = 2.5 ft   G2 = 2.0 ft   H2 = 1.5 ft   I2 = 1.3 ft

   Step 4: Add the paired offset lines.

   $$\Sigma E = E1 + E2 = 5.2 \text{ ft:}$$
   $$\Sigma F = F1 + F2 = 4 \text{ ft:}$$
   $$\Sigma G = G1 + G2 = 3 \text{ ft:}$$
   $$\Sigma H = H1 + H2 = 2.5 \text{ ft:}$$
   $$\Sigma I = I1 + I2 = 2.6 \text{ ft:}$$

Step 5: Subtract the sum of each offset line from the width to obtain the distance across the landscape feature.

$$E: \quad 10\,\text{ft} - 5.2\,\text{ft} = 4.8\,\text{ft}$$
$$F: \quad 10\,\text{ft} - 4\,\text{ft} = 6\,\text{ft}$$
$$G: \quad 10\,\text{ft} - 3\,\text{ft} = 7\,\text{ft}$$
$$H: \quad 10\,\text{ft} - 2.5\,\text{ft} = 7.5\,\text{ft}$$
$$I: \quad 10\,\text{ft} - 2.6\,\text{ft} = 7.4\,\text{ft}$$

Step 6: Add the lengths of the offset lines:

$$4.8\,\text{ft} + 6\,\text{ft} + 7\,\text{ft} + 7.5\,\text{ft} + 7.4\,\text{ft} = 32.7\,\text{ft}$$

Step 7: Multiply the sum of the offset lines by the distance between the offset lines.

$$32.7\,\text{ft} \times 2\,\text{ft} = 65.4\,\text{ft}^2$$

Solution

The small garden pond has a surface area of 65.4 ft^2.

Chapter 5

Practice Problem Set 5-1: How Much N, P, or K is in the Bag?

1. How many pounds of N, P, and K are there in a 20-lb box of a 10-10-10 garden fertilizer?

$$(20)(0.10) = 2\,\text{lb N}$$
$$(20)(0.10) = 2\,\text{lb P}_2\text{O}_5$$
$$(2)(0.44) = 0.88\,\text{lb P}$$
$$(20)(0.10) = 2\,\text{lb K}_2\text{O}$$
$$(2)(0.83) = 1.66\,\text{lb K}$$

2. How many pounds of N, P, and K are there in a 40-lb bag of a 20-3-9 fertilizer?

$$(40)(0.20) = 8\,\text{lb N}$$
$$(40)(0.03) = 1.2\,\text{lb P}_2\text{O}_5$$
$$(40)(0.09) = 3.6\,\text{lb K}_2\text{O}$$
$$(1.2\,\text{lb P}_2\text{O}_5)(0.44) = 0.53\,\text{lb P}$$
$$(3.6\,\text{lb K}_2\text{O})(0.83) = 3\,\text{lb K}$$

Practice Problem Set 5-2: Determining How Much Fertilizer to Apply When Rate Is Expressed as the Amount of Fertilizer Product per Unit Area

1. A 3,200 ft² garden bed is being prepared for a tulip bulb planting. If a 5-20-20 fertilizer product is to be applied at the rate of 3 lb/100 ft² of garden area, then how many pounds of 5-20-20 are required for this garden?

$$\frac{3 \text{ lb } 5\text{-}20\text{-}20}{100 \text{ ft}^2} = \frac{x \text{ lb } 5\text{-}20\text{-}20}{3,200 \text{ ft}^2}$$

$$100x = 3 \times 3,200$$

$$\frac{100x}{100} = \frac{9,600}{100}$$

$$x = \frac{9,600}{100}$$

$$x = 96 \text{ lb of } 5\text{-}20\text{-}20$$

2. A shrub border measuring 15 feet by 125 feet requires a starter fertilization using 10-10-10 at the rate of 1 lb product/100 ft² of garden area. How many pounds of 10-10-10 are needed for this application?

$$\text{Area of shrub border} = 15 \text{ ft} \times 125 \text{ ft} = 1,875 \text{ ft}^2$$

$$\frac{1 \text{ lb } 10\text{-}10\text{-}10}{100 \text{ ft}^2} = \frac{x \text{ lb } 10\text{-}10\text{-}10}{1,875 \text{ ft}^2}$$

$$100x = 1,875$$

$$\frac{100x}{100} = \frac{1,875}{100}$$

$$x = \frac{1,875}{100}$$

$$x = 18.75 \text{ lb of } 10\text{-}10\text{-}10$$

Practice Problem 5-3: Determining How Much Fertilizer to Apply When Rate Is Expressed as the Amount of N Required per Unit Area

How much 15-3-10 fertilizer would be needed to apply 1 lb N/1,000 ft² to 57,000 ft² of a soccer field and surrounding area?

$$(x)(0.15) = 1 \text{ lb N}$$

$$x = 1/0.15$$

$$x = 6.66 \text{ lb } 15\text{-}3\text{-}10$$

$$1,000x = (6.66)(57,000)$$

$$x = 379,620/1,000$$

$$x = 380 \text{ lb of } 15\text{-}3\text{-}10$$

Practice Problem Set 5-4: Determining How Much Fertilizer to Apply When Rate Is Expressed as the Amount of P or K Required Per Unit Area

1. How much 18-46-0 fertilizer would be needed to apply 1.5 lb P/1,000 ft² to a 70,000 ft² seed bed?

$$(x)(0.44) = 1.5$$
$$x = 1.5/0.44$$
$$x = 3.4 \text{ lb.P}_2\text{O}_5$$
$$(x)(0.46) = 3.4$$
$$x = 3.4/0.46$$
$$x = 7.4 \text{ lb } 18\text{-}46\text{-}0$$
$$1,000x = (7.4)(70,000)$$
$$x = 518,000/1,000$$
$$x = 518 \text{ lb } 18\text{-}46\text{-}0$$

2. How much 0-0-50 fertilizer would be needed to apply 1 lb of K/1,000 ft² to 80,000 ft² of sports fields?

$$(x)(0.83) = 1 \text{ lb K}$$
$$x = 1/0.83$$
$$x = 1.2 \text{ lb K}_2\text{O}$$
$$(x)(0.50) = 1.2 \text{ lb K}_2\text{O}$$
$$x = 1.2/0.50$$
$$x = 2.4 \text{ lb } 0\text{-}0\text{-}50/1,000 \text{ ft}^2$$
$$1,000x = (2.4)(80,000)$$
$$x = 192,000/1,000$$
$$x = 192 \text{ lb } 0\text{-}0\text{-}50 \text{ on } 80,000 \text{ ft}^2$$

3. A sports complex has 200,000 ft² of turf. A total of 3 lb N/1,000 ft² is to be applied during the season using a 22-2-8 fertilizer. The soil test indicates that 3 total lb of K should be applied to 1,000 ft²/year. How much *additional* potassium sulfate (0-0-50) will be needed to achieve the 3 lb K level?

$$(x)(0.22) = 3 \text{ lb N}$$
$$x = 3/0.22$$
$$x = 13.63 \text{ lb of } 22\text{-}2\text{-}8/1,000 \text{ ft}^2$$

How much K is applied per 1,000 ft²?

$$(13.6)(0.08) = 1.09 \text{ lb } K_2O$$

$$(1.09)(0.83) = 0.91 \text{ lb } K/1,000 \text{ ft}^2$$

3 lb K−0.91 lb K = 2.09 additional potassium needed for 1,000 ft²

How much 0-0-50 is needed to apply 1.42 lb K/1,000 ft²?

$$(x)(0.83) = 2.09 \text{ lb } K$$

$$x = 2.09/0.83$$

$$x = 2.52 \text{ lb } K_2O$$

$$(x)(0.50) = 2.52 \text{ lb } K_2O$$

$$x = 2.52/0.50$$

$$x = 5.04 \text{ lb } 0\text{-}0\text{-}50/1,000 \text{ ft}^2$$

How much is needed for 200,000 ft²?

$$1,000x = (5.04)(200,000)$$

$$x = 1,080,000/1,000$$

$$x = 1,008 \text{ lb } 0\text{-}0\text{-}50$$

Practice Problem Set 5-5: Determining How Much Fertilizer to Apply When Rate Is Expressed as the Amount of Oxide Form (P_2O_5 or K_2O) of a Nutritional Element Per Unit Area

1. How much 5-10-5 fertilizer would be needed to apply 2 lb P_2O_5/1,000 ft² to a 3,000 ft² perennial flower bed?

$$(x)(0.10) = 2 \text{ lb } P_2O_5$$

$$x = 2/0.10$$

$$x = 20 \text{ lb } 5\text{-}10\text{-}5/1,000 \text{ ft}^2$$

$$1,000x = (3,000)(20)$$

$$x = 60,000/1,000$$

$$x = 60 \text{ lb } 5\text{-}10\text{-}5$$

2. How many pounds N/1,000 ft² is applied with the application in question number one?

$$(20)(0.05) = 1 \text{ lb N}$$

336 Appendix E Solutions to Practice Problems

3. How much 0-0-60 fertilizer is needed to apply a rate of 2.0 lb K_2O/1,000 ft^2 to a 10,000 ft^2 lawn?

$$(x)(0.60) = 2 \text{ lb } K_2O$$
$$x = 2/0.60$$
$$x = 3.33 \text{ lb } K_2O$$
$$1,000x = (3.33)(10,000)$$
$$x = 33,300/1,000$$
$$x = 33.3 \text{ lb } 0\text{-}0\text{-}60$$

Practice Problem Set 5-6: Calculations Involving Liquid Fertilizer

1. How many gal of a 12-0-4 liquid fertilizer containing 1.2 lb N/gal and 0.33 lb K/gal will be needed to apply 0.5 lb N/1,000 ft^2 to a 60,000 ft^2 sports field?

$$(x)(1.2) = (1)(0.50)$$
$$x = 0.50/1.2$$
$$x = 0.42 \text{ gal needed to apply 0.5 lb N}/1,000\,\text{ft}^2$$
$$1,000x = (0.42)(60,000)$$
$$x = 25,200/1,000$$
$$x = 25.2 \text{ gal of 12-0-4 for } 60,000\,\text{ft}^2$$

2. How much potassium would be applied per 1,000 ft^2 using the product and product rate in question number one?

$$(1)(x) = (0.33)(0.42)$$
$$x = 0.14 \text{ lb K applied}/1,000\,\text{ft}^2$$

3. A hand-operated sprayer that holds 3 gal of spray is to be used to apply a liquid fertilizer to a flower bed. The fertilizer solution is an 8-1-3 that contains 0.74 lb N/gallon. The fertilizer is to be applied at a rate of 0.50 lb N/1,000 ft^2 with a total spray volume of 1 gal of water plus fertilizer. How much fertilizer should be placed in the sprayer to mix 3 gal of solution?

$$0.74x = (1)(0.50)$$
$$x = 0.50/0.74$$
$$x = 0.68 \text{ gal fertilizer}/1,000\,\text{ft}^2$$

Each gallon of spray will be 0.68 gal fertilizer and 0.32 gal water, and 3 gal will require 2.04 gal fertilizer and 0.96 gal water.

Practice Problem Set 5-7: Calculation Based on Rate Expressed in Metric Units

1. How many kilograms of a 15-3-10 fertilizer would be needed to apply 5 g N/m² to an 800 m² lawn?

$$(x)(0.15) = 5\,g$$

$$x = 5/0.15$$

$$x = 33.3\,g$$

$$(1)(x) = (33.3)(800)$$

$$x = 26{,}640/1$$

$$x = 26{,}640\,g \text{ of 15-3-10 or } 26.6\,kg$$

2. How many liters of a 14-2-3 liquid fertilizer with 168 g N/L would be needed to apply 0.50 kg N/100 m² to a 930 m² lawn?

$$0.50\,kg \text{ is } 500\,g$$

$$(168)(x) = (1)(500)$$

$$x = 500/168$$

$$x = 2.98\,L/100\,m^2$$

$$(100)(x) = (2.98)(930)$$

$$x = 2{,}771/100$$

$$x = 27.7\,L$$

Practice Problem 5-8: Fertilizer Economics: Comparing Fertilizer Products Based on Cost per Pound of Nitrogen

What is the cost per pound of nitrogen for a 50-lb bag of a 12-5-9 fertilizer that sells for $12.00/bag?

$$(50)(0.12) = 6\,lb\,N$$

$$\$12/6\,lb\,N = \$2/lb\,N$$

Chapter 6

Practice Problem Set 6-1: Calculations Involving Dry Pesticide Formulations

1. A 2G insecticide is to be applied at a rate of 1 lb a.i./ac to a 30,000 ft² lawn. How much of the insecticide will be needed for this application?

$$(x)(0.02) = 1$$
$$x = 1/0.02$$
$$x = 50 \text{ lb product/acre}$$
$$43{,}560x = (50)(30{,}000)$$
$$x = 1{,}500{,}000/43{,}560$$
$$x = 34.4 \text{ lb product}$$

2. A 50DF fungicide is to be applied at a rate of 0.25 lb a.i./ac to 5 ac of sports fields. How much of this fungicide will be needed for the application?

$$(x)(0.50) = 0.25 \text{ lb}$$
$$x = 0.25/0.50$$
$$x = 0.50 \text{ lb product/ac}$$

0.50 lb product on 1 ac will be $(0.50)(5) = 2.5$ lb on 5 ac

3. A 50WDG insecticide is to be applied at the rate of 0.25 oz product/1,000 ft^2 for insect control on 20,000 ft^2 planting of ornamental shrubs. How much product is needed to treat the 20,000 ft^2 planting?

$$1{,}000x = (0.25)(20{,}000)$$
$$x = 5{,}000/1{,}000$$
$$x = 5 \text{ oz of 50 WDG insecticide on } 20{,}000 \text{ ft}^2$$

4. In question number three, how much active ingredient was applied per acre?

$$1{,}000x = (0.25)(43{,}560)$$
$$x = 10{,}890/1{,}000$$
$$x = 10.9 \text{ oz product/ac}$$
$$(10.9)(0.50) = 5.45 \text{ oz a.i./ac}$$

5. In question number three, how much total active ingredient was applied to 20,000 ft^2?

$$(5 \text{ oz product})(0.50) = 2.5 \text{ oz a.i. on } 20{,}000 \text{ ft}^2$$

Practice Problem Set 6-2: Calculations Involving Liquid Pesticide Formulations

1. A 2EC herbicide is to be applied at 1.2 lb a.i./ac. How many gallons will be needed to treat 6 ac of turf?

Appendix E Solutions to Practice Problems **339**

$$2x = (1)(1.2)$$
$$x = 1.2/2$$
$$x = 0.60 \text{ gal/ac}$$
$$1x = (0.60)(6)$$
$$x = 3.6 \text{ gal for 6 ac}$$

2. If 1.6F insecticide is to be applied at 3.8 fl oz product/ac, how much is needed to treat 20,000 ft^2?

$$43,560x = (3.8)(20,000)$$
$$x = 76,000/43,560$$
$$x = 1.75 \text{ oz on } 20,000 \text{ ft}^2$$

3. In question number two, how much active ingredient in pounds per acre is being applied?

There are 128 oz/gal. The 3.8 oz/ac rate is $3.8/128 = 0.03$ gal/acre. One gallon has 1.6 lb a.i. and 0.03 gal will have $(1.6)(0.03) = 0.05$ lb a.i./ac.

Practice Problem Set 6-3: Making Economic Decisions

1. A fungicide is available in a 6AS for $75/gal, an 82.5WDG for $50/5-lb bag, and a 5G for $36/25 lb bag. What is the cost per pound of active ingredient for each material?

6AS has 6 lb a.i. in 1 gal.

$75/6 = $12.50/lb a.i.

82.5WDG: (5 lb)(0.825) = 4.13 lb a.i.

$50/4.13 = $12.10/lb a.i.

5G: (25 lb)(0.05) = 1.25 lb a.i.

$36/1.25 = $28.80/lb a.i.

2. A 50DF fungicide sells for $75/10 lb bag. The same active ingredient in a 2G sells for $40/50 lb bag. Which formulation is the most expensive?

$$(10)(0.50) = 5 \text{ lb a.i.}$$
$$\$75 = \$15/\text{lb a.i.}$$

340 Appendix E Solutions to Practice Problems

$$5 \text{ lb a.i.}$$

$$(50)(0.20) = 1 \text{ lb a.i.}$$

$$\frac{\$40}{16 \text{ a.i.}} = \$40/\text{lb a.i.}$$

$$1 \text{ lb a.i.}$$

Practice Problem 6-4: Metric Calculations for Dry Formulations

A 0.5G insecticide is to be applied to 0.22 kg a.i./ha for ornamental beds. How many grams are needed to treat a 200 m² bed?

$$0.22 \text{ kg is } 220 \text{ g}$$

$$(x)(0.005) = 0.22$$

$$x = 0.22/0.005$$

$$x = 44 \text{ kg/ha}$$

44 kg is 440 g and 1 ha is 10,000 m²

$$10,000x = (44)(200)$$

$$x = 8,800/10,0000$$

$$x = 0.88 \text{ kg or } 880 \text{ g needed to treat } 200 \text{ m}^2$$

Practice Problem Set 6-5: Metric Calculations for Liquid Formulations

1. A 250EC fungicide contains 250 g a.i./L. How many liters are needed to treat 20 ha at a rate of 0.125 kg a.i./ha?

   $$250 \text{ g is } 0.25 \text{ kg}$$

   $$0.25x = (1)(0.125)$$

   $$x = 0.125/0.25$$

   $$x = 0.50 \text{ L will provide } 0.125 \text{ kg on L ha}$$

   It will take the following:

   $$(0.50)(20) = 10 \text{ L to treat 20 ha}$$

2. A 480E herbicide that contains 480 g a.i./l is to be applied at 0.6 ml/m². At this rate of application, how many kilograms of active ingredient per hectare are being applied?

0.60 ml on one m² is $(0.6)(10{,}000) = 6{,}000$ ml on a ha.

6,000 ml is 6 L

There are 480 g or 0.48 kg a.i./L.

$$(6)(0.48) = 2.88 \text{ kg a.i./ha}$$

Chapter 7

Practice Problem 7-1 Calibration of a Drop Spreader

A 42 in. wide drop spreader is to be calibrated to apply 1 lb N/1,000 ft² using an 8-4-4 fertilizer. A 20 ft test strip will be used for the calibration. Place a plastic sheet over the test strip, operate the spreader over the plastic sheet, and collect the material by lifting the sheet. The spreader will be calibrated to apply 0.50 lb N/1,000 ft² and two passes at right angles to one another will be used to make the application. (Figure 7-5b) How many ounces of 8-4-4 fertilizer will be collected from the test strip when the spreader is properly calibrated at 0.5 lb N/1,000 ft²?

$$42 \text{ in.}/12 = 3.5 \text{ ft wide}$$
$$(3.5 \text{ ft})(20 \text{ ft}) = 70 \text{ ft}^2$$
$$(x)(0.08) = 0.50$$
$$x = 0.50/0.08$$
$$x = 6.25 \text{ lb 8-4-4}$$
$$1{,}000x = (6.25)(70)$$
$$x = 437.5/1{,}000$$
$$x = 0.4375 \text{ lb on 20 ft strip}$$
$$(0.4375 \text{ lb})(16 \text{ oz/lb}) = 7 \text{ oz}$$

Practice Problem 7-2 Drop-Spreader Calibration Using a Granular Herbicide

A 2G herbicide is to be applied to 2.0 lb a.i./ac using a 42 in. wide spreader. A 30 ft test strip will be used and the material will be collected in a catch tray attached to the base of the spreader. The spreader will be calibrated at 1 lb a.i./ac, and two passes will be used to make the 2 lb treatment.

How many ounces of material will be released on the test strip when the spreader is properly calibrated?

$$(x)(0.20) = 1$$
$$x = 1/0.20$$
$$x = 50 \text{ lb product}$$
$$42 \text{ in.}/12 = 3.5 \text{ ft}$$
$$(3.5 \text{ ft})(30 \text{ ft}) = 105 \text{ ft}^2$$

$$43,560x = (50)(105)$$
$$x = 5,250/43,5600$$
$$x = 0.12\,\text{lb}$$
$$(0.12\,\text{lb})(16\,\text{oz/lb}) = 1.92\,\text{oz of 2G on the 30 ft strip}$$

How many square feet will a 40 lb bag of the material cover when the spreader is properly calibrated?

The half rate was 50 lb product/ac and the full rate is 100 lb product/ac.
$$100x = (43,560)(40)$$
$$x = 1,742,400/100$$
$$x = 17,424\,\text{ft}^2$$

A 40 lb bag will cover 17,424 ft².

Practice Problem 7-3 Calibration of a Broadcast Spreader 1

A broadcast spreader that is designed to overlap from wheel track to wheel track is to be used to apply a 1 lb N/1,000 ft² using a 20-2-8 fertilizer. The application will be made in two passes at a right angle to one another. A test strip of 50 ft is to be used, and the effective spread width has been determined to be 12 ft. How much product should be released on the test strip when the spreader is properly calibrated?

Calibrate to apply 0.25 lb N/1,000 ft².
$$(x)(0.20) = 0.25$$
$$x = 0.25/0.20$$
$$x = 1.25\,\text{lb}$$
$$(12)(50) = 600\,\text{ft}^2$$
$$1,000x = (1.25)(600)$$
$$x = 750/1,000$$
$$x = 0.75\,\text{lb on test strip}$$

How many square feet will a 40 lb bag cover?

0.75 is one quarter of the rate, (4)(0.75) = 3 lb for the full rate
$$(3)(x) = (600)(40)$$
$$x = 24,000/3$$
$$x = 8,000\,\text{ft}^2$$

Practice Problem 7-4 Calibration of a Broadcast Spreader 2

A soccer complex that includes 180,000 ft² of turf is to be treated with a 2G insecticide at 1.6 lb a.i./ac. A broadcast spreader is to be calibrated to apply the material with overlap at the edge of the fan of material (Figure 7-6b). One-half of the material will be applied in each of two applications at right angles to one another (Figure 7-5b). A 100 ft test strip will be used. The effective spread width is 12 ft. How much material will be released on the test strip when the spreader is properly set?

$$(12\,\text{ft})(100\,\text{ft}) = 1,200\,\text{ft}^2$$

$$1.6/2 = 0.80\,\text{lb a.i. will be applied per acre}$$

$$(x)(0.02) = 0.80$$

$$x = 0.80/0.02$$

$$x = 40\,\text{lb material/acre}$$

$$43,560x = (40)(1,200)$$

$$x = 48,000/43,560$$

$$x = 1.1\,\text{lb on the test strip}$$

If the material comes in 50 lb bags, how many bags will be needed for the application?

$$43,560x = (80)(180,000)$$

$$x = 14,400,000/43,560$$

$$x = 331\,\text{lb}$$

$$331/50 = 6.6\,\text{bags will be needed}$$

Practice Problem 7-5 Calibration of a Boom Sprayer 1

A 15-ft-wide boom sprayer with 12 nozzles was timed to cover a 150 ft test strip in 21 sec. If each nozzle releases 27.5 oz in 21 sec, what is the application rate of this spreader in gallons/acre?

$$(15)(150) = 2,250\,\text{ft}^2$$

$$(27.5\,\text{oz/noz})(12\,\text{noz}) = 330\,\text{oz in 21 seconds}$$

$$2250x = (21)(43,560)$$

$$x = 914,760/2250$$

$$x = 406.6\,\text{sec/acre}$$

$$21x = (330)(406.6)$$

$$x = 134,178/21$$

$$x = 6,389\,\text{oz}$$

$$6,389/128\,\text{oz/gal} = 50\,\text{gal/acre}$$

Practice Problem 7-6 Calibration of a Boom Sprayer 2

An 18-ft-wide boom sprayer with 12 nozzles is to be operated at 7 mph. Each nozzle releases an average of 82 oz in 60 sec. What is the application rate in gallons/acre for this sprayer?

$$(7)(5280) = 36,960 \text{ ft}$$

$$(36,960 \text{ ft})(18) = 665,280 \text{ ft}^2$$

$$(82)(12) = 984 \text{ oz in one minute}$$

$$665,280x = (60)(43,560)$$

$$x = 2,613,600/665,280$$

$$x = 3.93/\text{acre}$$

$$x = (984)(3.93)$$

$$x = 3867 \text{ oz}$$

$$3,867/128 \text{ oz/gal} = 30 \text{ gal/acre}$$

Practice Problem 7-7 Calibration of a Small Lawn Sprayer

A small lawn and garden sprayer that is designed to be pulled behind a lawn tractor has a 5 gal tank and two nozzles on a rear boom that release spray over an area 5 ft wide. When operated over a 50 ft test strip, it was found to release 64 oz of spray material from the tank. How much spray material would be applied per 1,000 ft^2 using this sprayer, and how many square feet will a full sprayer cover?

$$(50)(5) = 250 \text{ ft}^2 \text{ test strip}$$

$$250x = (64)(1,000)$$

$$x = 64,000/250$$

$$x = 256 \text{ oz on } 1,000 \text{ ft}^2$$

$$256 \text{ oz}/128 \text{ oz/acre} = 2 \text{ gal}/1,000 \text{ ft}^2$$

If the sprayer releases 2 gal/1,000 ft^2, 5 gal will cover 2,500 ft^2.

If this sprayer were to be used to apply a 4EC herbicide to a lawn at 1.5 lb a.i./ac, how many ounces of the herbicide should be placed in the 5-gal tank as it is filled to capacity?

$$4x = (1)(1.5)$$

$$x = 1.5/4$$

$$x = 0.375 \text{ gal/ac}$$

$$(0.375 \text{ gal})(128 \text{ oz/gal}) = 48 \text{ oz/ac}$$

$$43,560x = (48)(2,500)$$

$$x = 120,000/43,560$$

$$x = 2.75 \text{ oz in the 5 gal tank will treat } 2,500 \text{ ft}^2$$

Chapter 8

Practice Problem Set 8-1 Establishment

1. Determine the pounds of PLS for the following seed lots:
   a. ton of tall fescue seed that contains 91% pure seed and 86% germination rate.

   Step 1: Determine the PLS for the seed lot.
   $$(0.91)(0.86) = 0.7826 \text{ PLS}$$

   Step 2: Determine the lb PLS for seed lot. The seed lot weighs 1 ton, which is equal to 2,000 lb.
   $$(0.7826)(2,000) = 1,565 \text{ lb of PLS}$$

Solution
A total of 1,565 lb PLS is in 1 ton of tall fescue seed.

   b. 100 lb of Kentucky bluegrass that contains 95% pure seed and 96% germination rate.

   Step 1: Determine the PLS for the seed lot.
   $$(0.95)(0.96) = 0.912 \text{ PLS}$$

   Step 2: Determine the lb PLS for the seed lot. The seed lot weighs 100 lb.
   $$(0.912)(100) = 91.2 \text{ lb of PLS}$$

Solution
A total of 91.2 lb PLS is in 100 lb of Kentucky bluegrass seed.

2. A turf site that is 1.5 ac needs to be overseeded with perennial ryegrass. How much of the following seed lots would need to be applied to deliver 1.5 lb of PLS/1,000 ft^2?

   First determine how many pounds of PLS is needed to seed the 1.5 acres.

   Step 1: Determine how many square feet are in 1.5 acres. If there are 43,560 ft^2 in 1 acre, how many (x) ft^2 are in 1.5 acres?
   $$\frac{43,560 \text{ ft}^2}{1 \text{ acre}} = \frac{x \text{ ft}^2}{1.5 \text{ acres}}$$
   $$(43,560 \text{ ft}^2)(1.5 \text{ acres}) = 65,340 \text{ ft}^2$$

   Step 2: Determine how much PLS will be needed for the area.
   $$\frac{1.5 \text{ lb PLS}}{1,000 \text{ ft}^2} = \frac{x \text{ lb PLS}}{65,340 \text{ ft}^2}$$
   $$x \text{ lb PLS} = \frac{(1.5 \text{ lb PLS})(65,340 \text{ ft}^2)}{1,000 \text{ ft}^2}$$
   $$x \text{ lb PLS} = \frac{98,010}{1,000}$$
   $$x = 98 \text{ lb PLS}$$

Solution

A total of 98 lb PLS is needed to deliver 1.5 lb PLS/1,000 ft^2.

a. Lot A: 98% pure seed and 92% germination
Step 1: Determine the PLS of Lot A.

$$(0.98)(0.92) = 0.9016 \text{ PLS}$$

Step 2: Determine how much seed needs to be purchased to apply 98 lb of PLS to the 65,340 ft^2 area, if the seed lot has a PLS of 0.9016 or 90.16 lb in 100 lb of seed.

$$\frac{100 \text{ lb seed}}{90.16 \text{ lb PLS}} = \frac{x \text{ lb seed}}{98 \text{ lb PLS}}$$

Cross multiply and divide to achieve the answer.

$$x \text{ lb} = \frac{(98 \text{ lb PLS})(100 \text{ lb seed})}{90.16 \text{ lb PLS}}$$

$$x = 108.7 \text{ lb of seed}$$

Solution

A total of 108.7 lb of seed in Lot A is required to deliver 1.5 lb of pure live seed per 1,000 ft^2 to 1.5 acres.

b. Lot B: 89% pure seed and 92% germination
Step 1: Determine the PLS of Lot B.

$$(0.89)(0.92) = 0.8188 \text{ PLS}$$

Step 2: Determine how much seed needs to be purchased to apply 98.01 lb of PLS to the 65,340 ft^2 area, if the seed lot has a PLS of 0.8188 or 81.88 lb in 100 lb of seed.

$$\frac{100 \text{ lb seed}}{81.88 \text{ lb PLS}} = \frac{x \text{ lb seed}}{98 \text{ lb PLS}}$$

Cross multiply and divide to achieve the answer.

$$x \text{ lb} = \frac{(98 \text{ lb PLS})(100 \text{ lb seed})}{81.88 \text{ lb PLS}}$$

$$x = 119.7 \text{ lb of seed}$$

Solution

A total of 119.7 lb of seed in Lot B is required to deliver 1.5 lb of pure live seed per 1,000 ft^2 to 1.5 acres.

c. Lot C: 85% pure seed and 90% germination
Step 1: Determine the PLS of Lot C.

$$(0.85)(0.90) = 0.765 \text{ PLS}$$

Step 2: Determine how much seed needs to be purchased to apply 98.01 lb of PLS to the 65,340 ft² area, if the seed lot has a PLS of 0.765 or 76.5 lb in 100 lb of seed.

$$\frac{100 \text{ lb seed}}{76.5 \text{ lb PLS}} = \frac{x \text{ lb seed}}{98 \text{ lb PLS}}$$

Cross multiply and divide to achieve the answer.

$$x \text{ lb} = \frac{(98 \text{ lb PLS})(100 \text{ lb seed})}{76.5 \text{ lb PLS}}$$

$$x = 128.1 \text{ lb of seed}$$

Solution
A total of 128.1 lb of seed in Lot C is required to deliver 1.5 lb of pure live seed per 1,000 ft² to 1.5 acres.

3. What is the cost per pound of PLS for the following seed lots?

   a. Lot A: 98% pure seed and 96% germination. Cost $120 for 50 lb
   Step 1: Determine the PLS of Lot A.

   $$(0.98)(0.96) = 0.9408 \text{ PLS}$$

   Step 2: Calculate the pounds of PLS by multiplying the percent PLS by the total pounds of seed.

   $$(0.9408 \text{ PLS})(50 \text{ lb}) = 47.04 \text{ lb of PLS}$$

   Step 3: To determine the cost of seed per pound of PLS, divide the cost (dollars) by pounds of PLS.

   $$x\$/\text{lb} = \frac{\$120}{47.04 \text{ lb PLS}}$$

   $$x = \$2.55/\text{lb PLS}$$

   b. Lot B: 90% pure seed and 90% germination. Cost $110 for 50 lb
   Step 1: Determine the PLS of Lot B.

   $$(0.90)(0.90) = 0.81 \text{ PLS}$$

   Step 2: Calculate the pounds of PLS by multiplying the percent PLS by the total pounds of seed.

   $$(0.81 \text{ PLS})(50 \text{ lb}) = 40.5 \text{ lb of PLS}$$

   Step 3: To determine the cost of seed per pound of PLS, divide the cost (dollars) by pounds of PLS.

   $$x \$/\text{lb} = \frac{\$110}{40.5 \text{ lb PLS}}$$

   $$x = \$2.72/\text{lb PLS}$$

348 Appendix E Solutions to Practice Problems

c. Lot C: 84% pure seed and 88% germination. Cost $105 for 50 lb
Step 1: Determine the PLS of Lot C.

$$(0.84)(0.88) = 0.7392 \text{ PLS}$$

Step 2: Calculate the pounds of PLS by multiplying the percent PLS by the total pounds of seed.

$$(0.7392 \text{ PLS})(50 \text{ lb}) = 36.96 \text{ lb of PLS}$$

Step 3: To determine the cost of seed per pound of PLS, divide the cost (dollars) by pounds of PLS.

$$x \, \$/\text{lb} = \frac{\$105}{36.96 \text{ lb PLS}}$$

$$x = \$2.84 \text{ lb/PLS}$$

4. A lawn that measures 6,400 ft² is to be sodded with sod slabs that measure 18 in. by 24 in.

  a. What is the square footage of each piece of sod?
  Step 1: Convert inches into feet.

  $$\frac{1 \text{ ft}}{12 \text{ in.}} = \frac{x \text{ ft}}{18 \text{ in.}}$$

  $$x = 1.5 \text{ ft}$$

  $$\frac{1 \text{ ft}}{12 \text{ in.}} = \frac{x \text{ ft}}{24 \text{ in.}}$$

  $$x = 2 \text{ ft}$$

  Step 2:

  $$\text{Area of 1 sod slab} = \text{Length} \times \text{Width}$$

  $$\text{Area} = (1.5 \text{ ft})(2 \text{ ft})$$

  $$\text{Area} = 3 \text{ ft}^2$$

Solution
Area of one slab is 3 ft².

  b. Assuming a 5% waste, how many sod pieces need to be purchased?
  Step 1: There is 6,400 ft² to be sodded. To calculate 5% waste, multiply by 0.05.

  $$x = (6{,}400 \text{ ft}^2)(0.05)$$

  $$x = 320 \text{ ft}^2 \text{ is equal to 5\% waste}$$

  Step 2: Calculate area plus 5% waste.

  $$x = 6{,}400 \text{ ft}^2 + 320 \text{ ft}^2$$

  $$x = 6{,}720 \text{ ft}^2 \text{ is the square footage of sod needed.}$$

Step 3: Calculate the number of sod slabs to purchase.

$$x = \frac{6{,}720 \text{ ft}^2}{3 \text{ ft}^2/\text{slab}}$$

$$x = 2{,}240 \text{ sod slabs}$$

Solution
A total of 2,240 sod pieces need to be purchased.

   c. If the sod cost $1.90/yd², what is the total cost of the sod?
     Step 1: There are 9 ft² in 1 yd². Set up a proportion to determine the total of square yards.

$$\frac{1 \text{ yd}^2}{9 \text{ ft}^2} = \frac{x \text{ yd}^2}{6{,}720 \text{ ft}^2}$$

$$x \text{ yd}^2 = \frac{6{,}720 \text{ ft}^2}{9 \text{ ft}^2}$$

$$x = 747 \text{ yd}^2$$

     Step 2: Calculate the cost of the sod based on $1.90/yd².

$$x = (747 \text{ yd}^2)(\$1.90/\text{yd}^2)$$

$$x = \$1{,}419.30$$

Solution
The cost to sod 6,400 ft² is $1,419.30.

5. A lawn that measures 5,000 ft² is to be established.

   a. How many 4 in. plugs would need to be purchased if planting on 10 in. centers?
     Step 1: Determine the number of plugs per square foot. A plug that is planted on 10 in. centers means that each plug occupies an area 10 in. by 10 in. or 100 in.².

$$\frac{1 \text{ plug}}{100 \text{ in.}^2} = x \text{ plug } 144 \text{ in.}^2$$

$$100 x = 144$$

$$x \text{ plug} = \frac{144 \text{ in.}^2}{100 \text{ in.}^2}$$

$$x = 1.44 \text{ plugs/ft}^2$$

     Step 2: If 1.4 plugs are planted per 1 ft², how many plugs are needed to plant 5,000 ft²?

$$\frac{1.44 \text{ plugs}}{1 \text{ ft}^2} = \frac{x \text{ plugs}}{5{,}000 \text{ ft}^2}$$

$$x = (1.44 \text{ plugs})(5{,}000 \text{ ft}^2)$$

$$x = 7{,}200 \text{ plugs}$$

Solution

A total of 7,200 plugs are needed to establish a 5,000 ft² area of lawn.

b. How many bushels of stolons would need to be purchased if sprigging at a rate of 3 bushels/1,000 ft²?

Step 1: Set up the proportion:

$$\frac{3 \text{ bushels}}{1{,}000 \text{ ft}^2} = \frac{x \text{ bushels}}{5{,}000 \text{ ft}^2}$$

Step 2: Isolate and solve for x.

$$1{,}000\, x = (3)(5{,}000)$$

$$x \text{ bushels} = \frac{(3 \text{ bushel})(5{,}000 \text{ ft}^2)}{1{,}000 \text{ ft}^2}$$

$$x = 15 \text{ bushels}$$

Solution

A total of 15 bushels of stolons are needed to sprig 5,000 ft² of lawn.

Practice Problem 8-2: Soil Modification

A 10,000 ft² lawn is to be established. The topsoil was stripped off prior to construction, and it needs to be replaced. It has been determined that 6 inches of amended topsoil will be placed on the lawn. How much amended topsoil needs to be ordered?

Step 1: Since the area has previously been determined, the first step is to convert depth into feet so that total soil volume can be calculated. If there are 12 inches in 1 foot, then there are (x) feet in 6 inches.

$$\frac{1 \text{ ft}}{12 \text{ in.}} = \frac{x \text{ ft}}{6 \text{ in.}}$$

$$x = 0.50 \text{ ft}$$

There are 0.50 feet in 6 inches.

Step 2: Determine the volume of modified soil that is needed for the lawn. The area is 10,000 ft² and the soil depth is 0.50 feet. To determine the volume of modified topsoil, multiply the area (ft²) by the depth (ft) to yield cubic feet (ft³).

$$(10{,}000 \text{ ft}^2)(0.50 \text{ ft}) = 5{,}000 \text{ ft}^3$$

A total of 5,000 ft³ of amended topsoil is needed for the lawn.

Step 3: Convert cubic feet to cubic yards. There are 27 ft³ in 1 yd³; therefore, it is necessary to divide the total volume of modified topsoil by 27 ft³.

$$x \text{ yd}^3 = \frac{5{,}000 \text{ ft}^3 \text{ of topsoil}}{27 \text{ ft}^3/\text{yd}^3}$$

$$x = 185.2 \text{ yd}^3$$

Solution
A total of 185.2 yd³ of modified topsoil needs to be ordered for the lawn establishment project.

Practice Problem Set 8-3: Topdressing

1. A landscape company is going to topdress a 6,000 ft² seedbed with compost amended soil to a depth of $\frac{1}{2}$ in. How many cubic yards of compost amended soil needs to be ordered?

    Step 1: To determine volume, convert the topdress depth from inches to feet. The topdress depth is $\frac{1}{2}$ in., which is the same as saying 0.50 in.

    $$\tfrac{1}{2} = 0.50 \text{ in.}$$

    Since there are 12 in. in a foot, convert 0.50 in. by dividing by 12.

    $$\frac{1 \text{ ft}}{12 \text{ in.}} = \frac{x \text{ ft}}{0.50 \text{ in.}}$$

    $$x \text{ ft} = \frac{0.50 \text{ in.}}{12 \text{ in./ft}}$$

    $$x = 0.04167 \text{ ft}$$

    Step 2: Volume is determined by multiplying the depth (in feet) by the square footage of the infield.

    $$\text{Volume} = (6,000 \text{ ft}^2)(0.04167 \text{ ft})$$

    $$\text{Volume} = 250 \text{ ft}^3$$

    Step 3: Topdress will be purchased in cubic yards; therefore, convert cubic feet to cubic yards. Set up the proportion by asking the following: If 1 yd³ = 27 ft³, then (x) yd³ = 250 ft³.

    $$\frac{1 \text{ yd}^3}{27 \text{ ft}^3} = \frac{x \text{ yd}^3}{250 \text{ ft}^3}$$

    $$x \text{ yd}^3 = \frac{(250 \text{ ft}^3)(1 \text{ yd}^3)}{27 \text{ ft}^3}$$

    $$x = 9.3 \text{ yd}^3$$

Solution
It will take 9.3 yd³ to topdress the infield to a $\frac{1}{2}$ in. depth.

2. A sports turf supply company has a quantity of topdress media that is stored in a cone-shaped pile. The pile has a base diameter of 30 ft and a height of 18 ft.

    a. How many cubic yards are in the pile of topdress (see Chapter 3 for the formula for determining the volume of a cone).

Step 1: Determine the surface area of the storage bin.
The formula for the area of the base of a cone is $B = \pi r^2$ where $r = \frac{1}{2} d$.

$$r = \tfrac{1}{2}\, 30$$
$$r = 15 \text{ ft}$$
$$B = \pi (15^2)$$
$$B = \pi \times 225 \text{ ft}^2$$
$$B = 706.9 \text{ ft}^2$$

Step 2: Determine the volume (x) of the bin of topdress material in cubic feet. The surface area of the storage bin is 707 ft² and the height of the topdress material is 18 ft. To calculate volume, multiply the base area by the height.

$$V = \tfrac{1}{3}(B \times h)$$
$$x = \tfrac{1}{3}(706.9 \text{ ft}^2 \times 18 \text{ ft})$$
$$x = \tfrac{1}{3}(12{,}724 \text{ ft}^3)$$
$$x = 4{,}241 \text{ ft}^3$$

Step 3: Convert 4,241 ft³ to yd³. Set up the proportion, as shown in Example 8-8.

$$\frac{1 \text{ yd}^3}{27 \text{ ft}^3} = \frac{x \text{ yd}^3}{4{,}241 \text{ ft}^3}$$

$$x \text{ yd}^3 = \frac{4{,}241 \text{ ft}^3}{27 \text{ ft}^3/\text{yd}^3}$$

$$x = 157.1 \text{ yd}^3$$

b. How many square feet will the pile cover if the topdress is $\frac{5}{16}$ in.?
The topdress depth is $\frac{5}{16}$ in., which is the same as saying 0.3125 in.

$$\frac{5}{16} = 0.3125 \text{ in.}$$

Since there are 12 in. in a foot, convert 0.3125 in. by dividing by 12.

$$\frac{1 \text{ ft}}{12 \text{ in.}} = \frac{x \text{ ft}}{0.3125 \text{ in.}}$$

$$x \text{ ft} = \frac{0.3125 \text{ in.}}{12 \text{ in.}/\text{ft}}$$

$$x = 0.02604 \text{ ft}$$

Volume for 1 ft² is determined by multiplying the area (1 ft²) by the height (0.02604 ft).

$$x = (0.02604 \text{ ft})(1 \text{ ft}^2)$$

$$x = 0.02604 \text{ ft}^2$$

Convert cubic feet to cubic yards.

$$\frac{1 \text{ yd}^3}{27 \text{ ft}^3} = \frac{x \text{ yd}^3}{0.02604 \text{ ft}^3}$$

$$x \text{ yd}^3 = \frac{0.02604 \text{ ft}^3/\text{ft}^2}{27 \text{ ft}^3/\text{yd}^3}$$

$$x = 0.000965 \text{ yd}^3$$

Step 5: Determine how much surface area the pile of topdress material will cover. Set up the proportion by asking the following: If 0.000965 yd³ is needed to cover 1 ft², then how many (x) ft² will be covered by 157.1 yd³?

$$\frac{1 \text{ ft}^2}{0.000965 \text{ yd}^3} = \frac{x \text{ ft}^2}{157.1 \text{ yd}^3}$$

$$x \text{ ft}^2 = \frac{157.1 \text{ yd}^3}{0.000965 \text{ yd}^3/\text{ft}^2}$$

$$x = 162,798 \text{ ft}^2$$

Solution

A cone-shaped pile of topdress material that contains 157 yd³ of topdress material will cover 162,798 ft² when applied at a $\frac{5}{16}$ in. depth.

Practice Problem Set 8-4: Water Use Calculations

1. A total of 125 ac of sod are to be irrigated with 2.5 in. of water over the next 2 weeks. How much water will be needed to make this application?

   Step 1: Calculate total water use for 1 ac.

   $$\frac{27,154.3 \text{ gal}}{1 \text{ in. H}_2\text{O}} = \frac{x \text{ gal}}{2.5 \text{ in. H}_2\text{O}}$$

   $$(27,154.3 \text{ gal})(2.5 \text{ in. H}_2\text{O}) = (x \text{ gal})(1 \text{ in. H}_2\text{O})$$

   $$x \text{ gal} = (27,154.3 \text{ gal})(2.5 \text{ in. H}_2\text{O})$$

   $$x = 67,886 \text{ gal to apply 2.5 in. of water to 1 ac}$$

354 Appendix E Solutions to Practice Problems

Step 2: Determine how much water would need to be applied to 125 ac.

$$x \text{ gal} = (67,886 \text{ gal/ac})(125 \text{ acres})$$
$$x = 8,485,750 \text{ gal}$$

Solution

A total of 8,485,750 gal water would need to be applied to irrigate 125 ac with 2.5 in. of water.

2. The department of natural resources is monitoring the amount of irrigation water used by a sports complex. The complex irrigates 16 ac with 1.3 in. of water/week. Total irrigation for the year is 15.6 in. of water. How many cubic feet of water does the sports complex use for the year?

Step 1: Calculate the total water use for 1 acre. There are 27,154.3 gal in 1 ac-in. How many (x) gal will be in 15.6 ac-in.?

$$\frac{27,154.3 \text{ gal}}{1 \text{ ac-in.}} = \frac{x \text{ gal}}{15.6 \text{ ac-in.}}$$

$$x = (27,154.3 \text{ gal})(15.6 \text{ ac-in.})$$

$$x = 423,607.1 \text{ gal}$$

A total of 423,607.1 gal is needed to irrigate 1 ac with 15.6 ac-in. of water.

Step 2: Calculate total water use for 16 ac. If 423,607.1 gal is needed to irrigate 1 ac, how many (x) gallons are needed to irrigate 16 ac?

$$\frac{423,607.1 \text{ gal}}{1 \text{ ac}} = \frac{x \text{ gal}}{16 \text{ ac}}$$

$$x = (423,607.1 \text{ gal})(16 \text{ ac})$$

$$x = 6,777,714 \text{ gal}$$

A total of 6,777,714 gal is needed to irrigate 16 ac with 15.6 ac-in. of water.

Step 3: Calculate the cubic feet of water used by the sports complex in one year. There are 7.480519 gal in 1 ft³. The question is: How many cubic feet are in 6,777,714 gal?

$$\frac{1 \text{ ft}^3}{7.480519 \text{ gal}} = \frac{x \text{ ft}^3}{6,777,714 \text{ gal}}$$

$$x \text{ ft}^3 = \frac{6,777,714 \text{ ft}^2/\text{gal}}{7.480519 \text{ gal}}$$

$$x = 906,048.6 \text{ ft}^3$$

Solution

A total of 906,048.6 ft³ of water is the projected volume needed by the sports complex in one year.

3. An office complex draws its irrigation water from a small retention pond. The surface area of the pond is 95,832 ft² and the depth is 10 ft. There are 5.5 ac of turf that needs to be irrigated. They want to determine how many inches of water can be applied to irrigate 5.5 ac during one season.

Step 1: Determine the surface area of the pond in acres. If 1 ac equals 43,560 ft², how many acres are in 500,000 ft²?

$$\frac{1 \text{ ac}}{43,560 \text{ ft}^2} = \frac{x \text{ ac}}{95,832 \text{ ft}^2}$$

$$x \text{ ac} = \frac{95,832 \text{ ft}^2}{43,560 \text{ ft}^2}$$

$$x = 2.2 \text{ ac of surface area}$$

Step 2: Determine the amount of water in the lake as acre-feet. The average depth of the lake has been determined to be 10 ft and the surface area is 2.2 ac. Therefore:

$$x \text{ ac-ft} = (2.2 \text{ ac})(10 \text{ ft})$$

$$x = 22 \text{ ac-ft}$$

There are 22 ac-ft of water in the lake.

Step 3: Convert acre-feet to gallons of water. There are 325,851.4 gal in 1 ac-ft. Therefore:

$$\frac{325,851.4 \text{ gal.}}{1 \text{ ac-ft}} = \frac{x \text{ gal.}}{22 \text{ ac-ft}}$$

$$x \text{ gal} = (325,851.4 \text{ gal/ac-ft})(22 \text{ ac-ft})$$

$$x = 7,168,731 \text{ gal}$$

There are 7,168,731 gallons of water in the pond.

Step 4: Calculate the gallons of water needed to irrigate 5.5 ac with 1 in. of water. There are 27,154.3 gal in 1 ac-in. This is done by multiplying gal/ac-in. by total acres.

$$x \text{ gal} = (27,154.3 \text{ gal/ac-in.})(5.5 \text{ ac})$$

$$x = 149,348.7 \text{ gal}$$

A total of 149,348.7 gal are needed to apply 1 in. of water to 5.5 ac of turfgrass.

Step 5: Calculate the total inches of water that can potentially be applied to 5.5 ac of turfgass. There are 7,168,731 gallons of water in the pond and irrigating the turfgrass with 1 in. of water delivers 149,348.7 gallons. Therefore:

$$x \text{ in.} = \frac{7,168,731 \text{ gal}}{149,348.7 \text{ gal}}$$

$$x = 48 \text{ in. of water}$$

356 Appendix E Solutions to Practice Problems

Solution:
The lake is capable of supplying 48 in. of water to the 5.5 ac without any recharge.

Chapter 9

Practice Problem Set 9-1

The Smiths plan to install a new paver patio adjacent to the north side of their house. The patio dimensions are 20 ft E-W and 35 ft N-S (see Figure 9-2). They have calculated a drop of 8.4 inches over 35 feet from the patio edge adjacent to the house to the north edge of the patio.

1. What percent slope are they using on this patio?

   Step 1: Convert 8.4 inches to feet

   $$\frac{8.4 \text{ in.}}{12 \text{ in./ft}} = 0.7 \text{ ft}$$

   Step 3:

   $$\text{Ratio} = \text{rise:run}$$

   $$0.7 \text{ ft}:35 \text{ ft}$$

   Step 4:

   $$\text{Percent slope} = \frac{\text{rise}}{\text{run}}$$

   $$\frac{0.7 \text{ ft}}{35 \text{ ft}} = 0.02$$

   $$(0.02)(100) = 2.00\% \text{ slope}$$

2. Is it within the slope guidelines for a seating area?
   Yes, the slope is within the guidelines for a seating area.

Practice Problem Set 9-2

The Schuhs plan to install a 10′ × 14′ paver patio using a complex paving pattern.
   Calculate the following:

1. The number of 4″ × 8″ pavers

   Step 1: Calculate the area covered by one paver.

   $$(\text{paver length})(\text{paver width}) = \text{area of one paver}$$

   $$(8 \text{ in.})(4 \text{ in.}) = 32 \text{ in.}^2/\text{paver}$$

   Step 2: Calculate the number of pavers per square foot, given there are 144 in.²/ ft².

   $$\frac{144 \text{ in.}^2/\text{ft}^2}{32 \text{ in.}^2/\text{paver}} = 4.5 \text{ pavers/ft}^2$$

Step 3: Calculate the area of the patio

$$(\text{patio length})(\text{patio width}) = \text{area of patio}$$
$$(10\,\text{ft})(14\,\text{ft}) = 140\,\text{ft}^2$$

Step 4: Calculate the number of pavers needed for the patio.

$$(\text{area of patio in ft}^2)(\text{number of pavers/ft}^2) = \text{total number of pavers}$$
$$(140\,\text{ft}^2)(4.5\,\text{pavers/ft}^2) = 630\,\text{pavers}$$

Step 5: Calculate the amount of overage to account for waste/breakage and using a complex paving pattern.

$$(\text{number of pavers})(\text{overage percentage}) = \text{additional pavers needed for waste/breakage}$$
$$(630\,\text{pavers})(0.08) = 50.4\,\text{pavers or 51 pavers}$$

Solution:

calculated amount + overage amount

= total amount

$630 + 51$

$= 681$ total pavers

681 pavers are required to install the Schuhs' new 140 ft² patio

2. The amount of crushed limestone required for a 4-inch layer

   Step 1:
   $$\frac{\text{total area}}{\text{area covered/yd}^3} = \text{amount of bulk material needed}$$
   $$\frac{140\,\text{ft}^2}{81\,\text{ft}^2/\text{yd}^3} = 1.73\,\text{yd}^3\,\text{crushed limestone}$$

   Step 2: Calculate the amount of overage using the higher number from Table 9-4 since the patio is medium sized.

   $$(\text{amount of limestone})(\text{overage percent}) = \text{amount of overage}$$
   $$(1.73\,\text{yd}^3)(0.15) = 0.259\,\text{yd}^3$$

   Step 3:
   calculated amount + overage amount = total amount
   $$1.73\,\text{yd}^3 + 0.259\,\text{yd}^3 = 1.99\,\text{or}\,2.0\,\text{yd}^3$$

Solution:

2.0 yd³ of crushed limestone are required for the Schuhs' new 140 ft² patio.

3. The amount of sand needed for the project

Step 1:

$$\frac{\text{total area}}{\text{area covered/yd}^3} = \text{amount of bulk material needed}$$

$$\frac{140 \text{ ft}^2}{324 \text{ ft}^2/\text{yd}^3} = 0.43 \text{ yd}^3 \text{ sand}$$

Step 2: Since sand is also needed to sweep into the joints between the pavers, it is a good idea to use the higher overage percentage.

$$(\text{amount of sand})(\text{overage percent}) = \text{amount of overage}$$

$$(0.43 \text{ yd}^3)(0.15) = 0.06 \text{ yd}^3$$

Step 3:

$$\text{calculated amount} + \text{overage amount} = \text{total amount}$$

$$0.43 \text{ yd}^3 + 0.06 \text{ yd}^3 = 0.49 \text{ yd}^3$$

In this case, round up to 0.50 yd^3 when ordering the sand.

Solution:

0.50 yd^3 of sand is required for the Schuhs' new 140-ft^2 patio.

Practice Problem 9-3

Determine the breakeven price and selling price for a 5-gallon rhododendron assuming a wholesale cost of $80.00, overhead markup of 60%, and profit of 10%.

Rhododendron (wholesale cost; 5 gal)
                $80.00

Overhead markup (60%)
    (percent overhead markup)(wholesale cost) = markup
    (0.60)($80.00) = $48.00 $48.00
Breakeven price (wholesale cost + markup) $128.00
Profit (10%)
    (percent profit)(breakeven price) = selling price
    (0.10)($128.00) = $12.80 $12.80
Total selling price $140.80

Practice Problem 9-4

Determine the materials selling price for the new foundation planting at the Shoemaker's residence (see Figure 9-4). The project includes: of 375 ft^2 bed area; adding a 4″ layer of conditioned topsoil and tilling it into the bed area; two 12″ B&B kousa dogwood trees (labeled A); three 3 gal-sized arrowwood viburnum shrubs (labeled B); sixteen 1 gal-sized dwarf barberry shrubs (labeled C); and a 350-ft^2 area covered with bugleweed groundcover spaced 10″ on center.

Appendix E Solutions to Practice Problems **359**

1. Area of the planting bed?   375 ft^2
2. How many yd^3 of conditioned topsoil are needed to cover the bed with a 4″ layer?

$$\frac{375 \text{ ft}^2}{81 \text{ ft}^2/\text{yd}^3} = 4.63 \text{ yd}^3 \text{ topsoil}$$

3. Estimate the number of ground cover plants needed assuming the groundcover area is 350 ft^2 including under the tree canopies, and calculate the plantings are on 10″ centers.

$$350 \text{ ft}^2 \times 1.4 \text{ plants/ft}^2 = 490 \text{ plants}$$

Description	Quantity or Unit (A)	Selling Price / unit (including wholesale cost, overhead & profit) (B)	Total selling price for materials C = (A)(B)
Conditioned topsoil	5 yd^3	$15.00	$ 75.00
Dwarf barberry (1 gal)	16	$16.00	$ 256.00
Arrowwood viburnum (3 gal)	4	$19.00	$ 76.00
Kousa dogwood (12″ B&B; by hand)	2	$91.52	$ 183.04
Bugleweed (4″ pot)	490	$ 5.00	$2,450.00
Materials Selling Price Sub-Total			$3,040.04

Practice Problem 9-5

Calculate the total hourly rate for an employee who earns $10.00/hour and works for a company where labor burden is 25%, labor overhead is 120%, and profit on labor is calculated as 10%.

Hourly labor rate	$10.00
Labor burden (20%)	
(percent labor burden)(hourly rate) = labor burden	
(0.20)($10.00) = $2.00	$2.00
Hourly labor cost (hourly labor rate + labor burden)	$12.00
Labor overhead(120%)	
(percent labor overhead)(hourly labor cost) = hourly labor overhead	
(1.20)($12.00) = $14.40	$14.40
Breakeven hourly labor cost (hourly labor cost + labor overhead)	$26.40
Profit(10%)	
(percent profit on labor) (breakeven hourly labor cost) = hourly labor profit	
(0.10)($26.40) = $2.64	$2.64
Total hourly labor rate (breakeven hourly labor cost + hourly profit)	$29.04

Appendix E Solutions to Practice Problems

Practice Problem 9-6

Determine the installation hours and associated labor price for the new foundation planting at the Shoemaker residence (see Practice Problem 9-4 and Figure 9-4). Use an hourly labor rate of $29.04.

Description	Quantity or Unit (A)	Production Rate (labor hours per unit)* (B)	Total hours required for job C = (A)(B)	Labor rate / hour (including labor, burden, labor overhead & profit) (D)	Total Labor Price E = (C)(D)
Spread soil	41.66 SY	0.051	2.12	$29.04	$ 61.56
Till soil	41.66 SY	0.008	0.33	$29.04	$ 9.58
Rake soil	0.375 MSF	0.800	0.30	$29.04	$ 8.71
Dwarf barberry (1 gal)	16	0.271	4.34	$29.04	$126.03
Arrowwood viburnum (3 gal)	4	0.571	2.28	$29.04	$ 66.21
Kousa dogwood (12" B&B; by hand)	2	1.23	2.46	$29.04	$ 71.44
Bugleweed (4" pot)	490	0.018	8.82	$29.04	$256.13
Clean up	0.120 MSF	0.267	0.03	$29.04	$ 0.87
Time and Labor Selling Price Sub-total			20.68	$29.04	$600.55

Practice Problem 9-7

Determine the total selling price for the installation of the new foundation planting at the Shoemaker residence. (See Practice Problem 9-4 and Figure 9-4.)

Description	Materials Price (A)	Labor Price (B)	Total Price A + B = C
Spread Soil	$ 75.00	$ 61.56	$ 136.56
Till soil	0	$ 9.58	9.58
Rake soil	0	$ 8.71	$ 8.71
Dwarf barberry (1 gal.)	$ 256.00	$126.03	$ 382.03
Arrowwood viburnum (3 gal.)	$ 76.00	$ 66.21	$ 142.21
Kousa dogwood (12" B&B; by hand)	$ 183.04	$ 71.44	$ 254.48
Bugleweed (4" pot)	$2,450.00	$256.13	$2,706.13
Clean up	0	$ 0.87	$ 0.87
Total Selling Price			$3,640.57

Practice Problem 9-8

1. Determine the annual hours for each landscape service performed for the Newberry property maintenance contract. Based on the information Table 9-13,

Newberry Property					
Maintenance Service	Unit or Quantity (A)	Production Rate (B)	Hours / Unit C = (A)(B)	Annual Frequency (D)	Total Annual Hours E = (C)(D)
Turf					
Mowing	1.50 MSF	0.167	0.2505	23	5.76
Fertilizing	1.50 MSF	0.067	0.1005	3	0.30
Shrubs (Deciduous)					
Corrective Pruning	15	0.067	1.005	1	1.01
Flowerbeds					
Herbicide: Pre-emergence	0.50 MSF	0.05	0.025	1	0.03
Mulching	0.50 MSF	0.50	0.25	1	0.25
				Total Hours	7.35

2. Calculate the labor price for the maintenance contract for the Newberry property assuming a total hourly labor rate of $25.00/hour.

Solution:

(total hours)(total labor rate/hour) = total labor price

(7.35 hours)($25.00/hour) = $183.75 total labor price

Practice Problem Set 9-9

Calculate the materials price for **1.** the pre-emergence herbicide and **2.** hardwood mulch for the Newberry property (Table 9-13).

1. A granular pre-emergence herbicide (Trifluralin) is applied to 500 ft^2 of flower bed area. The product price (including overhead and profit) is $64.25 for a 17.5 lb bag.

   Step 1: Calculate how many bags of herbicide are needed. According to the product label, the 17.5 lb bag of pre-emergence herbicide covers 2800 ft^2.

   $$\frac{\text{total flower bed area}}{\text{area covered by herbicide/bag}} = \text{total bags of herbicide}$$

   $$\frac{500 \text{ ft}^3}{2800 \text{ ft}^2 \text{ per bags}} = 0.179$$

   Step 2: Calculate the total amount of herbicide needed in pounds.

   (number of bags)(pounds/bag) = total pounds of herbicide needed

   (0.179 bags)(17.5 lb/bag) = 3.13 lb

Step 3: Calculate the price of the herbicide per pound.

$$\frac{\text{price of herbicide}}{\text{size of bag in pounds}} = \text{price/pound}$$

$$\frac{\$64.25}{17.5\,\text{lb}} = \$3.67/\text{lb}$$

Step 4: Calculate the price of the herbicide required for the job.

(herbicide price/pound)(number of pounds needed)

= price of herbicide needed

($3.67/lb)(3.13 lb) = $11.49

Solution:
The price for the pre-emergence herbicide is $11.49.

2. The flower bed needs a 3″ application of hardwood mulch. From Table 9-4 you can find that 1 yd³ of material, spread 3″ deep, covers 108 ft². The area to be covered with mulch is 500 ft². The mulch price is $27.00/yd³.

Step 1: Calculate how many yd³ of mulch is needed.

$$\frac{\text{total bed area}}{\text{area covered/yd}^3} = \text{total yd}^3 \text{ needed}$$

$$\frac{500\,\text{ft}^2}{108\,\text{ft}^2/\text{yd}^3} = 4.63\,\text{yd}^3$$

Step 2: Calculate the price of the mulch required for the job.

(total yd³ of mulch needed)(price/yd³) = total price for mulch

(4.63 yd³)($27.00/yd³) = $125.01

Solution:
The price for the hardwood mulch is $125.01.

Practice Problem 9-10

Calculate the price of the annual maintenance contract for the Newberry property.

The annual maintenance contract for the Newberry property requires 7.35 hours (calculated in Practice Problem 9-8).

Labor Price (7.35 hours)($25.00/hour)	$183.75
Materials Price Pre-emergence herbicide: $11.49 Hardwood mulch: $125.01	$136.50
Total	$320.25

Chapter 10

Practice Problem Set 10-1 Calculating Greenhouse Bench Efficiency

1. Calculate the bench efficiency of an older greenhouse that measures 24 ft wide by 100 ft long. It contains four ground beds designed for cut flower production. Each ground bed measures 4 ft wide by 95 ft long and they are oriented parallel to one another and to the long side of the greenhouse.

    Step 1: Determine the Total Greenhouse Area by multiplying the length of the greenhouse in feet by the width of the greenhouse in feet.
    $$100 \text{ ft} \times 24 \text{ ft} = 2,400 \text{ ft}^2$$

    Step 2: Determine the Total Growing Area by adding the surface area of all of the benches.
    $$4 \text{ beds} \times 4 \text{ ft} \times 95 \text{ ft} = 1,520 \text{ ft}^2$$

    Step 3: Calculate the bench efficiency.
    $$\frac{\text{Total Growing Area (ft)}^2}{\text{Total Greenhouse Area (ft)}^2} \times 100 = \text{Percent (\%) Bench Efficiency}$$

    $$\frac{1,520 \text{ ft}^2}{2,400 \text{ ft}^2} \times 100 = 63.3\% \text{ Bench Efficiency}$$

Solution:
This greenhouse has 63.3% bench efficiency.

2. Calculate the bench efficiency of a greenhouse that measures 41.5 ft wide and 144 ft long. This greenhouse contains 48 rolling benches that measure 5.5 ft by 18 ft each.

    Step 1: Determine the Total Greenhouse Area by multiplying the length of the greenhouse in feet by the width of the greenhouse in feet.
    $$144 \text{ ft} \times 41.5 \text{ ft} = 5,976 \text{ ft}^2$$

    Step 2: Determine the Total Growing Area by summing the surface area of all of the benches.
    $$48 \text{ benches} \times 5.5 \text{ ft} \times 18 \text{ ft} = 4,752 \text{ ft}^2$$

    Step 3: Calculate the bench efficiency.
    $$\frac{\text{Total Growing Area (ft)}^2}{\text{Total Greenhouse Area (ft)}^2} \times 100 = \text{Percent (\%) Bench Efficiency}$$

    $$\frac{4,752 \text{ ft}^2}{5,976 \text{ ft}^2} \times 100 = 79.5\% \text{ Bench Efficiency}$$

Solution:
This greenhouse has 79.5% bench efficiency.

3. Recalculate the bench efficiency of the greenhouse in question number two after adding hanging baskets over the 5.5 ft wide central aisle.

Step 1: Calculate the growing space gained by suspending baskets over the central aisle by multiplying the width of the central aisle by the length of the greenhouse.

$$5.5 \text{ ft} \times 144 \text{ ft} = 792 \text{ ft}^2$$

Step 2: Add the space gained by suspending baskets over the central aisle to the existing growing space.

$$792 \text{ ft}^2 + 4{,}752 \text{ ft}^2 = 5{,}544 \text{ ft}^2$$

Step 3: Calculate the bench efficiency.

$$\frac{\text{Total Growing Area (ft)}^2}{\text{Total Greenhouse Area (ft)}^2} \times 100 = \text{Percent (\%) Bench Efficiency}$$

$$\frac{5{,}544 \text{ ft}^2}{5{,}976 \text{ ft}^2} \times 100 = 92.8\% \text{ Bench Efficiency}$$

Solution:
By adding hanging baskets over the central aisle, bench efficiency is increased from 79.5% to 92.8%.

Practice Problem 10-2 Calculating the Surface Area of an Even-Span Greenhouse

Calculate the glazed surface area and the curtain wall surface area of a greenhouse that measures 30 ft wide and 96 ft long. The eave height is 8 ft, the curtain wall height is 2 ft, the rafter length is 18.75 ft, and the gable height is 11.25 ft.

Part One: Calculate the glazed surface area

Step 1: Determine the surface area of the roof.

Roof Surface Area = 2 Roof Sides × Rafter Length × Side Wall Length

Roof Surface Area = 2 × 18.75 ft × 96 ft

Roof Surface Area = 3,600 ft^2

Step 2: Determine the surface area of the gable.

Gable Surface Area = 2 Gables × $\frac{1}{2}$ (Gable End Width)(Gable Height)

Gable Surface Area = 2 × $\frac{1}{2}$ (30 ft)(11.25 ft)

Gable Surface Area = 337.5 ft^2

Step 3: Determine the glazed surface area of the gable end walls.

Gable End Wall Area = 2 Gable Ends[Gable End Width × (Eave Height − Curtain Wall Height)] ×

Gable End Wall Area = 2[30 ft(8 ft − 2 ft)]

Gable End Wall Area = 2 × 30 ft × 6 ft

Gable End Wall Area = 360 ft^2

Step 4: Determine the glazed surface area of the side walls.

Side Wall Surface Area = 2 Side Walls[Side Wall Length × (Eave Height − Curtain Wall Height)]

Side Wall Surface Area = 2[96 ft × (8 ft − 2 ft)]

Side Wall Surface Area = 2 × 96 ft × 6 ft

Side Wall Surface Area = 1,152 ft^2

Step 5: Add the surface areas of all the glazed portions of the greenhouse to determine the total glazed surface area.

Total Glazed Surface Area = Roof Area + Gable Area + Glazed Gable End Wall Area + Glazed Side Wall Area

Total Glazed Surface Area = 3,600 ft^2 + 337.5 ft^2 + 360 ft^2 + 1,152 ft^2

Total Glazed Surface Area = 5,449.5 ft^2

Solution: Part One

The total glazed surface area of this greenhouse is 5,449.5 ft^2.

Part Two: Calculate the surface area of the curtain wall

Step 1: Calculate the curtain wall area on the gable ends.

Gable End Curtain Wall Area = 2 Gable Ends × Gable End Width × Height of Curtain Wall

Gable End Curtain Wall Area = 2 × 30 ft × 2 ft

Gable End Curtain Wall Area = 120 ft^2

Step 2: Calculate the curtain wall area on the side walls.

Side Wall Curtain Wall Area = 2 Greenhouse Sides × Side Wall Length × Curtain Wall Height

Side Wall Curtain Wall Area = 2 × 96 ft × 2 ft

Side Wall Curtain Wall Area = 384 ft^2

Step 3: Add the surface areas of all the curtain walls.

Total Curtain Wall Area = Gable End Curtain Wall Area + Side Wall Curtain Wall Area

Total Curtain Wall Area = 120 ft² + 384 ft²

Total Curtain Wall Area = 504 ft²

Solution: Part Two

The total curtain wall surface area of this greenhouse is 504 ft².

Practice Problem 10-3: Calculating the Surface Area of an Arch-Top Greenhouse

Calculate the surface area of an arch-top greenhouse that measures 24 ft wide and 48 ft long. The eave height is 12 ft, the gable height is 12 ft, and the arch length is 37.7 ft long.

Step 1: Calculate the surface area of the roof.

Roof Surface Area = Arch Length × Greenhouse Length

Roof Surface Area = 37.7 feet × 48 feet

Roof Surface Area = 1,809.6 ft²

Step 2: Calculate the surface area of the gable.

Surface Area of the Gables = π × Gable Height × $\frac{1}{2}$ Greenhouse Width

Surface Area of the Gables = π × 12 ft × 12 ft

Surface Area of the Gables = 452.39 ft²

Step 3: Calculate the surface area of the gable end walls.

Surface Area of the Gable End Walls = 2 Gable Ends × Eave Height × Greenhouse Width

Surface Area of the Gable End Walls = 2 Gable Ends × 12 ft × 24 feet

Surface Area of the Gable End Walls = 576 ft²

Step 4: Calculate the surface area of the side walls.

Surface Area of the Side Walls = 2 Greenhouse Side Walls × Eave Height × Greenhouse Length

Surface Area of the Side Walls = 2 Greenhouse Side Walls × 12 ft × 48 ft

Surface Area of the Side Walls = 1,152 ft²

Step 5: Add the surface areas of the greenhouse components.

Total Surface Area of the Greenhouse = Roof Surface Area + Gable Surface Area + Gable End Wall Surface Area + Side Wall Surface Area

Total Surface Area of the Greenhouse = $1{,}809.6\,\text{ft}^2 + 452.39\,\text{ft}^2$
$+ 576\,\text{ft}^2 + 1{,}152\,\text{ft}^2$

Total Surface Area of the Greenhouse = $3{,}989.99\,\text{ft}^2$

Solution:
The surface area of this arch-top greenhouse is $3{,}989.99\,\text{ft}^2$.

Practice Problem Set 10-4: Calculating the Surface Area of Gutter-Connected Greenhouses

1. Calculate the glazed surface area and the curtain wall surface area of an even-span gutter-connected greenhouse with four bays that each measures 30 ft wide and 96 ft long. The eave height is 8 ft, the curtain wall height is 2 ft, the rafter length is 18.75 ft, and the gable height is 11.25 ft.

Part One: Calculate the glazed surface area

Step 1: Determine the surface area of the roof.

Roof Surface Area = 2 × Rafter Length × Side Wall Length
× Number of Greenhouse Bays

Roof Surface Area = 2 × 30 ft × 96 ft × 4 Greenhouse Bays

Roof Surface Area = $23{,}040\,\text{ft}^2$

Step 2: Determine the surface area of the gable.

Gable Surface Area = $2 \times \frac{1}{2}$ (Gable End Width)(Gable Height)
(Number of Greenhouse Bays)

Gable Surface Area = $2 \times \frac{1}{2}$ (30 ft)(11.25 ft)(4)

Gable Surface Area = $1{,}350\,\text{ft}^2$

Step 3: Determine the glazed surface area of the gable end walls.

Gable End Wall Area = 2 × [Gable End Width (Eave Height
− Curtain Wall Height)] ×
(Number of Greenhouse Bays)

Gable End Wall Area = 2 × [30 ft(8 ft − 2 ft)](4)

Gable End Wall Area = 2 × 30 ft × 6 ft × 4

Gable End Wall Area = $1{,}440\,\text{ft}^2$

Step 4: Determine the glazed surface area of the side walls.

Side Wall Surface Area = 2 × Side Wall Length (Eave Height
− Curtain Wall Height)

Side Wall Surface Area = 2 × 96 ft(8 ft − 2 ft)

Side Wall Surface Area = 2 × 96 ft(6 ft)

Side Wall Surface Area = 1,152 ft^2

Step 5: Add the surface areas of all the glazed portions of the greenhouse to determine total glazed surface area.

Total Glazed Surface Area = Roof Area + Gable Area
   + Glazed Gable End Wall Area
   + Glazed Side Wall Area

Total Glazed Surface Area = 23,040 ft^2 + 1,350 ft^2 + 1,440 ft^2 + 1,152 ft^2

Total Glazed Surface Area = 26,982 ft^2

Solution: Part One

The total glazed surface area of this greenhouse is 26,982 ft^2.

Part Two: Calculate the surface area of the curtain wall

Step 1: Calculate the curtain wall area on the gable ends.

Gable End Curtain Wall Area = 2 × Gable End Width
   × Height of Curtain Wall
   × Number of Greenhouse Bays

Gable End Curtain Wall Area = 2 × 30 ft × 2 ft × 4 Greenhouse Bays

Gable End Curtain Wall Area = 480 ft^2

Step 2: Calculate the curtain wall area on the side walls.

Side Wall Curtain Wall Area = 2 × Side Wall Length × Curtain Wall Height

Side Wall Curtain Wall Area = 2 × 96 ft × 2 ft

Side Wall Curtain Wall Area = 384 ft^2

Step 3: Add the surface areas of all the curtain walls.

Total Curtain Wall Area = Gable End Curtain Wall Area
   + Side Wall Curtain Wall Area

Total Curtain Wall Area = 480 ft^2 + 384 ft^2

Total Curtain Wall Area = 864 ft^2

Solution: Part Two

The total curtain wall surface area of this greenhouse is 864 ft^2.

2. Calculate the glazed surface area of a gutter-connected arch-top greenhouse with six bays that each measures 24 ft wide and 48 ft long. For each bay, the arch length is 37.7 ft, the gable height is 12 ft, and the eave height is 12 ft. The glazing extends from the eave to the ground so there is no curtain wall.

Step 1: Calculate the surface area of the roof.

Roof Surface Area = Number of Greenhouse Bays × Arch Length × Greenhouse Length

Roof Surface Area = 6 Greenhouse Bays × 37.7 feet × 48 feet

Roof Surface Area = 10,857.6 ft^2

Step 2: Calculate the surface area of the gable.

Surface Area of the Gables
= 6 Greenhouse Bays × 2 Gables
× $\left[\dfrac{\pi \times \text{Gable Height} \times \frac{1}{2} \text{Greenhouse Width}}{2} \right]$

Surface Area of the Gables
= 6 Greenhouse Bays × $\left[\pi \times \text{Gable Height} \times \frac{1}{2} \text{Greenhouse Width} \right]$

Surface Area of the Gables = 6 × π × 12 ft × 12 ft

Surface Area of the Gables = 2,714.34 ft^2

Step 3: Calculate the surface area of the gable end walls.

Surface Area of the Gable End Walls
= 6 Greenhouse Bays × 2 Gable Ends
× Eave Height × Greenhouse Width

Surface Area of the Gable End Walls
= 6 Greenhouse Bays × 2 Gable Ends × 12 ft × 24 ft

Surface Area of the Gable End Walls = 3,456 ft^2

Step 4: Calculate the surface area of the side walls.

Surface Area of the Side Walls
= 2 Greenhouse Side Walls × Eave Height
× Greenhouse Length

Surface Area of the Side Walls
= 2 Greenhouse Side Walls × 12 ft × 48 ft

Surface Area of the Side Walls = 1,152 ft^2

Step 5: Add the surface areas of the greenhouse components.

$$\text{Total Surface Area of the Greenhouse}$$
$$= \text{Roof Surface Area} + \text{Gable Surface Area}$$
$$+ \text{Gable End Wall Surface Area} + \text{Side Wall Surface Area}$$

$$\text{Total Surface Area of the Greenhouse}$$
$$= 10{,}857.6\,\text{ft}^2 + 2{,}714.34\,\text{ft}^2 + 3{,}456\,\text{ft}^2 + 1{,}152\,\text{ft}^2$$

$$\text{Total Surface Area of the Greenhouse} = 18{,}179.94\,\text{ft}^2$$

Solution:
The surface area of this arch-top greenhouse is 18,179.94 ft^2.

Practice Problem 10-5: Calculating the Surface Area of a Quonset Greenhouse

Calculate the surface area of a Quonset greenhouse that is 22 ft wide and 72 ft long. The gable height is 11 ft and the arch length is 34.6 ft long.

Step 1: Calculate the surface area of the roof.

$$\text{Surface Area of Roof} = \text{Arch Length} \times \text{Greenhouse Length}$$
$$\text{Surface Area of Roof} = 34.6\,\text{ft} \times 72\,\text{ft}$$
$$\text{Surface Area of Roof} = 2{,}491.2\,\text{ft}^2$$

Step 2: Calculate the surface area of the gable ends.

$$\text{Surface Area of the Gables} = \pi \times \text{Gable Height} \times \tfrac{1}{2}\,\text{Greenhouse Width}$$
$$\text{Surface Area of the Gables} = \pi \times 11\,\text{ft} \times \tfrac{1}{2}\,22\,\text{ft}$$
$$\text{Surface Area of the Gables} = \pi \times 11\,\text{ft} \times 11\,\text{ft}$$
$$\text{Surface Area of the Gables} = 380.13\,\text{ft}^2$$

Step 3: Add the surface area of the gables and the roof.

$$\text{Surface Area of Quonset Greenhouse} = \text{Roof Surface Area}$$
$$+ \text{Gable Surface Area}$$
$$\text{Surface Area of Quonset Greenhouse} = 2{,}491.2\,\text{ft}^2 + 380.13\,\text{ft}^2$$
$$\text{Surface Area of Quonset Greenhouse} = 2{,}871.33\,\text{ft}^2$$

Solution:
The total surface area of this Quonset greenhouse is 2,871.33 ft^2.

Practice Problem 10-6: Calculating the Volume of an Even-Span Greenhouse

Calculate the volume of an even-span greenhouse that measures 30 ft wide and 96 ft long. The eave height is 8 ft and the gable height is 11.25 ft.

Step 1: Calculate the area of the gable. The gable is triangular in shape so the basic formula for the area of a triangle is adapted to the greenhouse.

$$\text{Gable Area} = \tfrac{1}{2} \, (\text{Gable End Width}) \times (\text{Gable Height})$$

$$\text{Gable Area} = \tfrac{1}{2} \, (30 \, \text{ft}) \times (11.25 \, \text{ft})$$

$$\text{Gable Area} = 15 \, \text{ft} \times 11.25 \, \text{ft}$$

$$\text{Gable Area} = 168.75 \, \text{ft}^2$$

Step 2: Calculate the area of the gable end wall.

$$\text{Gable End Wall Area} = \text{Gable End Width} \times \text{Eave Height}$$

$$\text{Gable End Wall Area} = 30 \, \text{ft} \times 8 \, \text{ft}$$

$$\text{Gable End Wall Area} = 240 \, \text{ft}^2$$

Step 3: Calculate the area of the gable end of the greenhouse.

$$\text{Gable End Area} = \text{Gable Area} + \text{Gable End Wall Area}$$

$$\text{Gable End Area} = 168.75 \, \text{ft}^2 + 240 \, \text{ft}^2$$

$$\text{Gable End Area} = 408.75 \, \text{ft}^2$$

Step 4: Multiply the area of the gable end by the length of the greenhouse to determine greenhouse volume.

$$\text{Volume of Greenhouse} = \text{Gable End Area} \times \text{Greenhouse Length}$$

$$\text{Volume of Greenhouse} = 408.75 \, \text{ft}^2 \times 96 \, \text{ft}$$

$$\text{Volume of Greenhouse} = 39{,}240 \, \text{ft}^3$$

Solution:
The volume of this even-span greenhouse is $39{,}240 \, \text{ft}^3$.

Practice Problem 10-7: Calculating the Volume of an Arch-Top Greenhouse

Calculate the volume of an arch-top greenhouse that measures 24 ft wide and 48 ft long. The eave height is 12 ft and the gable height is 12 ft.

Step 1: Calculate the area of the gable. The gable is semicircular or semi-elliptical in shape, so the basic formula for the area of an ellipse is adapted to the greenhouse.

$$\text{Gable Area} = \left(\pi \times \text{Gable Height} \times \tfrac{1}{2} \, \text{Greenhouse Width}\right) \div 2$$

$$\text{Gable Area} = (\pi \times 12 \, \text{ft} \times 12 \, \text{ft}) \div 2$$

$$\text{Gable Area} = 226.19 \, \text{ft}^2$$

Step 2: Calculate the area of the gable end wall.

$$\text{Gable End Wall Area} = \text{Gable End Width} \times \text{Eave Height}$$
$$\text{Gable End Wall Area} = 24 \text{ ft} \times 12 \text{ ft}$$
$$\text{Gable End Wall Area} = 288 \text{ ft}^2$$

Step 3: Calculate the area of the gable end of the greenhouse.

$$\text{Gable End Area} = \text{Gable Area} + \text{Gable End Wall Area}$$
$$\text{Gable End Area} = 226.19 \text{ ft}^2 + 288 \text{ ft}^2$$
$$\text{Gable End Area} = 514.19 \text{ ft}^2$$

Step 4: Multiply the area of the gable end by the length of the greenhouse to determine greenhouse volume.

$$\text{Volume of Greenhouse} = \text{Gable End Area} \times \text{Greenhouse Length}$$
$$\text{Volume of Greenhouse} = 514.19 \text{ ft}^2 \times 48 \text{ ft}$$
$$\text{Volume of Greenhouse} = 24{,}681.12 \text{ ft}^3$$

Solution:

The volume of this arch-top greenhouse is 24,681.12 ft^3.

Practice Problem Set 10-8 Calculating the Volume of Gutter-Connected Greenhouses

1. Calculate the volume of a gutter-connected even-span greenhouse with four bays that each measures 30 ft wide and 96 ft long. The eave height is 8 ft and the gable height is 11.25 ft.

    Step 1: Calculate the area of the gable of one of the greenhouse bays. The gable is a triangle so:

    $$\text{Gable Area} = \tfrac{1}{2} \, (\text{Gable End Width}) \times (\text{Gable Height})$$
    $$\text{Gable Area} = \tfrac{1}{2} \, (30 \text{ ft}) \times (11.25 \text{ ft})$$
    $$\text{Gable Area} = 15 \text{ ft} \times 11.25 \text{ ft}$$
    $$\text{Gable Area} = 168.75 \text{ ft}^2$$

    Step 2: Calculate the area of the gable end wall of one of the greenhouse bays.

    $$\text{Gable End Wall Area} = \text{Gable End Width} \times \text{Eave Height}$$
    $$\text{Gable End Wall Area} = 30 \text{ ft} \times 8 \text{ ft}$$
    $$\text{Gable End Wall Area} = 240 \text{ ft}^2$$

Appendix E Solutions to Practice Problems **373**

Step 3: Calculate the area of the gable end of one of the greenhouse bays.

$$\text{Gable End Area} = \text{Gable Area} + \text{Gable End Wall Area}$$

$$\text{Gable End Area} = 168.75 \text{ ft}^2 + 240 \text{ ft}^2$$

$$\text{Gable End Area} = 408.75 \text{ ft}^2$$

Step 4: Calculate the volume of one of the greenhouse bays.

$$\text{Volume of Greenhouse Bay} = \text{Gable End Area} \times \text{Greenhouse Length}$$

$$\text{Volume of Greenhouse Bay} = 408.75 \text{ ft}^2 \times 96 \text{ ft}$$

$$\text{Volume of Greenhouse Bay} = 39{,}240 \text{ ft}^3$$

Step 5: Calculate the volume of the greenhouse by multiplying the number of bays by the volume of a single bay.

$$\text{Volume of Greenhouse} = \text{Number of Bays} \times \text{Volume of Greenhouse Bay}$$

$$\text{Volume of Greenhouse} = 4 \text{ Bays} \times 39{,}240 \text{ ft}^3$$

$$\text{Volume of Greenhouse} = 156{,}960 \text{ ft}^3$$

Solution:
The volume of this gutter-connected greenhouse is $156{,}960 \text{ ft}^3$.

2. Calculate the volume of a gutter-connected arch-top greenhouse with six bays that each measures 24 ft wide and 48 ft long. For each bay, the gable height is 12 ft and the eave height is 12 ft.

Step 1: Calculate the area of the gable of one of the greenhouse bays.

$$\text{Gable Area} = \left(\pi \times \text{Gable Height} \times \tfrac{1}{2} \text{ Greenhouse Width}\right) \div 2$$

$$\text{Gable Area} = (\pi \times 12 \text{ ft} \times 12 \text{ ft}) \div 2$$

$$\text{Gable Area} = 226.19 \text{ ft}^2$$

Step 2: Calculate the area of the gable end wall of one of the greenhouse bays.

$$\text{Gable End Wall Area} = \text{Gable End Width} \times \text{Eave Height}$$

$$\text{Gable End Wall Area} = 24 \text{ ft} \times 12 \text{ ft}$$

$$\text{Gable End Wall Area} = 288 \text{ ft}^2$$

Step 3: Calculate the area of the gable end of one of the greenhouse bays.

$$\text{Gable End Area} = \text{Gable Area} + \text{Gable End Wall Area}$$
$$\text{Gable End Area} = 226.19 \text{ ft}^2 + 288 \text{ ft}^2$$
$$\text{Gable End Area} = 514.19 \text{ ft}^2$$

Step 4: Calculate the volume of one of the greenhouse bays.

$$\text{Volume of Greenhouse Bay} = \text{Gable End Area} \times \text{Greenhouse Length}$$
$$\text{Volume of Greenhouse Bay} = 514.19 \text{ ft}^2 \times 48 \text{ ft}$$
$$\text{Volume of Greenhouse Bay} = 24{,}681.12 \text{ ft}^3$$

Step 5: Calculate the volume of the greenhouse.

$$\text{Volume of Greenhouse} = \text{Number of Bays} \times \text{Volume of Greenhouse Bay}$$
$$\text{Volume of Greenhouse} = 6 \text{ Bays} \times 24{,}681.12 \text{ ft}^3$$
$$\text{Volume of Greenhouse} = 148{,}086.72 \text{ ft}^3$$

Solution:
The volume of this greenhouse is 148,086.72 ft^3.

Practice Problem 10-9: Calculating the Volume of a Quonset Greenhouse

Calculate the volume of a Quonset greenhouse that is 22 ft wide and 72 ft long. The gable height is 11 ft.

Step 1: Calculate the area of the gable end.

$$\text{Gable End Area} = (\pi \times \text{Gable Height} \times \tfrac{1}{2} \text{ Greenhouse Width}) \div 2$$
$$\text{Gable End Area} = (\pi \times 11 \text{ ft} \times \tfrac{1}{2} \ 22 \text{ ft}) \div 2$$
$$\text{Gable End Area} = (\pi \times 11 \text{ ft} \times 11 \text{ ft}) \div 2$$
$$\text{Gable End Area} = 190.07 \text{ ft}^2$$

Step 2: Calculate the volume of the greenhouse.

$$\text{Greenhouse Volume} = \text{Gable End Area} \times \text{Greenhouse Length}$$
$$\text{Greenhouse Volume} = 190.07 \text{ ft}^2 \times 72 \text{ ft}$$
$$\text{Greenhouse Volume} = 13{,}685.04 \text{ ft}^3$$

Solution
The volume of this greenhouse is 13,685.04 ft^3.

Practice Problem Set 10-10: Fertilizer Applications Using Proportioning Equipment

1. A homeowner plans to apply a water-soluble fertilizer with a brass-siphon proportioner at the rate of 2.5 Tbsp/gal. The brass-siphon proportioner has been calibrated and delivers at a 1:15 proportion. The homeowner would like to prepare 4 gal of concentrated fertilizer stock solution. How many tablespoons of fertilizer are needed to prepare the 4 gal of stock solution? The dilute fertilizer solution is to be applied at the rate of 1 gal per 25 ft^2. How many square feet can be covered if all of the stock solution is used?

   Determine the number of gallons delivered when the entire 4 gal of stock tank solution is used up. If 1 gal of stock is used when a total of 16 gal is delivered, then how many gallons will be delivered when 4 gal of stock solution is used?

   Step 1: Set up a proportion.
   $$\frac{1 \text{ gallon of stock solution}}{16 \text{ gallons of dilute solution delivered}} = \frac{4 \text{ gallons of stock solution}}{x \text{ gallons of dilute solution delivered}}$$

   Step 2: Isolate and solve for x.
   $$x = 4 \times 16$$
   $$x = 64 \text{ gallons}$$

Solution

Sixty-four gallons of dilute solution are delivered when 4 gal of stock solution is used.

The amount of fertilizer to be added to the stock tank is equivalent to the amount of fertilizer required for 64 gal of solution. If 2.5 Tbsp of fertilizer is added to 1 gal of dilute solution, then how many tablespoons need to be added to produce 64 gal of dilute solution?

   Step 1: Set up a proportion.
   $$\frac{2.5 \text{ tablespoon of fertilizer}}{1 \text{ gallon of dilute solution}} = \frac{x \text{ tablespoons of fertilizer}}{64 \text{ gallons of dilute solution}}$$

   Step 2: Isolate and solve for x.
   $$x = 64 \times 2.5$$
   $$x = 160 \text{ Tbsp}$$

Solution

One hundred sixty tablespoons of fertilizer are needed to formulate 4 gal of stock solution.

If 1 gal of solution is used to fertilize 25 ft² of garden space, then how many square feet will 64 gal cover?

Step 1: Set up a proportion.

$$\frac{1 \text{ gallon of fertilizer solution}}{25 \text{ ft}^2 \text{ of garden area}} = \frac{64 \text{ gallons of fertilizer solution}}{x \text{ ft}^2 \text{ of garden area}}$$

Step 2: Isolate and solve for x.

$$x = 25 \times 64$$
$$x = 1{,}600 \text{ ft}^2$$

Solution

A volume of 64 gal of fertilizer solution will cover 1,600 ft² of garden area at the rate of 1 gal/25 ft².

2. A nursery manager would like to use a brass-siphon device to apply fertilizer solution to a small crop of potted nursery stock. The manager set up the delivery system and calibrated the device. A total of 17 cups of solution were collected when 1 cup of test stock solution was used. What is the proportioner ratio for this brass-siphon device?

Answer: 1:16

3. A nursery manager would like to fertilize a hydrangea crop with 13-0-22 at a rate of 150 ppm N. The injector ratio is 1:200 and the stock tank volume is 100 gallons. How much 13-0-22 is needed to prepare stock solution in the 100 gal tank? How many parts per million of potassium are delivered with the nitrogen?

Step 1: Determine how many ounces of dry fertilizer product are required to make the desired parts per million solution in 100 gal of water.

$$\frac{\text{Desired ppm} \div 75}{\text{Decimal Fraction of the Desired Fertilizer Element}}$$

$$\frac{150 \text{ ppm N} \div 75}{0.13} = 15.38 \text{ ounces}$$

Solution

Dissolve 15.38 oz of 13-0-22 in 100 gal of water to produce a 150 ppm N solution.

Step 2: Determine the number of ounces of dry fertilizer to place in a stock tank.

$$\frac{\text{Volume of Stock Tank (gallons)} \times \text{Total Gallons Delivered When 1 Gallon of Stock is Injected} \times \text{Solution to Step 1}}{100}$$

$$\frac{100 \times 201 \times 15.38}{100} = 3{,}091.38 \text{ oz}$$

Step 3: For ease of measuring, convert ounces to pounds plus ounces by setting up a proportion.

$$\frac{16 \text{ ounces}}{1 \text{ pound}} = \frac{3{,}091.38 \text{ ounces}}{x \text{ pounds}}$$

Step 4: Isolate and solve for x.

$$3{,}091.38 = 16x$$

$$x = \frac{3{,}091.38}{16}$$

$$x = 193.21 \text{ lb}$$

Step 5: Convert 0.21 to ounces by setting up a proportion and solving for x.

$$\frac{0.21 \text{ pounds}}{x \text{ ounces}} = \frac{1 \text{ pound}}{16 \text{ ounces}}$$

$$x = 0.21 \times 16$$

$$x = 3.36 \text{ ounces}$$

Solution:
Dissolve 193 lb plus 3.4 oz of 13-0-22 in 100 gal of water to produce a stock solution that delivers 150 ppm N when using a 1:200 proportioner.

Determine how much potassium (K) is being delivered.

Step 1: Begin with determining the number of ounces of 13-0-22 needed to be dissolved in 100 gal of water to produce a 150 ppm N solution. (This is the first step in the injector short-cut equations.)

$$\frac{\text{Desired ppm} \div 75}{\text{Decimal Fraction of the Desired Fertilizer Element}}$$

$$\frac{150 \div 75}{0.13} = 15.38 \text{ oz}$$

Step 2: Use the following equation to find how many parts per million of K_2O is being delivered in this solution.

Number of ounces of fertilizer product/100 gallon from the original problem $\times$ 75 $\times$ decimal fraction of K_2O in the fertilizer product

$= \text{ppm of } K_2O$

15.38 ounces $\times$ 75 $\times$ 0.22 K_2O = 253.77 ppm of K_2O

Step 3: How many parts per million of potassium are found in 253.77 ppm of K_2O? Chapter 5 indicates that there is 83% K found in the K_2O molecule; therefore, 83% of 250 ppm of K_2O is K or:

$$0.83 \times 253.77 \text{ ppm } K_2O = 210.6 \text{ ppm K}$$

Solution

There is 210.6 ppm K found in a solution containing 150 ppm N made using a 13-0-22 fertilizer product.

4. A greenhouse manager would like to fertilize a poinsettia crop using a combination of potassium nitrate (13-0-44) at 100 ppm N and calcium nitrate (15.5-0-0) at 100 ppm N. The injector ratio is 1:100 and the stock tank volume is 50 gal. How much 13-0-44 and 15.5-0-0 are needed to prepare the stock solution in the 50 gal tank? How many parts per million of potassium are delivered with both sources of nitrogen?

   Step 1 for 13-0-44: Determine the number of ounces of dry fertilizer product required to make a desired parts per million solution in 100 gal of water.

   $$\frac{\text{Desired ppm} \div 75}{\text{Decimal Fraction of the Desired Fertilizer Element}}$$

   $$\frac{100 \text{ ppm N} \div 75}{0.13} = 10.26 \text{ oz}$$

   Step 2 for 13-0-44: Determine the number of ounces of dry fertilizer to place in a stock tank.

   $$\frac{\text{Volume of Stock Tank (gallons)} \times \text{Total Gallons Delivered When 1 Gallon of Stock is Injected} \times \text{Solution to Step 1}}{100}$$

   $$\frac{50 \times 101 \times 10.26}{100} = 518.13 \text{ oz}$$

   Step 3 for 13-0-44: For ease of measuring, convert ounces to pounds plus ounces by setting up a proportion.

   $$\frac{16 \text{ ounces}}{1 \text{ pound}} = \frac{518.13 \text{ ounces}}{x \text{ pounds}}$$

   Step 4 for 13-0-44: Isolate and solve for x.

   $$518.13 = 16x$$

   $$x = \frac{518.13}{16}$$

   $$x = 32.38 \text{ lb}$$

   Step 5 for 13-0-44: Convert 0.38 to ounces by setting up a proportion and solving for x.

   $$\frac{0.38 \text{ pounds}}{x \text{ ounces}} = \frac{1 \text{ pound}}{16 \text{ ounces}}$$

   $$x = 0.38 \times 16$$

   $$x = 6.08 \text{ oz}$$

Solution for 13-0-44
Dissolve 32 lb plus 6 oz of 13-0-44 in 50 gal of water to produce a stock solution that delivers 100 ppm N when using a 1:100 proportioner.

Step 1 for 15.5-0-0: Determine the number of ounces of dry fertilizer product required to make the desired parts per million solution in 100 gal of water.

$$\frac{\text{Desired ppm} \div 75}{\text{Decimal Fraction of the Desired Fertilizer Element}}$$

$$\frac{100 \text{ ppm N} \div 75}{0.155} = 8.6 \text{ oz}$$

Step 2 for 15.5-0-0: Determine how many ounces of dry fertilizer to place in a stock tank.

$$\frac{\text{Volume of Stock Tank (gallons)} \times \text{Total Gallons Delivered When 1 Gallon of Stock Is Injected} \times \text{Solution to Step 1}}{100}$$

$$\frac{50 \times 101 \times 8.6}{100} = 434.3 \text{ oz}$$

Step 3 for 15.5-0-0: For ease of measuring, convert ounces to pounds plus ounces by setting up a proportion.

$$\frac{16 \text{ ounces}}{1 \text{ pound}} = \frac{434.3 \text{ ounces}}{x \text{ pounds}}$$

Step 4 for 15.5-0-0: Isolate and solve for x.

$$434.3 = 16x$$
$$x = \frac{434.3}{16}$$
$$x = 27.14 \text{ lb}$$

Step 5 for 15.5-0-0: Convert 0.14 to ounces by setting up a proportion and solving for x.

$$\frac{0.14 \text{ pounds}}{x \text{ ounces}} = \frac{1 \text{ pound}}{16 \text{ ounces}}$$
$$x = 0.14 \times 16$$
$$x = 2.24 \text{ oz}$$

Solution for 15.5-0-0
Dissolve 27 lb plus 2.2 oz of 15.5-0-0 in 50 gal of water to produce a stock solution that delivers 100 ppm N when using a 1:100 proportioner.

Determine how much potassium is being delivered.
For 15.5-0-0, no potassium is being delivered.

Calculation for 13-0-44:

Step 1: Begin with determining the number of ounces of 13-0-44 needed to be dissolved in 100 gal of water to produce a 100 ppm N solution. (This is the first step in the injector short-cut equations.)

$$\frac{\text{Desired ppm} \div 75}{\text{Decimal Fraction of the Desired Fertilizer Element}}$$

$$\frac{100 \div 75}{0.13} = 10.26 \text{ oz}$$

Step 2: Use the following equation to find how many K_2O parts per million is being delivered in this solution.

\# ounces of fertilizer product/100 gallon from the original problem

$\times$ 75 decimal fraction of K_2O in the fertilizer product = ppm of K_2O

10.26 ounces $\times$ 75 $\times$ 0.44 K_2O = 338.58 ppm of K_2O

Step 3: How many parts per million of potassium are found in 338.58 ppm of K_2O?

Chapter 5 indicates that there is 83% K found in the K_2O molecule; therefore, 83% of 250 ppm of K_2O is K or:

$$0.83 \times 338.58 \text{ ppm } K_2O = 281 \text{ ppm K}$$

Solution

There is 281 ppm K found in a solution containing 100 ppm N made using a 13-0-44 fertilizer product.

Practice Problem Set 10-11: Calculating the Volume of a Growing Container

1. Calculate the volume of a cube-shaped free-standing planter that measures 2 ft².

$$\text{Volume} = s^3$$
$$\text{Volume} = 2 \text{ ft} \times 2 \text{ ft} \times 2 \text{ft}$$
$$\text{Volume} = 8 \text{ ft}^3$$

2. Calculate the volume of an L-shaped planter in a shopping mall interior that measures 1.5 ft deep and 4 ft wide. One arm of the L measures 15 ft long and the other arm of the L measures 25 ft long.

$$\text{Volume} = l \times w \times h$$
$$\text{Volume} = (15 \text{ ft} \times 1.5 \text{ ft} \times 4 \text{ ft}) + (25 \text{ ft} \times 1.5 \text{ ft} \times 4 \text{ ft})$$
$$\text{Volume} = 90 \text{ ft}^3 + 150 \text{ ft}^3$$
$$\text{Volume} = 240 \text{ ft}^3$$

Appendix E Solutions to Practice Problems **381**

3. Cylindrical free-standing containers that measure 8 ft in diameter and 3.5 ft deep are found in an indoor courtyard. What is the volume of each of these containers?

$$\text{Volume} = Bh$$
$$\text{Volume} = (\pi r^2)h$$
$$\text{Volume} = (\pi 4^2 \text{ ft})(3.5 \text{ ft})$$
$$\text{Volume} = \pi \times 16 \text{ ft}^2 \times 3.5 \text{ ft}$$
$$\text{Volume} = 175.9 \text{ ft}^3$$

4. Calculate the volume of a round, tapered container that measures 10 in. in diameter at the top, 7 in. in diameter at the bottom, and 5.5 in. tall.

$$\text{Volume} = [(B_{\text{top}} + B_{\text{bottom}}) \div 2]h$$
$$\text{Volume} = [(\pi r^2)_{\text{top}} + (\pi r^2)_{\text{bottom}}] \div 2]h$$
$$\text{Volume} = [(\pi 5^2)_{\text{top}} + (\pi 3.5^2)_{\text{bottom}}] \div 2]5.5$$
$$\text{Volume} = [(78.54 + 38.48) \div 2]5.5$$
$$\text{Volume} = (58.51)(5.5)$$
$$\text{Volume} = 321.81 \text{ in.}^3$$

5. Calculate the volume of a square, tapered container that measures 6.5 in. square on the top, 5 in. square on the bottom, and 6.5 in. tall.

$$\text{Volume} = [(B_{\text{top}} + B_{\text{bottom}}) \div 2]h$$
$$\text{Volume} = [(s^2)_{\text{top}} + (s^2)_{\text{bottom}}] \div 2]h$$
$$\text{Volume} = [(6.5^2)_{\text{top}} + (5^2)_{\text{bottom}}] \div 2]6.5$$
$$\text{Volume} = [(42.25 + 25) \div 2]6.5$$
$$\text{Volume} = (33.625)(6.5)$$
$$\text{Volume} = 218.56 \text{ in.}^3$$

Practice Problem Set 10-12: Calculating Root Medium Volume Requirements

1. A nursery manager is planning a crop of 3,000 garden mums grown in 2.9 L mum pans. How many 3.8 ft³ compressed bales that yield 7.5 ft³ each need to be ordered for this crop?

   Step 1: Determine the total volume of root media needed in liters by multiplying the number of pots by the volume of each pot.

$$3{,}000 \text{ pots} \times 2.9 \frac{\text{liters}}{\text{pot}} = 8{,}700 \text{ liters}$$

Solution

A volume of 8,700 L of root medium is needed to fill 3,000 mum pans.

> **Step 2:** Convert liters of root medium to cubic feet of root medium. Set up a proportion.
>
> $$\frac{28.316846592 \text{ liters}}{1 \text{ cubic foot}} = \frac{8,700 \text{ liters}}{x \text{ cubic feet}}$$
>
> Isolate and solve for x.
>
> $$28.316846592x = 8,700$$
>
> $$x = \frac{8,700}{28.316846592}$$
>
> $$x = 307.24 \text{ ft}^3$$

Solution

A volume of 307.24 ft³ of root medium is needed to fill 3,000 mum pans.

> **Step 3:** Determine how many 3.8 ft³ compressed bales are equivalent to 307.24 ft³.
> Note that each compressed bale loosens to a volume of 7.5 ft³.
> Set up a proportion.
>
> $$\frac{1 \text{ compressed bale}}{7.5 \text{ ft}^3} = \frac{x \text{ compressed bales}}{307.24 \text{ ft}^3}$$
>
> Isolate and solve for x.
>
> $$7.5x = 307.24$$
>
> $$x = \frac{307.24}{7.5}$$
>
> $$x = 40.9 \text{ compressed bales}$$

Solution

A total of 41 compressed bales are required for this crop.

2. A greenhouse manager is planning a crop of 17,500 flats of petunias grown in flat inserts that hold 0.1667 ft³ of root media each. How many 55 ft³ loose bulk bags are needed for this crop?

> **Step 1:** Determine the total volume of root media needed in cubic feet by multiplying the number of flats by the volume of each flat.
>
> $$17,500 \text{ flats} \times 0.1667 \frac{\text{ft}^3}{\text{flat}} = 2,917.25 \text{ ft}^3$$

Solution

A volume of 2,917.25 ft³ of root medium is needed to fill 17,500 flats.

Appendix E Solutions to Practice Problems **383**

Step 2: Determine how many 55 ft³ loose bulk bags are equivalent to 2,917.25 ft³.

Set up a proportion.

$$\frac{1 \text{ bulk bag}}{55 \text{ ft}^3} = \frac{x \text{ bulk bags}}{2,917.25 \text{ ft}^3}$$

Isolate and solve for x.

$$55x = 2,917.25$$

$$x = \frac{2,917.25}{55}$$

$$x = 53 \text{ loose bulk bags}$$

Solution
A total of 53 loose bulk bags are required for this crop.

3. An interior landscape manager needs to fill 17 round planters measuring 6 ft in diameter and 3 ft tall with root media. How many 60 ft³ bulk bags of root media are needed for this project?

Step 1: Determine the volume of each planter.

$$\text{Volume} = Bh$$

$$\text{Volume} = (\pi r^2)h$$

$$\text{Volume} = (\pi 3^2)3$$

$$\text{Volume} = 84.82 \text{ ft}^3$$

Step 2: Determine the volume of 17 planters.

$$17 \text{ planters} \times 84.82 \text{ ft}^3/\text{planter} = 1,441.94 \text{ ft}^3$$

Step 3: Determine how many 60 ft³ bulk bags are required. Set up a proportion.

$$\frac{1 \text{ bulk bag}}{60 \text{ ft}^3} = \frac{x \text{ bulk bags}}{1,441.94 \text{ ft}^3}$$

Isolate and solve for x.

$$60x = 1,441.94$$

$$x = \frac{1,441.94}{60}$$

$$x = 24 \text{ bulk bags}$$

Solution
A total of 24 bulk bags are needed for this project.

4. A club house manager at a country club needs to fill 25 planter boxes on a patio deck that measure 18 in. by 18 in. by 6 ft long. How many 3 ft³ bags of root medium are required to fill all of the planter boxes?

Step 1: Determine the volume of each planter box in cubic feet.
18 in. is equivalent to 1.5 ft

$$\text{Volume} = l \times w \times h$$

$$\text{Volume} = 6 \text{ ft} \times 1.5 \text{ ft} \times 1.5 \text{ ft}$$

$$\text{Volume} = 13.5 \text{ ft}^3$$

Step 2: Determine the volume of 25 planter boxes.

$$25 \times 13.5 \text{ ft}^3 = 337.5 \text{ ft}^3$$

Step 3: Determine how many 3 ft³ bags of root medium are equivalent to 337.5 ft³.
Set up a proportion.

$$\frac{1 \text{ bag}}{3 \text{ ft}^3} = \frac{x \text{ bags}}{337.5 \text{ ft}^3}$$

Isolate and solve for x.

$$3x = 337.5$$

$$x = \frac{337.5}{3}$$

$$x = 112.5 \text{ bags}$$

Solution
A total of 113 bags are required for this project.

Practice Problem 10-13: Formulating Root Media

A greenhouse manager is planning a crop of 450 pots of hydrangeas. They are to be grown in 8-inch azalea pots (0.1429 ft³/pot) filled with a root medium containing 2 parts vermiculite, 2 parts peat moss, and 1 part perlite.

1. What volume of root medium is needed for this crop?
Determine the total number of cubic feet of root medium required.

$$450 \text{ pots} \times 0.1429 \text{ ft}^3/\text{pot} = 64.305 \text{ ft}^3$$

2. How many cubic feet of each root medium component are required?

Step 1: Determine how many units are found in the formulation by adding all the parts of the formula.

$$2 \text{ vermiculite} + 2 \text{ peat moss} + 1 \text{ perlite} = 5 \text{ units in the formula}$$

Appendix E Solutions to Practice Problems **385**

Step 2: Determine the volume of each unit by dividing the total volume of mix by the number of units.

$$64.305 \text{ ft}^3 \text{ of mix} \div 5 \text{ units} = 12.86 \frac{\text{ft}^3}{\text{unit}}$$

Step 3: Calculate the volume of each component required for this mix.

$$2 \text{ units vermiculite} \times 12.86 \text{ ft}^3/\text{unit} = 25.72 \text{ ft}^3 \text{ vermiculite}$$

$$2 \text{ units peat moss} \times 12.86 \text{ ft}^3/\text{unit} = 25.72 \text{ ft}^3 \text{ peat moss}$$

$$1 \text{ unit perlite} \times 12.86 \text{ ft}^3/\text{unit} = 12.86 \text{ ft}^3 \text{ perlite}$$

3. How much will each component cost (use Table 10-2)?

Step 1: Calculate the cost of each component by first determining the cost per cubic foot of each component.

Step 1A: Vermiculite is sold in 6 ft^3 bags and costs $17.39/bag.

How much does each cubic foot of vermiculite cost?

Set up a proportion.

$$\frac{6 \text{ ft}^3}{\$17.39} = \frac{1 \text{ ft}^3}{\$x}$$

Isolate and solve for x.

$$6x = 17.39$$

$$x = \frac{17.39}{6}$$

$$x = 2.90$$

Each cubic foot of vermiculite costs $2.90.

Step 1B: Peat moss is sold in 3.8 ft^3 compressed bales with a compression ratio of 2:1. That means each bale yields 7.6 ft^3. Confirm this by setting up a proportion.

$$\frac{3.8 \text{ ft}^3}{x \text{ ft}^3} = \frac{1}{2}$$

Isolate and solve for x.

$$x = 3.8 \times 2$$

$$x = 7.6$$

If 7.6 ft^3 of peat moss costs $9.35, then how much does 1 ft^3 of peat moss cost?

Set up a proportion.

$$\frac{7.6 \text{ ft}^3}{\$9.35} = \frac{1 \text{ ft}^3}{\$x}$$

Isolate and solve for x.

$$7.5x = 9.35$$

$$x = \frac{9.35}{7.6}$$

$$x = 1.23$$

Each cubic foot of peat moss costs $1.23.

Step 1c: Perlite is sold in 6 ft³ bags and costs $16.49/bag. How much does each cubic foot of perlite cost?

Set up a proportion.

$$\frac{6\,\text{ft}^3}{\$16.49} = \frac{1\,\text{ft}^3}{\$x}$$

Isolate and solve for x.

$$6x = 16.49$$

$$x = \frac{16.49}{6}$$

$$x = 2.75$$

Each cubic foot of perlite costs $2.75.

Step 2: Calculate the cost of each component required for this mix.

$$25.72\,\text{ft}^3\,\text{vermiculite} \times \$2.90/\text{ft}^3 = \$74.59$$

$$25.72\,\text{ft}^3\,\text{peat moss} \times \$1.23/\text{ft}^3 = \$31.64$$

$$12.86\,\text{ft}^3\,\text{perlite} \times \$2.75/\text{ft}^3 = \$35.37$$

4. What is the total cost of materials for the root medium for this crop?

Add the cost of each component to calculate the total cost of the mix.

$$\$74.59 + \$31.64 + \$35.37 = \$141.60$$

Solution

The cost of this mix is $141.60.

Practice Problem Set 10-14: Incorporation of Fertilizers into Root Media During Formulation

1. A greenhouse manager plans to produce 500 flats of bedding plants using a 1206 flat insert. Each 1206 flat insert holds 0.1087 ft³ of mix. How much soilless root medium is required? How much dolomitic limestone and Uni-Mix® are required to add to this volume of root medium?

Step 1: Determine how many cubic yards of medium are needed for this crop.

$$500\,\text{flats} \times 0.1087\,\text{ft}^3/\text{flat} = 54.35\,\text{ft}^3$$

Convert cubic feet to cubic yards by using a proportion.

$$\frac{27 \text{ ft}^3}{1 \text{ yd}^3} = \frac{54.35 \text{ ft}^3}{x \text{ yd}^3}$$

Isolate and solve for x.

$$27x = 54.35$$
$$x = \frac{54.35}{27}$$
$$x = 2 \text{ yd}^3$$

Solution
The grower needs to prepare 2 yd^3 of root medium for this crop.

Step 1A: Go to Table 10-3 and find the recommended rate of dolomitic limestone required.

10 lb of dolomitic limestone are required for every cubic yard of root medium.

Step 2A: Set up a proportion to find out how many pounds of dolomitic limestone are required for 2 yd^3 of root medium.

$$\frac{10 \text{ lb dolomitic limestone}}{1 \text{ yd}^3 \text{ of medium}} = \frac{x \text{ lb of dolomitic limestone}}{2 \text{ yd}^3 \text{ of medium}}$$

Step 3A: Isolate and solve for x.

$$x = 20 \text{ lb}$$

Solution A
A total of 20 lb of dolomitic limestone is required for 2 yd^3 of root medium.

Step 1B: Go to Table 10-4 and find the recommended rate of Uni-Mix® required.

2 lb of Uni-Mix® are required for every cubic yard of root medium.

Step 2B: Set up a proportion to find out how many pounds of Uni-Mix® are required for 2 yd^3 of root medium.

$$\frac{2 \text{ lb Uni-Mix}®}{1 \text{ yd}^3 \text{ of medium}} = \frac{x \text{ lb of Uni-Mix}®}{2 \text{ yd}^3 \text{ of medium}}$$

Step 3B: Isolate and solve for x.

$$x = 2 \times 2$$
$$x = 4 \text{ lb}$$

Solution B
A total of 4 lb of Uni-Mix® is required for 2 yd^3 of root medium.

388 Appendix E Solutions to Practice Problems

2. A nursery manager is preparing 8 yd³ of root medium. How much dolomitic limestone, superphosphate, micronutrients LG (1.5 lb/yd³ rate), calcium nitrate, and potassium nitrate need to be added to this volume of root medium?

   Step 1A: Go to Table 10-3 and find the recommended rate of dolomitic limestone required.

   10 lb of dolomitic limestone are required for every cubic yard of root medium.

   Step 2A: Set up a proportion to find out how many pounds of dolomitic limestone are required for 8 yd³ of root medium.

   $$\frac{10 \text{ lb dolomitic limestone}}{1 \text{ yd}^3 \text{ of medium}} = \frac{x \text{ lb of dolomitic limestone}}{8 \text{ yd}^3 \text{ of medium}}$$

   Step 3A: Isolate and solve for x.

   $$x = 80 \text{ lb}$$

Solution A

A total of 80 lb of dolomitic limestone are required for 8 yd³ of root medium.

   Step 1B: Go to Table 10-3 and find the recommended rate of superphosphate required.

   2.25 lb of superphosphate are required for every cubic yard of root medium.

   Step 2B: Set up a proportion to find out how many pounds of superphosphate are required for 8 yd³ of root medium.

   $$\frac{2.25 \text{ lb superphosphate}}{1 \text{ yd}^3 \text{ of medium}} = \frac{x \text{ lb of superphosphate}}{8 \text{ yd}^3 \text{ of medium}}$$

   Step 3B: Isolate and solve for x.

   $$x = 2.25 \times 8$$
   $$x = 18 \text{ lb}$$

Solution B

A total of 18 lb of superphosphate are required for 8 yd³ of root medium.

   Step 1C: Review the question to determine how many pounds of Micronutrients LG® are required per cubic yard of root medium.

   A total of 1.5 lb of Micronutrients LG® are required for every cubic yard of root medium.

   Step 2C: Set up a proportion to find out how many pounds of Micronutrients LG® are required for 8 yd³ of root medium.

   $$\frac{1.5 \text{ lb Micronutrients LG}^®}{1 \text{ yd}^3 \text{ of medium}} = \frac{x \text{ lb Micronutrients LG}^®}{8 \text{ yd}^3 \text{ of medium}}$$

Step 3c: Isolate and solve for x.

$$x = 1.5 \times 8$$

$$x = 12 \text{ lb Micronutrients LG}^{\circledR}$$

Solution C

A total of 12 lb of Micronutrients LG$^{\circledR}$ is required for 8 yd^3 of root medium.

Step 1d: Go to Table 10-3 and find the recommended rate of calcium nitrate and potassium nitrate required.

One pound each of calcium nitrate and potassium nitrate are required for every cubic yard of root medium.

Step 2d: Set up a proportion to find out how many pounds of calcium nitrate or potassium nitrate are required for 8 yd^3 of root medium.

$$\frac{1 \text{ lb calcium nitrate or potassium nitrate}}{1 \text{ yd}^3 \text{ of medium}} = \frac{x \text{ lb calcium nitrate or potassium nitrate}}{8 \text{ yd}^3 \text{ of medium}}$$

Step 3d: Isolate and solve for x.

$$x = 8 \text{ lb}$$

Solution D

A total of 8 lb each of calcium nitrate and potassium nitrate is required for 8 yd^3 of root medium.

Practice Problem 10-15: Monitoring Crop Nutritional Status: Electrical Conductivity

Fill in Table 10-5 by converting the electrical conductivity measurements.

mmhos/cm	mhos × 10^{-5}/cm	mS/cm	µmhos/cm	dS/m
4.12 mmhos/cm	412 mhos × 10^{-5}/cm	4.12 mS/cm	4,120 µmhos/cm	4.12 dS/m
1.83 mmhos/cm	183 mhos × 10^{-5}/cm	1.83 mS/cm	1,830 µmhos/cm	1.83 dS/m
2.35 mmhos/cm	235 mhos × 10^{-5}/cm	2.35 mS/cm	2,350 µmhos/cm	2.35 dS/m
2.50 mmhos/cm	250 mhos × 10^{-5}/cm	2.50 mS/m	2,500 µmhos/cm	2.50 dS/m
1.95 mmhos/cm	195 mhos × 10^{-5}/cm	1.95 mS/cm	1,950 µmhos/cm	1.95 dS/m

Practice Problem Set 10-16: Monitoring Light Intensity

Convert each of the following light intensity values.

1. Convert 54 klx to footcandles.

390 Appendix E Solutions to Practice Problems

Step 1: Set up a proportion using equivalents.
$$\frac{10.764 \text{ klux}}{1,000 \text{ fc}} = \frac{54 \text{ klux}}{x \text{ fc}}$$

Step 2: Isolate and solve for x.
$$10.764x = 1,000 \times 54$$
$$x = \frac{(1,000)(54)}{10.764}$$
$$x = 5,016.7 \text{ fc}$$

Solution
A light level of 5,016.7 fc is equivalent to 54 klx.

2. Convert 6,000 fc to kilolux.

Step 1: Set up a proportion using equivalents.
$$\frac{10.764 \text{ klx}}{1,000 \text{ fc}} = \frac{x \text{ klx}}{6,000 \text{ fc}}$$

Step 2: Isolate and solve for x.
$$1,000x = 10.764 \times 6,000$$
$$x = \frac{(10.764)(6,000)}{1,000}$$
$$x = 64.6 \text{ klx}$$

Solution
A light level of 6,000 fc is equivalent to 64.6 klx.

3. Convert 48 klx to lux.

Step 1: Set up a proportion using equivalents.
$$\frac{1 \text{ klx}}{1,000 \text{ lux}} = \frac{48 \text{ klx}}{x \text{ lux}}$$

Step 2: Isolate and solve for x.
$$x = 1,000 \times 48$$
$$x = 48,000 \text{ lux}$$

Solution
A light level of 48 klx is equivalent to 48,000 lux.

4. Convert 50 fc to lux.

Step 1: Set up a proportion using equivalents.
$$\frac{10.764 \text{ lux}}{1 \text{ fc}} = \frac{x \text{ lux}}{50 \text{ fc}}$$

Step 2: Isolate and solve for x.

$$x = 10.764 \times 50$$

$$x = 538.2 \text{ lux}$$

Solution
A light level of 50 fc is equivalent to 538.2 lux.

5. Convert 3,000 lx to footcandles.

   Step 1: Set up a proportion using equivalents.

   $$\frac{1 \text{ lux}}{0.0929 \text{ fc}} = \frac{3,000 \text{ lux}}{x \text{ fc}}$$

   Step 2: Isolate and solve for x.

   $$x = 0.0929 \times 3,000$$

   $$x = 278.7 \text{ fc}$$

Solution
A light level of 3,000 lux is equivalent to 278.7 fc.

Practice Problem Set 10-17: Calculations Involving Pesticides and Growth Regulators

1. A greenhouse manager needs to make a drench application of an insecticide to a crop of 1,800 poinsettias grown in 8 in. pots. A drench volume of 8 fl oz is required, and 12 oz of dry insecticide product are dissolved in every 100 gal of drench solution. How many gallons of drench solution are required for this application? How many ounces of dry insecticide product are needed for this application?

   Step 1A: Determine how many fluid ounces of drench are required to treat 1,800 pots at a rate of 8 fl oz/pot by setting up a proportion.

   $$\frac{8 \text{ fluid ounces of drench}}{1 \text{ pot}} = \frac{x \text{ fluid ounces of drench}}{1,800 \text{ pots}}$$

   Isolate and solve for x.

   $$x = 8 \times 1,800$$

   $$x = 14,400 \text{ fl oz of drench}$$

   Step 2A: Convert fluid ounces to gallons by setting up a proportion.

   $$\frac{1 \text{ gallon}}{128 \text{ fluid ounces}} = \frac{x \text{ gallons}}{14,400 \text{ fluid ounces}} =$$

Isolate and solve for x.

$$x \times 128 = 14{,}400$$
$$x = \frac{14{,}400}{128}$$
$$x = 112.5 \text{ gal}$$

Solution A

A total of 112.5 gal of drench solution is required to treat 1,800 pots of poinsettia. A greenhouse manager will prepare 113 gal of solution.

Step 1B: Determine how many ounces of growth regulator product are required to prepare 113 gal of solution when 12 oz are needed to prepare every 100 gal of solution. Set up a proportion.

$$\frac{12 \text{ ounces of product}}{100 \text{ gallon of solution}} = \frac{x \text{ ounces of product}}{113 \text{ gallons of solution}} =$$

Isolate and solve for x.

$$100\,x = 12 \times 113$$
$$x = \frac{12 \times 113}{100}$$
$$x = 13.56 \text{ oz of product}$$

Solution B

To prepare 113 gal of dilute plant growth regulator solution for a drench, 13.6 oz of dry insecticide product are needed.

2. If the dry insecticide product used in question 1 is a 25% active ingredient formulation, how many ounces of active ingredient are used in this application?

Step 1: Determine the number of ounces of active ingredient in 13.6 oz of product by using a proportion. If 25% or 25 oz in 100 oz are active ingredient, then the proportion is:

$$\frac{25 \text{ ounces of active ingredient}}{100 \text{ ounces of product}} = \frac{x \text{ ounces of active ingredient}}{13.6 \text{ ounces of product}}$$

Isolate and solve for x.

$$25 \times 13.6 = 100 \times x$$
$$\frac{25 \times 13.6}{100} = \frac{100 \times x}{100}$$
$$x = \frac{25 \times 13.6}{100}$$
$$x = 3.4 \text{ oz a.i.}$$
$$\text{or } 0.25 \times 13.6 = 3.4 \text{ oz a.i.}$$

Solution
There are 3.4 oz of a.i. in 13.6 oz of this insecticide product.

3. A greenhouse manager needs to make a foliar application of a growth regulator to 25,000 ft² of a young bedding plant crop. The label indicates that 1 qt of spray volume for every 100 ft² of crop is appropriate, and a rate of 4.7 ml of product in every gallon of spray is recommended for this crop. How many gallons of spray need to be prepared for this application? How many fluid ounces of product are required to prepare this spray solution?

Step 1A: Determine the volume of spray solution needed for this crop if the rate is 1 qt/100 ft² and there are 25,000 ft² of the crop. Set up a proportion.

$$\frac{1 \text{ quart spray solution}}{100 \text{ ft}^2} = \frac{x \text{ quarts spray solution}}{25{,}000 \text{ ft}^2}$$

Isolate and solve for x.

$$100x = 25{,}000$$
$$x = 250 \text{ quarts}$$

250 quarts of spray solution are needed for this crop.

Step 2A: Convert the volume from quarts to gallons by setting up a proportion.

$$\frac{1 \text{ gallon}}{4 \text{ quarts}} = \frac{x \text{ gallons}}{250 \text{ quarts}}$$

Isolate and solve for x.

$$4x = 250$$
$$x = 62.5 \text{ gal}$$

Solution A
A total of 62.5 gallons of spray solution is needed for this crop.

Step 1B: Determine how much product is required to prepare 62.5 gal of spray solution, if 4.7 ml of product are needed for each gallon of spray solution. Set up a proportion.

$$\frac{4.7 \text{ ml of product}}{1 \text{ gallon of spray solution}} = \frac{x \text{ ml of product}}{62.5 \text{ gallons of spray solution}}$$

Isolate and solve for x.

$$x = 4.7 \times 62.5$$
$$x = 293.75 \text{ ml}$$

Solution B
A total of 293.75 ml of growth regulator product are needed for this application.

4. How many grams of active ingredient are used in this application, if 0.12 g of active ingredient are found in 1 fl oz of product?

Step 1: Convert 293.75 ml to fluid ounces using a proportion.

$$\frac{1 \text{ fluid ounce}}{29.57353 \text{ ml}} = \frac{x \text{ fluid ounces}}{293.75 \text{ ml}}$$

Isolate and solve for x.

$$29.57353x = 293.75$$

$$x = \frac{293.75}{29.57353}$$

$$x = 9.9 \text{ oz}$$

Step 2: Use a proportion to determine the number of grams of active ingredient in 9.9 oz of product.

$$\frac{1 \text{ fluid ounce}}{0.12 \text{ g active ingredient}} = \frac{9.9 \text{ fluid ounces}}{x \text{ g active ingredient}}$$

$$x = 0.12 \times 9.9$$

$$x = 1.19 \text{ g}$$

Solution

A total of 1.19 g of active ingredient was applied in this application.

INDEX

Accuracy and precision
  definitions of, 32
  factors affecting, 33
  impact of measuring instrument, 32–33
Acre-foot
  definition of, 185
Acre-inch
  definition of, 185
Active ingredient (a.i.)
  calculating the amount applied, 133–134, 136–138
  cost of, 145–146
  definition of, 127
  percent by weight, 127–128
Area
  circle, 57
  ellipse, 60, 63
  irregular figures, 64–71
    modified offset method, 65–66
    offset method, 64–65
    Simpson's rule, 67–70
  landscape features, 79–93
    calculations using the geometric method, 79–86
      composite geometric forms, 81–86
      simple geometric forms, 80–81
    calculations using the modified offset method, 88–91
    calculations using the offset method, 86–88
    calculations using Simpson's rule, 91–92
  parallelogram, 50
  rectangle, 48–49
  regular figures, 47–63
  square, 48
  trapezoid, 50–51
  triangle, 52–54
  units of, 40, 288–289
    metric, 289
    U.S. Customary, 288

Base material calculations
  crushed limestone, 200
  sand, 201
Breakeven price, 202
Bulk materials, 197

Calibration
  spreader
    broadcast, 158–162
    drop, 151–158
  sprayer
    boom, 163–168
    small lawn, 168
Capacity (Liquid volume)
  units of, 41, 291–292
    metric, 40, 122, 148, 149, 292
    U.S. Customary, 291
Circle
  area of, 57
  circumference of, 57
  definition of, 56
  diameter of, 58
  example problems, 59–60
  radius of, 58
Circumference
  definition of, 57
Concentration
  units of, 41, 293
Cone
  volume of, 73–74
Containers
  volume of, 256–261
Conversion factors
  table of, 297–305
  using, 45

Conversions
  unit, 43–45
  using conversion factors, 45
  using equivalents, 44
Counting numbers
  definition of, 36
Cross multiplication
  definition of, 21
Cube
  volume of, 72–73
Cylinder
  volume of, 72–73

Decimal equivalents
  of commonly used fractions, 11
Decimal fraction
  definition of, 8
Decimal point
  definition of, 7
Decimal number
  converting to fractions, 9–10
  definition of, 7
Decisiemens, 272–275, 294
Diameter
  definition of, 58
Dry pesticides, 128, 130

Electrical conductivity
  definition of, 272
  meters, 272
  monitoring greenhouse crop nutritional status, 272–275
  units of, 273, 294
Ellipse
  area of, 60
  center of, 60–61
  definition of, 60
  focus points, 61
  major axis, 61
  major radius, 62

396 Index

Ellipse (*continued*)
  minor axis, 62
  minor radius, 62–63
Equation
  definition of, 18
Equipment costs, 210
Equivalents
  definition of, 36
  tables of, 287–294
  using, 44
Estimating
  landscape installation costs, 194–218
  landscape maintenance costs, 211–221
Exact numbers
  definition of, 36
Exponent
  definition of, 12
  in scientific notation 13
  negative, 13
Exponentiation
  of base two, 12
  of base ten, 14

Fertilizer
  analysis, 95–96
  cost, 125, 126, 217, 218
  formulation, 96
  rate expression, 96–98
  ratio, 96, 98, 129
  terminology, 95–98
Fertilizer analysis
  converting to decimal format, 99
  converting to ratio format, 100
  definition of, 96
Fertilizer applications
  using proportioning equipment, 246–255
    brass-siphon devices, 246–248
    commercial scale, 246, 248–255
Fertilizer calculations
  amount of product for landscape application, 102–119
  amount of N, P, and K in a bag, 100–102
  cost calculations, 125–126, 217–218
  greenhouse fertilizer proportioning devices, 246–255
  greenhouse root media, 269–272
  liquid, 119–120
  using metric units, 122–125
  using ppm, 246–255
    nitrogen delivery, 248–253
    phosphorous delivery, 253–254
    potassium delivery, 254–255
    short-cut equations, 251–253
Fertilizer product label, 99
Footcandle, 275–277, 295
Formulation
  fertilizer product, 96
  pesticide and pgr product, 128
Fractions
  adding, 5–6
  converting to decimals, 10
  converting to decimal fractions, 9
  definition of, 1
  dividing, 6
  improper, 3
  multiplying, 6
  proper fraction, 2
  reducing, 2
  subtracting, 6
Frequency
  of landscape maintenance service, 214

Greatest common factor (GCF), 2
Greenhouse bench arrangement
  aisle-eliminator system, 224–225
  floor system, 224–225
  parallel, 223–224
  peninsular, 223–224
  rolling, 224–225
Greenhouse bench efficiency
  definition of, 225
  example of, 226–227
  formula for, 225
Greenhouse crop
  container volume, 256–261
  fertilizer calculations, 246–255
  growth regulator calculations, 277–283
  monitoring light intensity, 275–277
  monitoring nutritional status, 272–275
  pesticide calculations, 277–283
  root media
    calculating cost, 266–268
    calculating volume requirements, 261–264
    formulation, 264–269
Greenhouse surface area, 227–239
  arch-top, 230–233
  even-span, 227–230
  gutter-connected, 233–237
    arch-top, 235–237
    even-span, 233–235
  quonset, 237–239
Greenhouse volume, 239–246
  arch-top, 241–242
  definition, 239
  even-span, 240–241
  general formula, 239
  gutter-connected, 242–245
    arch-top, 244–245
    even-span, 242–243
  Quonset, 245–246

Hardgoods, 197
Hardscape, 197
Hemisphere
  volume of, 76–78
Herbicide
  cost calculations, 218, 219
Heron's formula, 52–54

Improper fraction
  conversion to mixed number, 4
Irradiance
  equivalents for, 276
  units of, 275–276, 295

K, *see* potassium
Kilolux, 275–277, 295

Labor costs, 204
  hourly labor rate, 205, 215–216
Lawn, *see* turfgrass
Length
  units of, 40, 287–288
    metric, 288
    U.S. Customary, 287
Liquid fertilizer, 119, 120
Liquid pesticides, 128, 130
Lux, 275–277, 295

Index

Maintenance contract, 215
Materials costs
   installation, 195–204
   maintenance, 217–220
Mathematical constants
   definition of, 36
Measure
   units of, 40–43
Measurement, 31
Metric system, 40, 122, 148, 149
   prefixes, 286
Mhos, 272–275, 294
Micromhos, 272–275, 294
Millimhos, 272–275, 294
Millisiemens, 272–275, 294
Mixed number
   conversion to improper fraction, 4
   definition of, 3
Mulch calculations, 201, 219

Nitrogen, 95–126

Offset method
   definition of, 64
   example of, 64–65
   formula for, 64
   modified
      definition of, 65–66
      formula for, 65–66
      example of, 66–67
On center (o.c.) spacing, 196
Overage, 197
Overhead, 211
   direct job overhead, 211
   general overhead, 211
Overhead markup, 202, 203

P, *see* phosphorus
Parallelogram
   area of, 50
   definition of, 50
Parts per million, 41, 293
   greenhouse fertilizer applications, 246–255
Paver calculations, 197–199
Percent
   definition of, 27
   solving problems with percent, 27–29

Pesticide and plant growth regulator calculations for specific application methods
   banded spray applications, 279–280
   drench applications, 280–281
   dry formulations applied dry, 130–135
   dry formulations applied as a spray, 135–140
   foliar sprays with recommended spray volumes, 277–279
   liquid formulations applied as a spray, 140–144
   calculations in metric units, 147–149
   cost calculations, 145–147
   formulations
      dry, 130
      liquid, 130
      table of, 128
   product label, 129–130
   terminology, 127–129
Phosphorus, 95–126
   percent in P_2O_5, 101
Pi
   definition of, 58
Place value
   definition of, 8
   table, 8
Plant growth regulators, *see* pesticide and plant growth regulator
Plugs
   definition of, 176
Potassium, 95–126
   percent in K_2O, 101
Power
   see exponent, 12
Production rate, 206
   installation production rates, 207
   maintenance production rates, 212
Proper order of mathematical operations (PEMDAS)
   definition of, 16
   examples of, 16–17
Proportion
   definition of, 21
   use in solving problems, 23–25
   use in unit conversions, 22

Proportioning equipment
   brass-siphon devices, 246–248
   commercial scale, 246, 248–255
   fertilizer, 246–255
Pure live seed (PLS)
   definition of, 169
Pyramid
   volume of, 73–74
Pythagorean Theorem
   definition of, 54–55
   example of, 15
   use in squaring gardens and garden structures, 307–308
   use when determining a right triangle, 55–56
   use when determining the length of the hypotenuse, 15, 307–308

Quadrilateral
   definition of, 47

Radius
   definition of, 58
Rate
   definition of, 25
   solving problems with rate, 25–27
Ratio
   definition of, 20
   use in solving problems, 23–25
   use in unit conversions, 22
Rational number
   definition of, 1
Rectangle
   area of, 48–49
   definition of, 48
Rectangular prism
   volume of, 72–73
Rounding numbers
   application to horticulture, 34–36
   rules of, 38–39

Seeding rate, 197
Selling price, 197
Scientific notation
   definition of, 13
   used to express very large and small numbers, 13–14

Significant digits
  application to horticulture, 34–36
  definition of, 36
  rules, 36–37
Site grading, 191
Simpson's rule
  definition of, 67–68
  example of, 68–70
  formula for, 68
Slope
  calculating, 191–192
  definition of, 191
  landscape feature guidelines, 193
Sod, 174
Soil modification
  definition of, 178
  volume calculations for, 179–181
Solute
  definition of, 41–42
Solution
  definition of, 42
Solvent
  definition of, 42
Spacing
  annual, perennial and groundcover, 196
Sphere
  volume of, 76–78
Sports field, 114, 122
Sprayer calibration, 163–168
Spreader calibration, 151–163
Sprigging, 177
Square
  area of, 48
  definition of, 47
Square root
  definition of, 14
Squaring-up gardens and garden structures, 307–308
Stolons, 177

Temperature
  converting from Celsius to Fahrenheit, 42–43
  converting from Fahrenheit to Celsius, 42–43
  units of, 42
Time
  units of, 43, 294
Topdressing
  definition of, 178
  determining volume in storage, 183–184
  volume calculations for, 181–184
Trapezoid
  area of, 50–51
  definition of, 50
Triangle(s)
  area of, 52–54
  defining types, 52
  definition of, 51
Triangular prism
  volume of, 72–73
Turfgrass establishment
  plug, 176–177
  seed, 169–174
    cost calculations, 173–174
    pure live seed (PLS), 169
    seeding rates, 173
  sod, 174–177
    cost calculations, 175
  stolon, 177
Turfgrass seed label, 170
Turfgrass soil modification and topdressing
  definitions of, 178–179
  calculations for soil modification, 179–181
  calculations for topdressing, 181–184
Turfgrass water use calculations, 185–190

U.S. customary system, 40

Volume
  cone, 73–74
  figures with parallel bases and equal cross-sections
    definition of, 72
    example of, 73
    formula for, 72–73
  figures with pointed tops
    definition of, 73
    example of, 74
    formula for, 74
  figures with tapered sides
    definition of, 74–75
    example of, 75–76
    formula for, 75
  hemisphere
    definition of, 76
    example of, 78
    formula for, 77
  liquid, see Capacity
  pyramid, 73–74
  sphere
    definition of, 76
    example of, 77
    formula for, 76–77
  units of, 41, 289

Water use
  acre-foot, 185
  acre-inch, 185
  determining cost of water, 187–188
  determining storage pond capacity, 189–190
  total water use, 186–188
Weight
  units of, 41, 292–293
  metric, 293
  U.S. customary, 292
Whole number
  definition of, 1

x
  solving for, 18
  solving for in equations with addition and subtraction, 18
  solving for in equations with combined operations, 19–20
  solving for in equations with multiplication and division 19

Printed in the United States of America
ED-09-19-11